GIVEN TO T. HASKINS
25 MARCH 882

Pump Application Engineering

Pump Application Engineering

Tyler G. Hicks, P.E.

International Engineering Associates
Member: American Society of Mechanical Engineers
Institute of Electrical and Electronics Engineers
United States Naval Institute
International Oceanographic Foundation

T. W. Edwards, P.E.

Member: American Society of Mechanical Engineers
American Nuclear Society
Institute of Electrical and Electronics Engineers

McGRAW-HILL BOOK COMPANY

New York   St. Louis   San Francisco   Düsseldorf   London
Mexico   Panama   Sydney   Toronto

Sponsoring Editor   William G. Salo
Director of Production   Stephen J. Boldish
Editing Supervisor   Stanley E. Redka
Designer   Naomi Auerbach
Editing and Production Staff   Gretlyn Blau,
   Teresa F. Leaden, George E. Oechsner

PUMP APPLICATION ENGINEERING

07-028741-4

1234567890   MAMM   754321

Preface

THIS BOOK is a completely revised and updated version of "Pump Selection and Application." The wide acceptance of this work throughout the world in several languages led the authors to broaden its coverage. Thus, the present volume has been titled "Pump Application Engineering" to reflect the broad scope and depth of coverage.

Since "Pump Selection and Application" was published in 1957, many spectacular technological advances have caught the public eye—space exploration, moon landings, heart transplants, nuclear power, to name but a few. But, while it has not been so obvious to everyone, all of these advances have been dependent upon a rapidly developing pump technology. This really is not a surprising or unjustified statement to make when one remembers that, next to the electric motor, pumps are the most widely used machines in the world. There is hardly a single piece of major equipment that does not in some way depend upon pumps for its successful operation.

In "Pump Application Engineering" we have updated many of those areas of pump technology that have supported and contributed to the numerous advances of the 1960s. At the same time, however, we have retained the detailed information on correct pump selection and application that contributed so much to the value of "Pump Selection and Application." This combination provides the engineer in industry with the complete and up-to-date information he needs to select correctly and apply the pumping equipment required to perform a specific job.

Some recent technological advances have affected all types of pumping equipment. These advances include improved materials, better pump design, and superior quality-assurance techniques. Such advances permit successful pumping equipment application in increasingly severe services. These advances also prolong pump life and increase operating reliability.

Other recent advances pertain more specifically to certain pump types or services. Included are such items as increased equipment standardization, the wide application of so-called twin volutes, improved mechanical-seal designs, order-of-magnitude size increases, and the general pushing back of temperature limitations at both the low and high ends of the scale.

As a good example, take the pumping equipment needs of the nuclear-power industry. Here we have a whole new industry that has developed on a commercial basis since "Pump Selection and Application" was published in 1957. While pumps for nuclear power are numerous and diverse, probably the most severe and exacting service is for the primary circulators. In pressurized-water-reactor systems, these giant pumps handle capacities in the range of 100,000 gpm. Although developed heads are quite moderate, suction heads are high and the water pumped in extremely hot. Leakage limitations—or actual prohibition in some designs—require sophisticated sealing arrangements whose reliability must be of the highest order. Nuclear-power safety requirements are extremely stringent; this means that all pumps in the system are subjected to repeated design and quality-assurance checks. Reliability and safety are even further enhanced by built-in equipment redundancy and complex monitoring and control systems.

In gathering data and information for "Pump Application Engineering," the authors are indeed grateful to the many pump manufacturers who so generously supplied up-to-date illustrations and freely offered their informed suggestions for improvements in the text. As was the case when "Pump Selection and Application" was published, the Hydraulic Institute has again permitted wide use of its latest "Standards" material—an authoritative consensus of the vast store of pump application data developed over the years by the Institute's member companies. In addition, the Bureau of Labor Statistics of the United States Department of Labor provided invaluable information for updating economic evaluation problems.

The authors hope that this new book will assist engineers everywhere who must select and apply pumps of all types. Any suggestions for the improvement of future editions will be gladly welcomed.

Tyler G. Hicks
T. W. Edwards

Contents

                                                          vii

PART ONE

Pump Classes and Types

CHAPTER ONE

Centrifugal Pumps

PUMPS ARE COMMONLY DESIGNATED by two different kinds of terms: (1) those taking the hydraulic or liquid-moving characteristics into consideration, and (2) those based on the type or specific application for which the pump is intended. Use of these two methods of designating pumps causes much confusion among the uninitiated, and even among some old-timers.

Classes and Types Figure 1-1 is designed to clear up much of the mystery surrounding pump *classes* and *types*. It might be called a road map to the world of pumps. Based on often-used standard classifications, it incorporates a number of useful facts helpful in pump selection and application.

Three classes of pumps find use today—*centrifugal, rotary,* and *reciprocating.* Note that these terms apply only to the mechanics of moving the liquid, not to the service for which the pump is designed. This is important, because many pumps are built and sold for a specific service, and in the complex problem of finding which has the best design details the basic problems of class and type may be overlooked. Each class is further subdivided into a number of different types (Fig. 1-1). For example, under the rotary classification are cam, screw, gear, and vane pumps, to mention but a few. Each is a particular type of rotary pump. To go one step further, take a brief look at a fuel-oil pump in wide use

3

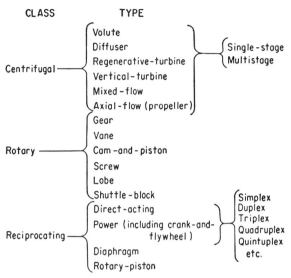

Fig. 1-1 Pump classes and types.

today. It is a rotary three-screw type available with rotors of a number of different materials and four means of balancing axial thrust.

Details The last two items in the above description are important *details* in pump application; the first two are keys to classification of the unit. The Hydraulic Institute recommends that the standard classification be considered as applying to type only, leaving it to the builder to use the details he has developed or standardized for that type of pump. So in selecting a pump it is often necessary to compare, detail for detail, a number of makes. The broad breakdown in Fig. 1-1 then comes in handy.

The Hydraulic Institute in its Standards classifies centrifugal pumps by the number of stages—single or multistage; casing type—volute, circular, or diffuser; position of shaft—horizontal, vertical (dry-pit type or submerged type); suction—single or double suction.

In terms of materials of construction the Hydraulic Institute uses the following designations: (1) bronze-fitted, (2) all-bronze, (3) specific-composition bronze, (4) all-iron, (5) stainless-steel fitted, and (6) all-stainless-steel pumps.

Bronze-fitted pumps have a cast-iron casing, bronze impeller, and bronze casing rings and shaft sleeves (if used). In an all-bronze pump every part in direct contact with the liquid is made of the manufacturer's standard bronze. The same is true of designation (3) except that the parts are made of a bronze composition suitable for the pump application. All-iron pumps have ferrous-metal parts contacting the liquid pumped. In a stainless-steel fitted pump the casing is a material suitable for the service, while the impellers, impeller rings, and shaft sleeves (if used) are made of a corrosion-resistant steel suitable for the liquid handled. In an all-stainless-steel pump, parts contacting the liquid are made of corrosion-resistant steel suitable for the application, while the shaft is a corrosion-resistant steel of a grade equal to the other parts of the pump. See Chap. 6 for a discussion of pump materials.

General Characteristics The next consideration is a general statement of the usual characteristics of a given class of pumps. Table 1-1 does just this. For example, to find a pump to handle relatively small capacities of clean clear liquids at high head, refer to the table. In any problem of this type, remember that the suction lift should not exceed the recommended maximum limit. Capacity in gallons per minute (gpm) determines pump size and influences the class of unit chosen. The nature of the fluid is also involved, as is pump construction. Head is a major factor, too.

TABLE 1-1 Characteristics of Modern Pumps

	Centrifugal		Rotary	Reciprocating		
	Volute and diffuser	Axial flow	Screw and gear	Direct-acting steam	Double-acting power	Triplex
Discharge flow...	Steady	Steady	Steady	Pulsating	Pulsating	Pulsating
Usual max suction lift, ft..	15	15	22	22	22	22
Liquids handled.	Clean, clear; dirty, abrasive; liquids with high solids content		Viscous, nonabrasive	Clean and clear		
Discharge pressure range...	Low to high		Medium	Low to highest produced		
Usual capacity range	Small to largest available		Small to medium	Relatively small		
How increased head affects:						
Capacity......	Decrease		None	Decrease	None	None
Power input...	Depends on specific speed		Increase	Increase	Increase	Increase
How decreased head affects:						
Capacity......	Increase		None	Small increase	None	None
Power input...	Depends on specific speed		Decrease	Decrease	Decrease	Decrease

Table 1-1 shows that a reciprocating pump is suitable for the general conditions of small capacity, high head, and clean and clear liquid. Then, depending on job needs, a piston- or plunger-type, direct-acting, crank-and-flywheel, or power-type pump may be chosen. It may be simplex, duplex, triplex, or have a larger number of cylinders.

Once these items are settled, study the pump valve details, construction materials, drives, etc. In general, it will be found that pump details are greatly influenced by job requirements. Thus the particular arrangement of a centrifugal pump may depend as much on piping, space, and working conditions as on any other existing factors. The drive chosen for the pump may be fixed by the pump speed, plant heat balance, power supply available, or cost of a particular fuel in the area. But again, these are details, to be decided after finding a pump suitable for the hydraulic conditions that must be met. And the key to meeting the hydraulic requirements is the right class and type of pump.

Where two or more units meet the hydraulic needs, the study must go one step further to determine which pump is best for the installation. The plant may require low first cost for the unit, long life, or maximum operating economy. Normally all these are not found in one package. So it must be decided which is most important for the installation being considered.

CENTRIFUGAL-PUMP ACTION

Volute-type Pumps   Here (Fig. 1-2) the impeller discharges into a progressively expanding spiral casing, proportioned to reduce the liquid

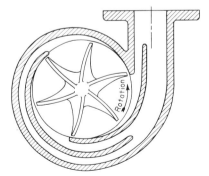

Fig. 1-2   Single pump volute converts velocity energy of the liquid into static pressure.

Fig. 1-3   Double (or twin) pump volute hydraulically acts the same as a single pump volute; mechanically, however, it balances radial shaft loads, minimizing shaft bending, especially at low flows.

velocity gradually. By this means some of the velocity energy of the liquid is converted into static pressure. Double (or twin) pump volutes (Fig. 1-3) produce near radial symmetry in high-pressure pumps and in pumps designed for low flow operation. The volutes balance the hydraulic radial loads on the pump shaft so that the loads cancel each other, materially reducing shaft loading and resultant bending.

Diffuser-type Pumps Stationary guide vanes (Fig. 1-4) surround the runner or impeller in a diffuser-type pump. These gradually expanding passages change the direction of liquid flow and convert velocity energy to pressure head.

Fig. 1-4 Diffuser changes flow direction and aids in converting velocity to pressure.

Turbine-type Pumps Also known as vortex, periphery, and regenerative pumps, liquid in this type is whirled by the impeller vanes at high velocity for nearly one revolution in an annular channel in which the impeller turns. Energy is added to the liquid in a number of impulses (Fig. 1-5). Deepwell diffuser-type pumps are often called turbine

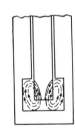

Fig. 1-5 Turbine pump adds energy to the liquid in a number of impulses.

pumps. However, they do not resemble the regenerative turbine pump in any way and should not be confused with it.

Mixed-flow and Axial-flow Types Mixed-flow pumps develop their head partly by centrifugal force and partly by the lift of the vanes on the liquid (Fig. 1-6). The discharge diameter of the vanes is greater than the inlet diameter. Axial-flow pumps develop their head by the propelling or

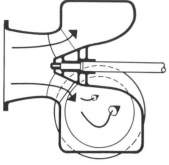

Fig. 1-6 Mixed-flow pumps use both centrifugal force and lift of vanes on liquid.

Fig. 1-7 A propeller pump develops most of its head by vane action on liquid.

lifting action of the vanes on the liquid (Fig. 1-7). The diameter of the impeller is the same at the suction and discharge sides. A propeller pump is a type of axial-flow pump.

Specific Speed This is an index of pump type, using the capacity and head obtained at the point of maximum efficiency. It determines the general profile or shape of the impeller. *In numbers, specific speed is the speed in revolutions per minute at which an impeller would run if reduced in size to deliver* 1 *gpm against a total head of* 1 *ft.* Impellers for high heads usually have low specific speed; impellers for low heads usually have high specific speed.

As Fig. 1-8 shows, each impeller design has a specific speed range for which it is best adapted. These ranges are approximate, without clean-cut divisions between them. Figure 1-8 gives the general relations between impeller shape, efficiency, and capacity. Suction limitations of different pumps bear a relation to the specific speed. These will be discussed later for various operating conditions.

PUMP AND SYSTEM CURVES

Characteristic Curves Unlike positive-displacement pumps (rotary and reciprocating), a centrifugal pump operated at constant speed delivers any capacity from zero to a maximum, depending on the head, design, and suction conditions. Characteristic curves (Fig. 1-9) show the interrelation of pump head, capacity, power, and efficiency for a specific impeller diameter and casing size. It is usual to plot head, power, and efficiency against capacity at constant speed, as in Fig. 1-9. But in special cases it is possible to plot any three variables against a fourth.

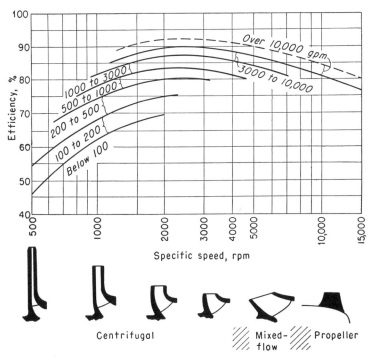

Fig. 1-8　Approximate relation of specific speed, impeller shape, and efficiency. (*Worthington Corp.*)

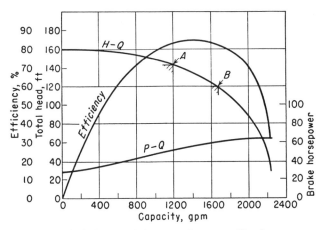

Fig. 1-9　Typical characteristic curves for a centrifugal pump.

The head-capacity curve, labeled *H–Q* (Fig. 1-9), shows the relation between capacity and total head, and may be rising, drooping, steep, or flat, depending on the impeller type and its design.   At *A* in Fig. 1-9 the head developed by the pump is 144 ft of liquid, capacity 1,200 gpm. At 120-ft head *B* the capacity of the pump rises to 1,680 gpm.

Variable Speeds   Where a pump is operated at more than one speed, a plot (Fig. 1-10) can be made to show the complete performance for a

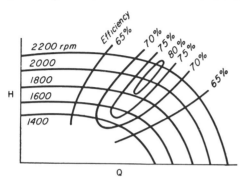

Fig. 1-10 Variable-speed head-capacity curves.

given suction lift. To plot this type of chart, *HQ* curves are drawn for the speeds being considered. Then curves of points having the same efficiency are superimposed. These constant-efficiency curves, also called *isoefficiency* curves, permit finding the required speed and the efficiency for any head-capacity condition within the limits of the chart.

Impeller Diameter   The first set of characteristic curves (Fig. 1-9) show pump performance for a specific impeller diameter, usually the maximum diameter. But impellers of more than one diameter can usually be fitted in a given casing. The curves in Fig. 1-11 show the per-

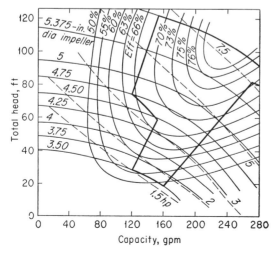

Fig. 1-11 Pump characteristics when impeller diameter is varied within same casing.

formance of a given pump with impellers of various diameters. The heavy line in Fig. 1-11 encloses the area of practical application for this particular design.

However, where a complete line of pumps of one design is available, the area outside the heavy outline is usually covered by other sizes. Then a plot (Fig. 1-12) called a *composite rating chart* can be used to give a complete picture of the available head and capacity when using a given line. It is common practice to refer to a line of pumps by figure or model number. The heavy outline in Fig. 1-12 denotes the pump size shown in Fig. 1-11.

While centrifugal pumps can be and often are selected from rating tables (Table 1-2, made by choosing certain points from the characteristic curves) performance curves give a much clearer picture of the characteristics of the unit at a given speed. Efficiency curves are usually omitted from a composite rating chart (Fig. 1-12) because they are impractical to plot. However, for purposes of easy selection, such charts give the data normally required—capacity, head, pump size, and motor horsepower. Once the size of pump has been chosen, a curve like that in Fig. 1-11 can be referred to for impeller diameter, efficiency, and other details.

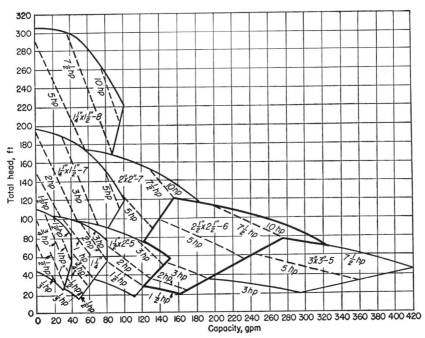

Fig. 1-12 Composite rating chart for a typical centrifugal pump. (*Goulds Pumps, Inc.*)

Note that the curves in Figs. 1-9 through 1-12 are general and apply only to a particular pump or line of pumps. Each design has its own characteristic curves, which may or may not resemble those shown. Characteristic curves for a large number of different types of centrifugal pumps are discussed in Parts Two and Three of this book.

System-head Curves   These (Fig. 1-13) are obtained by combining the friction-head curve of the system with the system static head and any pressure differences that exist.   A friction-head curve is a plot of the

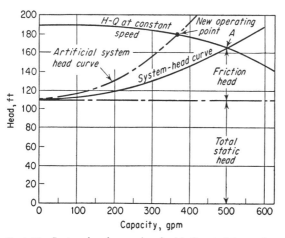

Fig. 1-13   System-head curve is valuable for studying a given hookup.

relation between flow and friction in the piping, valves, and fittings in the suction and discharge lines.   Since friction head varies roughly as the square of the flow, a plot of it is usually parabolic.   Static head is the difference in elevation between the liquid levels of the suction and discharge.

CLASSIFICATION BY APPLICATION

Earlier it was seen how modern pumps are classified and typed.   Now a closer look can be taken at another widely used method of classification —the specific application for which the pump was designed and built.

Application   While not all centrifugal pumps are classified by a term designating their ultimate application, a large number do carry some term related to their service.   Thus centrifugal pumps are termed boiler-feed, general-purpose, sump, deepwell, refinery (hot-oil), condensate, vacuum (heating), process, sewage, trash, circulating, ash, tailwater, etc.   In general, each has specific features of design and materials recommended by the builder for the particular service.

TABLE 1-2 Typical Centrifugal-pump Rating Table*

Size, gpm	Total head, ft				
	10	15	20	25	30
2C:					
100	....	....	1,000-0.8	1,060-1.0	1,150-1.2
150	....	....	1,070-1.2	1,150-1.5	1,240-1.7
200	....	....	....	1,290-2.1	1,360-2.4
3CS:					
150	750-0.53	850-0.78	950-1	1,030-1.2	1,100-1.5
200	....	950-1.1	1,010-1.4	1,100-1.7	1,170-2
250	....	....	1,170-1.9	1,190-2.3	1,260-2.6
300	....	....	....	....	1,400-3.5
3CL:					
200	690-0.63	800-0.95	910-1.3	1,010-1.6	1,110-2.05
300	870-1.2	950-1.6	1,000-1.9	1,100-2.4	1,170-2.8
400	....	....	1,200-3.1	1,230-3.7	1,290-4.1
500	....	....	....	....	1,490-5.8
4C:					
400	750-1.3	850-1.8	940-2.4	1,040-3	1,120-3.7
600	....	....	1,080-4	1,170-4.6	1,210-5.5
800	....	....	....	....	1,400-8.4
1¼D:					
25	....	617-0.21	707-0.03	778-0.40	845-0.51
50	....	680-0.37	760-0.49	865-0.63	900-0.76
75	....	....	856-0.78	916-0.94	980-1.1
100					
125					
150					
2DL:					
150	....	820-0.93	850-1.1	930-1.35	990-1.6
200	....	....	970-1.8	1,040-2.1	1,080-2.3
250					
300					

Example: 1,080-4 indicates pump speed is 1,080 rpm; actual input required to operate pump is 4 hp.
*Condensed from data of Goulds Pumps, Inc.

Another subdivision grows out of broad structural features. Thus terms like horizontal and vertical units, close-coupled designs, single- and double-suction impellers, horizontally split casings, barrel casings, etc., are used. Correct evaluation of all these variations is one of the big jobs in selecting a pump for a given application.

TYPICAL STANDARD PUMP DESIGNS

Standard designs for specific services make the job of pump selection easier because many of the usual problems that arise have already been solved by the manufacturer. However, this does not relieve the pumping-system designer of the responsibility of checking a given design in terms of his application. Nor does it eliminate the need for an economic analysis when more than one unit can be used to serve a given set of conditions.

General-purpose Pumps These (Fig. 1-14) are usually built to handle clear cool liquids at ambient or moderate temperatures. Often single-stage, these units may be split-case and are standard-fitted pumps equally good for a number of services. Some are multistage, while others handle liquids containing solids.

Fig. 1-14 Single-stage general-purpose pump has horizontally split casing. (*Peerless Pump Division, FMC Corp.*)

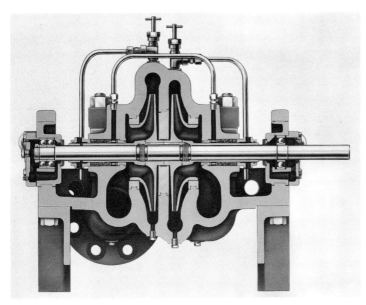

Fig. 1-15 Two-stage horizontally split pump with opposed impellers. (*Goulds Pumps Inc.*)

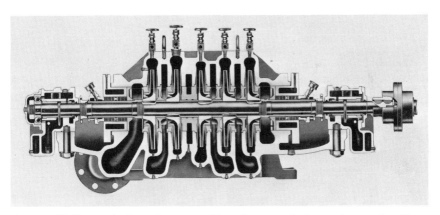

Fig. 1-16 Multistage single-suction opposed-impeller pump for heavy-duty service. (*Pennsylvania Pump and Compressor Co.*)

Multistage Pumps   Horizontal units of this design (Figs. 1-15 and 1-16) are built with both horizontally split and barrel-type casings. The barrel casing is most common in high-pressure designs with four or more stages, while the split casing is used for low to moderately high pressures with any number of stages.

Close-coupled Pumps   These combine the pump and its driving motor in a single unit, giving a compact, rugged, and efficient pump (Fig. 1-17a).   One specialized design (Fig. 1-17b) is sealless, has no stuffing boxes, and does not need a lubricant.   It is designed for handling a wide variety of chemical liquids.

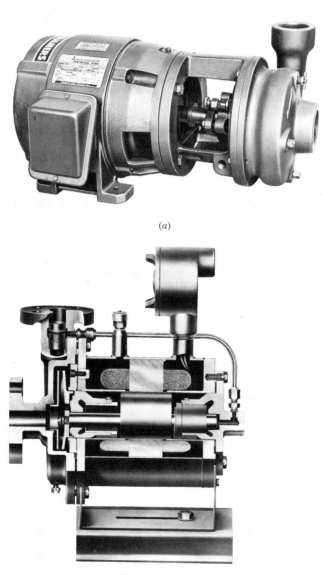

(a)

(b)

Fig. 1-17   (a) General-purpose, close-coupled pump has special motor shaft extension to mount pump impeller directly. (*Allis-Chalmers.*) (b) Sealless leakproof close-coupled chemical pump. (*Crane Co.*)

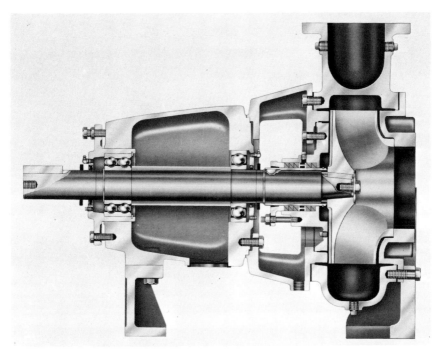

Fig. 1-18 Nonclogging pump with two-bladed impeller. (*Aurora Pump, A Unit of General Signal Corp.*)

Nonclogging Pumps Fitted with or without bladed impellers, these units handle sewage, paper-mill process liquids, slurries, and other similar solids-containing liquids.   The pump in Fig. 1-18 is a typical bladed-impeller design.

Slurry Pumps Slurry pumps are built of highly abrasion-resistant materials (Fig. 1-19).   Despite this special construction, wear still takes place and external adjustment is often provided to relocate the shaft horizontally, restoring internal clearances and original operating conditions. Also, the internal-impeller arrangement is sometimes reversed to keep the stuffing box under suction pressure at all times.   This permits injection of clean water into the stuffing box to constantly flush out solids and preserve the packing.

Fire Pumps Centrifugal fire pumps (Fig. 1-20) are now more extensively applied than any other type.   Early units were either rotary or direct-acting steam pumps.   Special fittings (Fig. 1-20) are usually supplied as part of the pump because the entire unit must meet underwriters' approval.   Standard sizes listed by underwriters and insurance laboratories are 500, 750, 1,000, 1,500, 2,000, and 2,500 gpm.   Special fire-service pumps are rated at 200, 300, and 450 gpm.   These are

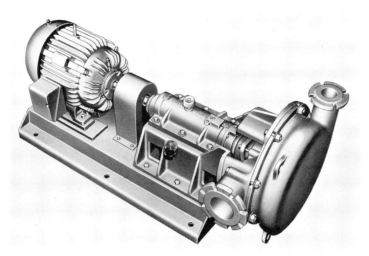

Fig. 1-19   Slurry pumps have impeller for stuffing-box operation at suction pressure permitting positive injection of clean water. (*Morris Machine Works, Inc.*)

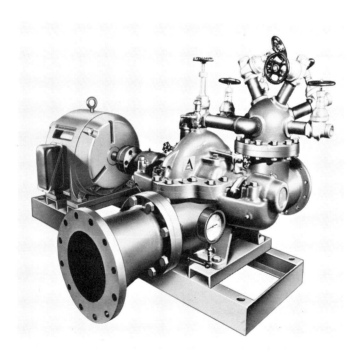

Fig. 1-20   Single-stage centrifugal fire pump, complete with motor and fittings. (*Allis-Chalmers.*)

limited to 130 per cent of their maximum rated capacity and use motors of 30 hp or less.

Vacuum Heating Service   Special designs (Fig. 1-21) for this type of service are built to give good efficiency over a wide range of heating loads.   An enclosed-type impeller on a stainless-steel shaft ensures longer life because the nonoverloading design protects the driving motor while the shaft material withstands corrosion and erosion by the liquid handled.

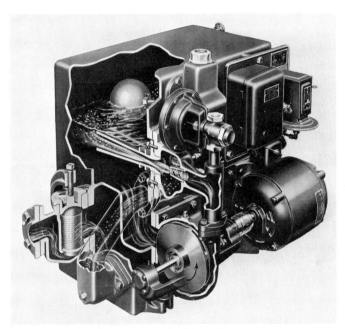

Fig. 1-21   Enclosed-type nonoverload impeller on a vacuum pump. (*Economy Pump Corp.*)

Hot-water Circulation   For heating and similar services, a close-coupled unit (Fig. 1-22a) is a popular choice.   The pump shown is fitted with sleeve bearings and a mechanical seal.

Vertical Pumps   A large number of designs of vertical pumps are available; those shown in Figs. 1-22b to 1-25 are but typical.   The sump pump in Fig. 1-22b is an example of a common design for this service, being fitted with a single-stage semiopen impeller, ball bearings, and sleeve shaft bearings.   The oil-lubricated vertical pump in Fig. 1-23 is an example of a type often used for deepwell service and a variety of other industrial pumping jobs.   This particular unit has closed impellers and sleeve-type line-shaft bearings.

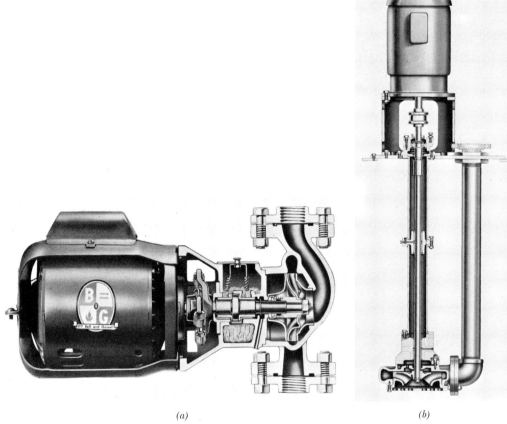

(a)   (b)

Fig. 1-22   (*a*) Single-stage pump for hot-water services. (*Bell & Gosset.*) (*b*) Vertical single-stage sump pump. (*The Deming Co.*)

Mixed-flow vertical pumps (Fig. 1-24) are generally applied for large-capacity jobs with low to moderate heads, though low-capacity units of this design are also available. Units like this are common in water-supply, irrigation, sewage, flood-control, dry-dock, condenser-circulating, and similar applications.

Vertical close-coupled pumps (Fig. 1-25) are common in process service. Called *in-line* pumps, they connect directly into a straight pipeline and are supported by it. Back-head impeller removal permits dismantling the pump without disturbing the piping connections. Thus, no baseplate or offset-piping arrangement is required.

Regenerative Turbine Pumps   These (Fig. 1-26) have fairly definite head and capacity limitations beyond which they cannot economically compete with the usual centrifugal pump. However, within their applicable range, they have definite advantages, including good suction-lift characteristics, a steep head-capacity characteristic, and good efficiency.

Other Designs   A large number of other designs of centrifugal pumps are regularly manufactured, plus a number of special designs. Many of these designs are discussed in the later chapters under the general heading of Pump Application.

Impellers   Besides being classified according to specific speed, an impeller is also typed as to how the liquid enters, its vane details, and the use for which it is intended.

Open impellers (Fig. 1-27A) have vanes attached to a central hub with relatively small shrouds. Semiopen impellers B have a shroud, or wall, on only one side. Closed impellers C and D have shrouds on both sides to enclose the liquid passages. Single- or end-suction units C have the liquid inlet on one side; in the double-suction type D, liquid enters both

Fig. 1-23 Oil-lubricated vertical pump having closed impellers. (*Layne & Bowler Corp.*)

Fig. 1-24 Two-stage open-impeller mixed-flow vertical pump. (*Johnston Pump Co.*)

Fig. 1-25 Vertical *in-line* process pump is supported by the pipeline itself. (*Allis-Chalmers.*)

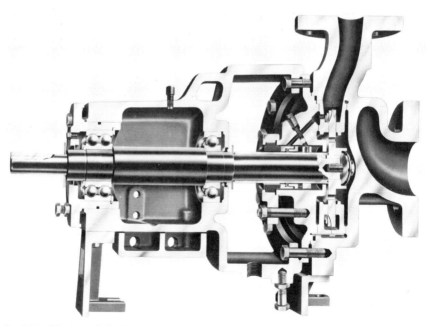

Fig. 1-26 Horizontal single-stage regenerative-turbine pump. (*Aurora Pump, A Unit of General Signal Corp.*)

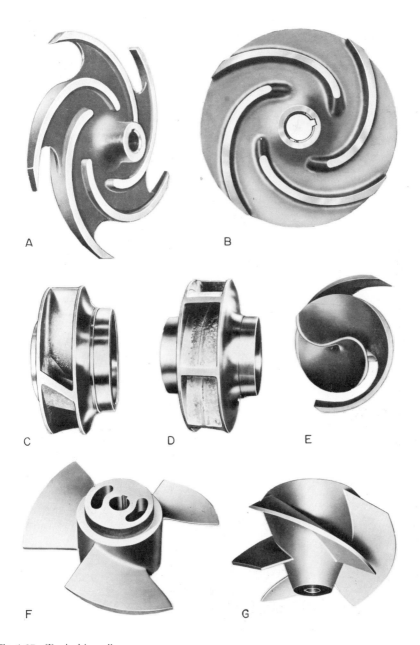

A

B

C

D

E

F

G

Fig. 1-27   Typical impellers.

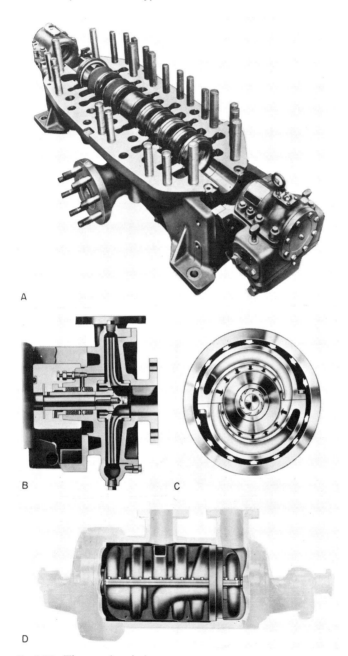

A

B

C

D

Fig. 1-28 Three casing designs.

sides. Shown at E, F, and G are paper-stock, propeller, and mixed-flow designs.

Casings Centrifugal-pump casings may be split horizontally (Fig. 1-28A), vertically (Fig. 1-28B), or diagonally (at an angle other than 90 deg). Horizontally split casings are also termed axially split. Both suction and discharge nozzles are normally in the lower half of the casing; the upper half lifts for easy inspection. Vertically split casings are also called radially split. They are used in close-coupled or frame-mounted end-suction designs. Barrel casings C and D are used on high-pressure diffuser and volute pumps. The inner casing fits in the outer barrel. Discharge pressure acting on the inner case provides the seal force required to hold the casing halves together.

Wearing Rings To prevent costly wear of the casing and impeller at the running joint, wearing rings, also called casing or sealing rings, are installed. Where these rings are removable, as they usually are, they can be replaced at a fraction of the cost of a new impeller or pump casing that might otherwise be needed. In Fig. 1-29, seal a is a plain flat joint. The similar joint b has a flat ring mounted on the pump casing. At c the ring fits into a casing groove; impeller may have a similar ring. In designs d, e, and f, rings are fitted to both the casing and the impeller. The form varies with the pump discharge pressure, service, etc.

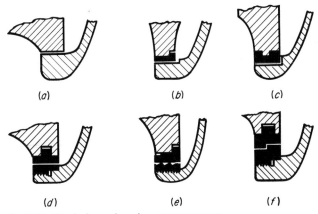

Fig. 1-29 Typical wearing-ring arrangements.

Bearings Practically every type of bearing has been used in centrifugal pumps. Today, antifriction, sleeve, and Kingsbury bearings find most common use. Many pumps are available with more than one type of bearing to meet different needs. Antifriction bearings (Fig. 1-30a) may be of the single- or double-row type. Spherical roller bearings are widely used for large pump shafts. Sleeve bearings *b* and *c* may be either horizontal or vertical. In the latter, water is often the lubricant. Kingsbury thrust bearings *d* find use in larger pumps. The design resembles that used in other rotating machinery.

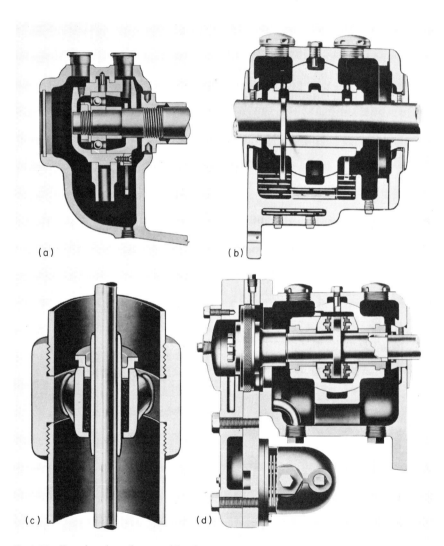

(a)

(b)

(c)

(d)

Fig. 1-30   Four bearings for centrifugal pumps.

Shaft Sleeves   Sleeves (Fig. 1-31) protect the shaft against corrosion, erosion, and wear. Many forms are used on large pumps but on small ones the sleeve is often left off to reduce the hydraulic and stuffing-box losses. The shaft is then made of a metal that is sufficiently corrosion- and wear-resistant for satisfactory life. Interstage sleeves guard multi- stage pump shafts. In some, a long hub on the impeller replaces the interstage sleeve.

Fig. 1-31 Shaft-sleeve assembly.

Lantern Rings Sometimes called seal cages, these (Fig. 1-32) are used to prevent air leakage into the pump when running with a suction lift and to distribute the sealing liquid uniformly around the annular space between the box bore and the shaft-sleeve surface. They receive liquid under pressure from the pump or an independent source. Grease sometimes serves as the sealing medium when clear liquid is not available or cannot be used (sewage pumps).

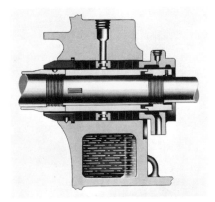

Fig. 1-32 Two lantern-ring arrangements.

Stuffing Boxes The stuffing box prevents air from leaking into the casing when the pressure in it is below atmospheric; it holds leakage out of the casing to a minimum when the pressure is above atmospheric. Figure 1-33 shows a solid-packed box, which has no lantern rings, and two injection designs, which do have lantern rings. On pumps handling hot liquids, or having high stuffing-box pressures, the box is often water- jacket-cooled. In some, the coolant and pumped liquid mix.

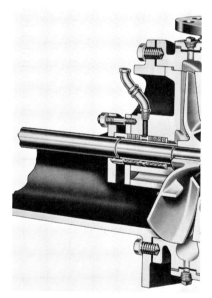
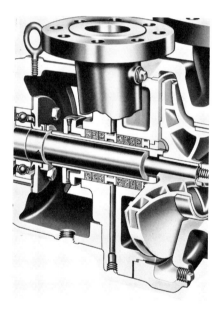

Fig. 1-33 Stuffing-box designs.

For many high-suction-pressure and high-temperature services, especially where these pressures and temperatures vary widely and proper stuffing-box adjustment is difficult, so-called *packless boxes* (Fig. 1-34) solve troublesome operating problems. In the box, a close-running, serrated breakdown bushing controls outward leakage, which is collected and piped away. On hot services, cold water at higher than suction pressure is injected part way through the breakdown bushing. This cold water dilutes the leakage, reduces its temperature, and

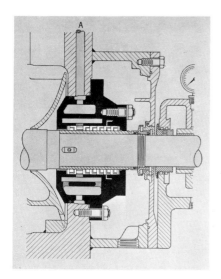

Fig. 1-34 Packless stuffing box for high-pressure high-temperature service has cold-water injection to prevent flashing.

prevents flashing which could set up shaft vibration within the close-running clearances.

Shaft Seals   Mechanical seals in wide variety serve where leakage around the shaft is objectionable.   They also find use where stuffing boxes cannot provide adequate leak protection.   The sealing surfaces are perpendicular to the pump shaft and usually comprise two polished lubricated parts running on each other.   Though not guaranteed leak-proof, leakage is usually nil.   The outside type (Fig. 1-35a) is used where

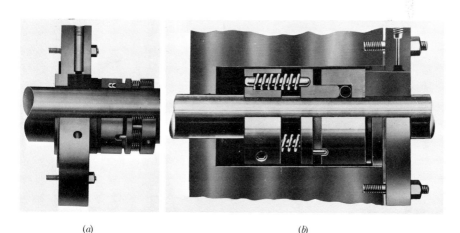

(a)   (b)

Fig. 1-35   Outside and inside mechanical seals.

gritty liquids or leakage retained in the stuffing box would be undesirable. The inside type (*b*) finds much use for volatile liquids. For high-pressure service where the pressure differential across a single seal face is excessive, two mechanical seals back-to-back, called a *double seal*, each take half of the pressure differential (Fig. 1-36).

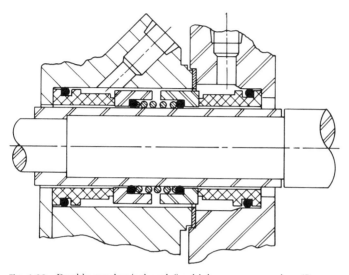

Fig. 1-36   Double mechanical seal for high-pressure service. (*Durame-tallic Corp.*)

CHAPTER TWO

Rotary Pumps

Usually positive-displacement units, rotary pumps consist of a fixed casing containing gears, vanes, pistons, cams, segments, screws, etc., operating with minimum clearance. Instead of "throwing" the liquid as in a centrifugal, a rotary pump traps it, pushing it around the closed casing, much like the piston of a reciprocating pump. But unlike a piston pump, a rotary pump discharges a smooth flow. Often thought of as viscous-liquid pumps, rotaries are by no means confined to this service alone. They will handle almost any liquid that is free of hard and abrasive solids. And hard solids can be present in the liquid if steam jacketing around the casing of the pump will keep them in a fluid condition.

Rotary-pump drive arrangement for multishaft designs are of two types. The pumping element on the driven shaft can drive its mating element on the idler shaft. But where abrasives in the liquid pumped would cause excessive wear or where flexible pumping elements are involved, external timing gears drive the idler shafts. This permits close-running clearances without hard contact.

TYPES OF ROTARY PUMPS

Cam-and-piston Pumps Also called rotary-plunger pumps, the cam-and-piston type consists of an eccentric and a slotted arm at its top (Fig.

2-1).   Rotation of the shaft causes the eccentric to trap liquid in the casing.   As rotation continues, liquid is forced from the casing through the slot to the pump outlet.

External-gear Pumps   These are the simplest rotary type.   As the gear teeth separate on the suction side of the pump (Fig. 2-2), liquid fills the

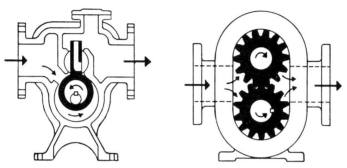

Fig. 2-1   Cam-and-piston rotary pump.

Fig.   2-2   External-gear   rotary pump.

spaces between them.   Then it is carried around and squeezed out as the teeth mesh.   The gears may have single-, double-helical, or spur teeth. Some designs have radial fluid holes in the idler gear from the tops and roots of the teeth to the internal bore.   These permit liquid to bypass from one tooth to the next, preventing build-up of excessive pressure that would overload the bearings and cause noisy operation.

Internal-gear Pumps   This type (Fig. 2-3) has one rotor with internally

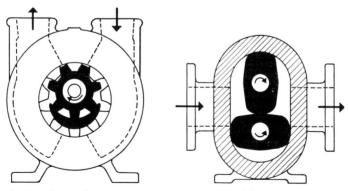

Fig. 2-3   Internal-gear   rotary pump.

Fig. 2-4   Two-lobe rotary pump.

cut teeth meshing with an externally cut gear idler.   A crescent-shaped partition (Fig. 2-3) may be used to prevent liquid from passing back to the suction side of the pump.

Lobular Pumps   These resemble the gear-type pumps in action and have two or more rotors cut with two, three, four, or more lobes on each rotor (Figs. 2-4 to 2-6).   The rotors are synchronized for positive rotation by external gears.   Because liquid is delivered in a smaller number

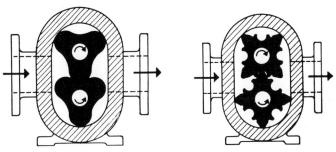

Fig. 2-5 Three-lobe rotary pump.

Fig. 2-6 Four-lobe rotary pump.

of larger quantities than in the gear pump, flow from the lobular type is not quite so constant as from the gear type.   Combination gear-and-lobe pumps are also available.

Screw Pumps   These (Figs. 2-7 to 2-9) have one, two, or three suitably threaded screws turning in a fixed casing.   A large number of designs are available for various applications.

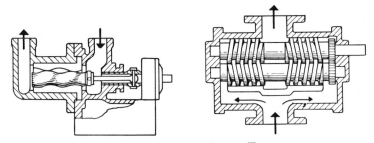

Fig. 2-7 Single-screw pump.

Fig. 2-8 Two-screw pump.

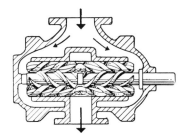

Fig. 2-9 Three-screw pump.

Single-screw pumps have a spiraled rotor turning eccentrically in an internal-helix stator or liner. The rotor is metal while the helix is hard or soft rubber, depending on the liquid handled.

Two- and three-screw pumps have one or two idlers, respectively. Flow is between the screw threads, along the axis of the screws. Opposed screws may be used to eliminate end thrust in the pump.

Vane Pumps   Swinging-vane pumps (Fig. 2-10) have a series of hinged vanes which swing out as the rotor turns, trapping liquid and forcing it

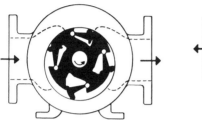

Fig. 2-10   Swinging-vane pump.   **Fig. 2-11**   Sliding-vane pump.

out the discharge pipe of the pump. Sliding-vane pumps (Fig. 2-11) use vanes that are held against the casing bore by centrifugal force when the rotor is turned. Liquid trapped between two vanes is carried around and forced out the pump discharge.

Other Designs   Shuttle-block pumps (Fig. 2-12) have a cylindrical rotor turning in a concentric casing. Inside the rotor is a shuttle block and piston reciprocated by an eccentrically located idler pin, producing suction and discharge.

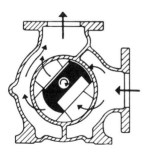
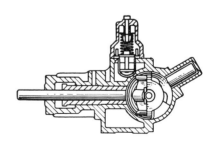

Fig. 2-12   Shuttle-block pump.   **Fig. 2-13**   Universal-joint pump.

The universal-joint pump (Fig. 2-13) has a stub shaft in the free end of the rotor supported in a bearing at about 30 deg with the horizontal. The opposite end of the rotor is fixed to the drive shaft. When the rotor revolves, four sets of flat surfaces open and close for a pumping action of four discharges per revolution.

An eccentric in a flexible chamber (Fig. 2-14) produces pumping action by squeezing the flexible member against the pump housing to

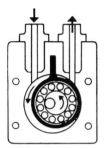

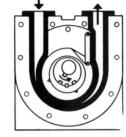

Fig. 2-14 Pump using an eccentric in a flexible chamber.

Fig. 2-15   Flexible-tube pump.

force liquid out the discharge. Flexible-tube pumps (Fig. 2-15) have a rubber tube that is squeezed by a compression ring on an adjustable eccentric. The pump shaft, attached to the eccentric, revolves it. Pumps of this design are built with one or two stages. Other designs of flexible-tube pumps are also available.

ROTARY-PUMP CHARACTERISTICS

Neglecting leakage, rotary pumps deliver almost constant capacity against variable discharge pressures. So the usual *HQ* curve is nearly a horizontal line (Fig. 2-16). Displacement of a rotary pump varies directly as the speed, except as the capacity may be affected by viscosity and other factors. Thick viscous liquids may limit pump capacity at higher speeds because the liquid cannot flow into the casing rapidly enough to fill it completely.

The slip or loss in capacity through clearances between the casing and rotating element, assuming a constant viscosity, varies as the discharge pressure increases. For example, in Fig. 2-16, at 600 rpm and 0 psi discharge pressure, capacity is 108 gpm. But at 300 psi and the same speed, capacity is 92 gpm. The difference, 16 gpm, is the slip or loss.

Power input to a rotary pump, *HQ* characteristic curve, increases with the liquid viscosity (Fig. 2-17). Efficiency decreases with a rise in viscosity. This may also be true, of course, with other classes of pumps. But since rotaries find wide use for viscous liquids, it is important to keep

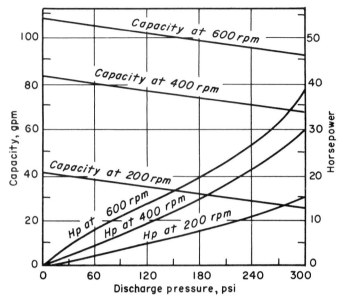

Fig. 2-16 Capacity and horsepower characteristics of an external-gear pump. (*Viking Pump Co.*)

these characteristics in mind. Figure 2-17 shows typical *HQ* and *PQ* curves for an internal-gear-type rotary pump.

Rating Tables As with centrifugal pumps, rating tables are often used to give pertinent data on pump capacity, power input, and head. Many rating tables for rotary pumps contain viscosity listings, showing the effect of increased or decreased viscosity on pump performance. Table 2-1 gives typical performance data for a three-screw rotary pump.

Classification The broad term *rotary pump* is used almost exclusively with this class. Few manufacturers classify their pumps in terms of ultimate use. Instead they list a number of possible applications for a given type. This practice is opposite that for centrifugal pumps where more stress is often given to application than to pump class, type, or construction.

Most builders of rotary pumps emphasize unit type in addition to class—for example, an internal-gear rotary pump. Such an identification is a useful guide during the initial stages of pump selection. Of course, practices in this respect vary somewhat from one manufacturer to another.

Construction Materials Rotary pumps are classed by the Hydraulic Institute as (1) all-iron, (2) bronze-fitted, and (3) all-bronze units. In an all-iron pump, every part of the unit in direct contact with the liquid handled is made of ferrous metal. In a bronze-fitted pump, the casing is

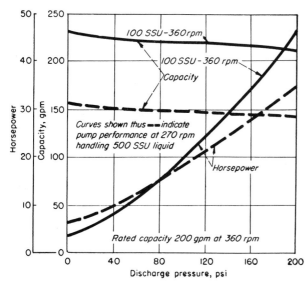

Fig. 2-17 Capacity and horsepower characteristics of an internal-gear pump. (*Viking Pump Co.*)

made of ferrous metal, and certain wearing parts such as rotors, vanes, and other moving parts are made of bronze. The shaft may be steel or a nonferrous metal. All-bronze pumps have every part of the unit in direct contact with the liquid handled made of the individual pump manufacturer's standard bronze, except the shaft, which may be stain-

TABLE 2-1   Typical Performance Data—Three-screw Rotary Pump*

Capacity at 150 SSU, 100 psig discharge pressure, gpm	Bhp at 50 psi			Bhp at 100 psi		
	200 SSU	500 SSU	Max. lift, in. Hg	200 SSU	500 SSU	Max. lift, in. Hg
3	0.2	0.3	24	0.3	0.4	24
8	0.5	0.7	22	0.8	1.0	22
17	0.9	1.1	20	1.5	1.7	20
25	1.2	1.4	20	2.1	2.3	20
40	2.0	2.5	20	3.3	3.8	20
63	3.0	3.8	20	5.1	5.8	20
95	4.5	5.6	19	7.6	8.7	19
130	6.0	7.5	18	10.1	11.6	18
191	9.0	11.1	16	14.8	16.9	16
270	12.3	15.4	13	20.8	23.9	13
405	18.3	22.9	9	31.0	35.9	9

* Condensed from De Laval Turbine Inc. data.

less steel or a nonferrous metal. Steel and ductile-iron externals are finding increased use in rotary pumps for refineries and high-temperature services. Also, 300-series stainless steel is gaining increased use in the chemical and food industries.

Pump Applications Most rotary pumps are self-priming and will, if necessary, handle entrained gas or air. Typical applications include transfer, recirculation, and metering of liquids of all viscosities, chemical processes, food handling, marine-cargo unloading, tank loading and unloading, fire protection, hydraulic power transmission, pressure lubrication, paint spray, machine-tool coolants, oil-burner service, grease handling, liquefied gases (propane, butane, ammonia, Freon, etc.), and a large number of other industrial services. Where liquids above a temperature of 180 F are to be pumped, consult the manufacturer for his recommendations.

CHAPTER THREE

Reciprocating Pumps

RECIPROCATING PUMPS are *positive-displacement* units—they discharge a definite quantity of liquid during piston or plunger movement through the stroke distance. However, not all the liquid may reach the discharge pipe because leaks or bypass arrangements may prevent this. Neglecting these, the volume of liquid displaced during one stroke of the piston or plunger equals the product of the piston area and stroke length.

TYPES OF RECIPROCATING PUMPS

Basically, there are two types of reciprocating pumps—direct-acting steam-driven units and power pumps. But there are many modifications of these basic designs, built for specific services in different fields. Some are classified as rotary pumps by their manufacturers but actually utilize the reciprocating motion of pistons or plungers to secure their pumping action. The more general term—*reciprocating*—will be used for these pumps in this book.

Direct-acting Pumps In this type, a common piston rod connects a steam piston and liquid piston (Fig. 3-1) or plunger (Fig. 3-3). Direct-acting pumps are built *simplex* (one steam and one liquid piston, respectively) and *duplex* (two steam and two liquid pistons).

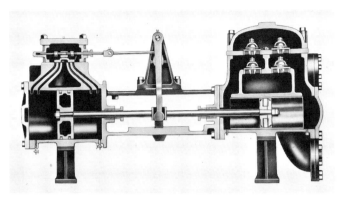

Fig. 3-1 Direct-acting horizontal duplex piston pump. Steam end is at left; liquid end is at right. (*Warren Pumps Inc.*)

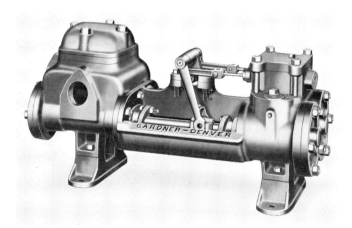

Fig. 3-2 Duplex steam slush pump. (*Gardner-Denver Co.*)

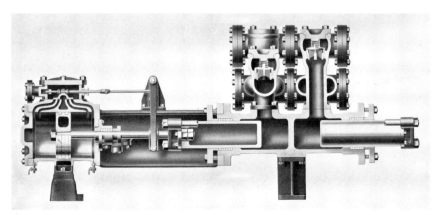

Fig. 3-3 Horizontal duplex outside-end-packed pot-valve-type plunger pump; tie rods attach the plungers. (*Worthington Corp.*)

Horizontal and vertical simplex and duplex direct-acting pumps have long been outstandingly successful in a large number of services, including low- to medium-pressure boiler feed, mud, grout, and slush handling, oil and water pumping, and many others. Characterized by easily adjustable head, speed, and capacity, they have good efficiency over wide capacity ranges (Tables 3-1). Plunger pumps (Fig. 3-3) are generally used for higher pressures than piston types (Figs. 3-1 and 3-2). As with all reciprocating pumps, direct-acting units have a pulsating discharge flow.

TABLE 3-1   Approximate Efficiencies, %, of Direct-acting Pumps

Stroke, in.	5	8	10	20	30	40	50
Crank-and-flywheel pump...				87	88	90	92
Piston pump	60	70	74	84	86	88	90
High-pressure pump	55	64	67	76	78	80	81

Power Pumps   These (Fig. 3-4) have a crankshaft driven from an outside source—commonly an electric motor, a belt, or a chain. Gears are often used between the drive and the crankshaft to reduce the output speed of the driver.

When driven at constant speed, power pumps deliver nearly constant capacity over a wide range of heads, and have good efficiency (Table 3-2). The liquid end, which may be either the piston or plunger type, will develop a high pressure when the discharge valve is closed. For this reason, it is common practice to fit a discharge relief valve to protect the pump and its piping. Direct-acting pumps will stall when the total force on the water piston equals that on the steam piston; power pumps develop a high pressure before they stall. The stalling pressure is several times the normal discharge pressure of power pumps.

TABLE 3-2   Approximate Efficiencies of Power Pumps

Water hp	3	5	10	20	30	50	75	100	200
Efficiency, %	55	65	72	77	80	83	85	86	88

Power pumps are particularly well adapted for high-pressure services and find some uses in boiler feeding, pipeline pumping, petroleum processing, and similar applications.

Crank-and-flywheel-type power pumps of early design were often

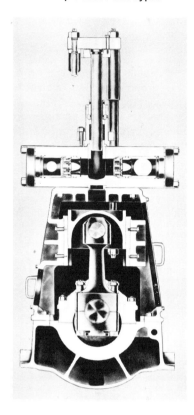

Fig. 3-4 Inverted triplex vertical plunger power pump for high-pressure applications. (*Worthington Corp.*)

steam-driven. Today, however, motor or engine drive (Fig. 3-4) is more common because this arrangement gives a more economical and compact installation requiring less maintenance. High-pressure plunger-type power pumps may be either horizontal or vertical. They are often used for hydraulic presses, petroleum processing, and similar services. But other designs find some use for the same services. Large high-pressure power pumps (Fig. 3-5) are often vertical but are also built as horizontal units.

Low-capacity Power-type Pumps These units (Figs. 3-6 through 3-12) are also known as variable-capacity, controlled-volume, and proportioning pumps. Their principal use is for controlling the flow of small quantities of liquids being fed to boilers, process equipment, and similar units. As such they occupy an important place in many industrial operations in all types of plants.

The capacity of these pumps may be varied by changing the stroke length (Fig. 3-6). The unit in Fig. 3-7 uses a diaphragm to pump the liquid being handled, but the diaphragm is actuated by a plunger which

Fig. 3-5   Large high-pressure plunger pump. (*Aldrich Division, Ingersoll-Rand Co.*)

displaces oil within the pump chamber.   Changing the length of the plunger stroke varies the diaphragm displacement.   A number of other designs of variable-displacement pumps are also available.   These will be discussed in later chapters on pump application.

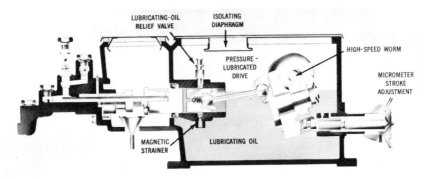

Fig. 3-6   Controlled-volume plunger pump has screw adjustment of stroke length. (*Milton Roy Co.*)

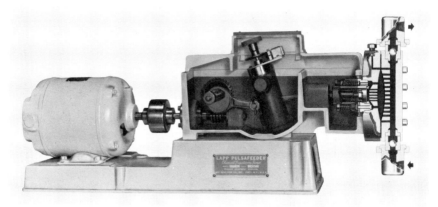

Fig. 3-7   Piston-diaphragm unit for controlled-volume pumping has oil for actuation of diaphragm which pumps the liquid. (*Lapp Insulator Co., Inc.*)

Diaphragm-type Pumps   The combined piston-diaphragm pump in Fig. 3-7 is generally used only for smaller capacities.   Diaphragm pumps (Figs. 3-8 and 3-9) are used for larger flows of clear or solids-containing liquids.   They are also suitable for thick pulps, sewage, sludge, acids or alkaline solutions, and mixtures of water and gritty solids.   A diaphragm of flexible nonmetallic material can better withstand corrosive or erosive

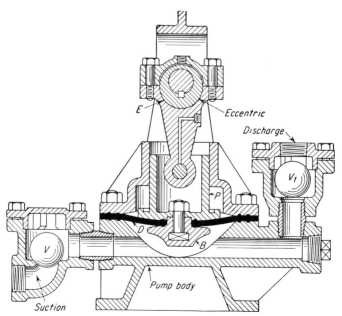

Fig. 3-8   Power-driven pressure-type diaphragm pump has its ball suction and discharge valves inside the pumping chamber.

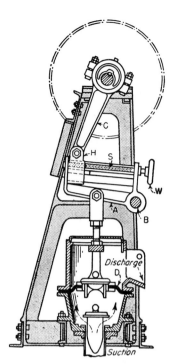

Fig. 3-9 Output of this pump may be adjusted while it is running.

Discharge

D

Suction

Fig. 3-10 Short-stroke high-speed diaphragm spray pump operates at 60 to 80 psi and handles chemicals.

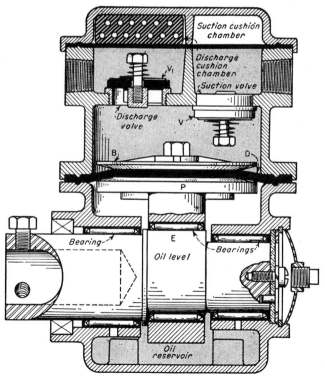

Suction cushion chamber

Discharge cushion chamber

Suction valve

Discharge valve

V

B

D

P

Bearing

E

Oil level

Bearings

Oil reservoir

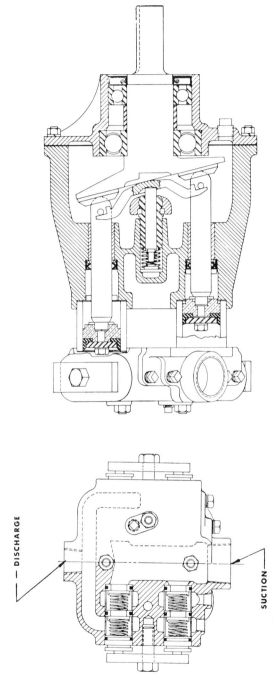

Fig. 3-11 Rotary-plunger single-acting pump unit has multiple plungers in a circle. (*John Bean Division, FMC Corp.*)

DISCHARGE

SUCTION

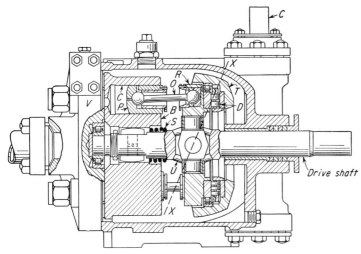

Fig. 3-12 Parallel-piston variable-displacement pump; the cylinder barrel and its pistons rotate with the drive unit.

action than metal parts of some reciprocating pumps. The short-stroke high-speed diaphragm spray pump in Fig. 3-10 is fitted with disk-type suction and discharge valves. It is designed for handling chemicals.

Other Designs   A large number of other types of reciprocating pumps, designed for specialized services, are also available. Many are used in industrial hydraulic, lubrication, chemical-feed, and similar systems. Their capacity and discharge pressure vary with the application for which the unit was designed.

The rotary-plunger single-acting unit in Fig. 3-11 has multiple plungers in a circle. Each plunger is connected to a common wobbler plate rotated by the driver. Rotation of the plate produces reciprocating motion of the plungers, giving suction and discharge action. Discharge flow from this pump is smooth. The parallel-piston pump in Fig. 3-12 has a tilting box T moved by control C to vary the liquid output of the unit. In another design (Fig. 3-13) the cylinder barrel and its pistons swing at an angle to the drive shaft to adjust the discharge flow. The horizontal-piston pump in Fig. 3-14 has a wobbler plate W which does not rotate. Pistons P give the desired pumping action.

CHARACTERISTICS OF RECIPROCATING PUMPS

The discharge flow from centrifugal and most rotary pumps is steady. But in reciprocating pumps the flow pulsates, with the character of the

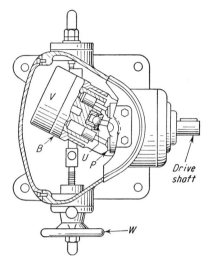

Fig. 3-13 The cylinder barrel and its pistons in this pump swing at an angle to adjust discharge rate.

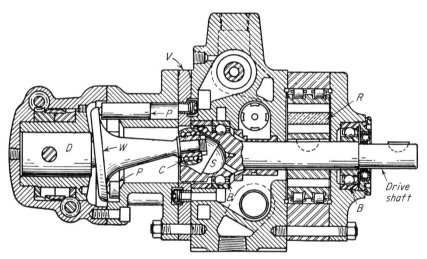

Fig. 3-14 Variable-displacement wobbler-plate pump in which the plate *W* does not rotate with the drive shaft.

pulsation depending on pump type and whether or not it has a cushion chamber.

Simplex Direct-acting Pumps Steam pumps operating at normal speed have a discharge curve like that shown in Fig. 3-15a. Flow is steady until the end of the stroke, where the liquid piston stops and reverses. Without a cushion chamber, flow theoretically ceases when the piston stops. But an air chamber prevents this, giving the loops shown. Duplex direct-acting steam pumps generally have the discharge of one cylinder displaced half a stroke from that of the other. The two add together to give the solid line in Fig. 3-15b having twice as many valleys as the simplex pump, but the low points are never below those of a simplex direct-acting pump.

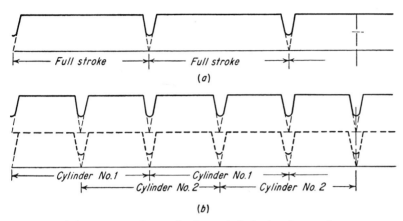

Full stroke ⟶ ⟵ Full stroke ⟶

(a)

Cylinder No.1 ⟶ ⟵ Cylinder No.1 ⟶
Cylinder No.2 ⟶ ⟵ Cylinder No.2 ⟶

(b)

Fig. 3-15   Discharge curves for (a) simplex and (b) duplex direct-acting pumps.

Power Pumps   Discharge curves for power pumps take the form of sine waves (Fig. 3-16) because the pistons or plungers are crank-driven. The discharge flow does not change so abruptly as with direct-acting pumps. The simplex double-acting power pump, for which the curve is plotted in Fig. 3-16a, has a maximum flow rate of 60 per cent above its average flow rate; the minimum flow below average is 100 per cent. This means that, at some point during each pumping cycle, flow from the pump is zero. But flow from the discharge line may be nearly constant, depending on the piping layout and the amount and kind of cushioning capacity used.

The duplex double-acting pump whose curve is plotted in Fig. 3-16b has a maximum flow rate 26.7 per cent above its average flow; the minimum is 21.6 per cent below the average flow. Thus there is always flow in the discharge pipe while the pump is operating. A triplex single-acting pump further smooths the discharge curve (Fig. 3-16c). Maximum flow rate above average for this unit is 6.64 per cent; minimum below the average flow is 18.4 per cent. With any reciprocating pump, the difference between the maximum discharge and the average discharge is stored in the cushion chamber until the discharge drops below the average.

Quintuplex and septuplex power pumps smooth the discharge curve even more, to give practically constant discharge-line flow. However,

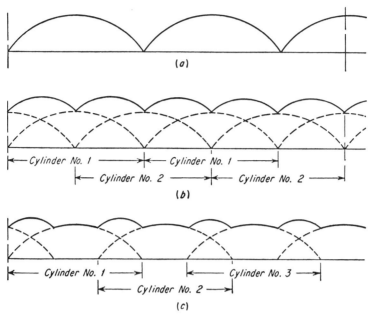

Fig. 3-16 Discharge curves for three types of power pumps. (*a*) Simplex double-acting. (*b*) Duplex double-acting. (*c*) Triplex single-acting.

the ultimate test of the suitability of a discharge curve is the job itself. Large pulsations may be of minor importance in one installation, yet in another they may be a major consideration.

Note that the flow percentages given for the units in Fig. 3-16 apply only to the curves shown. Pump design, crank angle, and a number of other factors will change the flow from one unit to the next. Values given, however, represent current practice and the variation from one manufacturer to another usually is not great.

Low-capacity Power-type Pumps   To a certain extent these units resemble their bigger brothers when it comes to the discharge curve. But the fact that most of these units are variable-capacity pumps alters the appearance of the curves somewhat. Figure 3-17 shows a series of discharge curves for typical variable-capacity units.

A simplex single-acting pump has the pulsating flow shown in Fig. 3-17*a* with no discharge during the suction stroke. When the capacity is decreased, the sine-type curve becomes flatter, as shown. Duplex single-acting designs deliver twice as much liquid and have discharge curves similar to simplex units (Fig. 3-17*b*). When one piston stops delivering liquid, the other starts, if the pistons are 180 deg apart, as shown. So there are no lengthy periods during which flow from the pump is zero.

Where constant delivery is desired without pulsation, a discharge curve like that shown in Fig. 3-17c can be obtained by using a special pump design for this service.   Note how the curve is flat throughout, whether the pump runs at full stroke, half, or less.

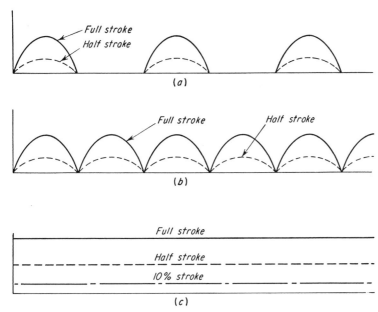

Fig. 3-17 Discharge curves for three small variable-capacity pumps. (*a*) Simplex single-acting. (*b*) Duplex single-acting. (*c*) Constant-delivery.

Once again, it is important to remember that the discharge curve of a reciprocating pump is not a measure of its efficiency, but simply a graphical representation of what is happening in the pump discharge.   The final measure of the importance of the curve shape is the job requirements for liquid flow in the line.

Capacity and Speed   Like other pumps, reciprocating pumps do not suck in liquids; they reduce pressure in the suction chamber, and external pressure, usually atmospheric, pushes the liquid into the pump.   For any pump with a given size suction line, capacity or maximum speed is fixed by the existing net positive suction head (npsh); see Chap. 4.

As the speed of a reciprocating pump increases, so does its capacity, provided nothing interferes with flow into and out of the pump.   The curves in Fig. 3-18 show the basic speeds for direct-acting and power pumps.   For the former, speed is expressed in feet per minute of piston motion; for the latter, as either piston speed or revolutions per minute.

In recent years the speeds of power pumps have increased measurably

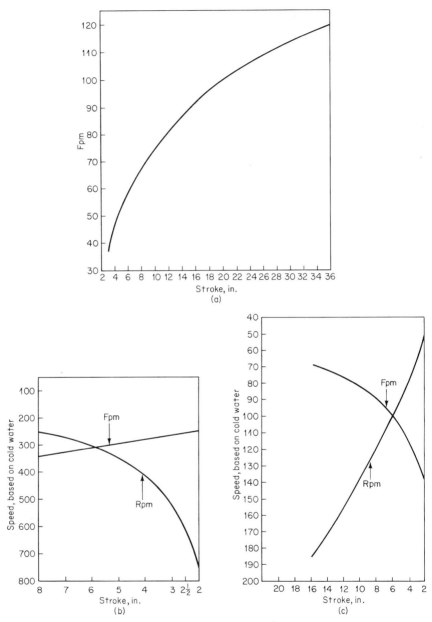

Fig. 3-18 Basic speeds for (*a*) simplex and duplex direct-acting steam pumps, (*b*) triplex and multiplex plunger pumps, and (*c*) duplex piston pumps. (*Hydraulic Institute.*)

for certain specific applications. The curves in Fig. 3-18, being for basic speeds, may not reflect the higher speeds in use today. However, high-speed units for general application throughout industry are still in the development stage. It appears that it will be some time before all the problems being met, especially those related to liquid valves, will be solved. Therefore the curves in Fig. 3-18 are valid for a large number of power pumps in use today, and for many existing designs on the market.

Liquid Viscosity and Water Temperature Both these variables affect maximum pump speed and capacity. Thus, as viscosity of the liquid changes from 250 to 5,000 SSU (Table 3-3) pump speed falls from rated to about 65 per cent of rated. As water temperature rises from 70 to 250 F, there should be a speed decrease to 62 per cent of rated. Semisolids, like acid sludge, molasses, and sirup, are handled in recipro-cating pumps designed to run without a suction valve. Either disk or ball discharge valves are used for these units.

TABLE 3-3 Speed-correction Factors

Liquid viscosity, SSU.........	250	500	1,000	2,000	3,000	4,000	5,000
Speed reduction, %..........	0	4	11	20	26	30	35
Water temp, F	70	80	100	125	150	200	250
Speed reduction, %..........	0	9	18	25	29	34	38

See Chap. 6 for an example showing the use of these factors.

When considering a given pump, check with its maker to determine the exact effects of viscosity, npsh, temperature, and design on capacity. Generalizations given above, while helpful guides, must not be used too freely. Table 3-3 summarizes speed-correction factors for viscosity and temperature recommended by the Hydraulic Institute. To use, simply multiply the rated speed by the percentage reduction, expressed as a decimal.

Liquid and Steam Ends Reciprocating-pump liquid ends are built in a large number of designs for various liquids, service conditions, and pres-sures. Figures 3-19 through 3-25 show several typical arrangements for modern pumps. The direct-acting steam pump in Fig. 3-19 has a valve-plate liquid end with removable discharge-valve decks. The cup-type packing used for the liquid piston is also shown. The valve-pot-type liquid end in Fig. 3-20 has valve chambers closed by individual covers, while the outside-packed plunger pump in Fig. 3-21 has its valves in pots and its packing glands arranged so all leakage is external, where it can be seen.

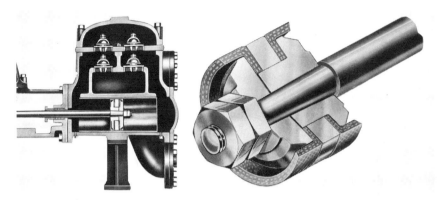

Fig. 3-19   Valve-plate-type liquid end.

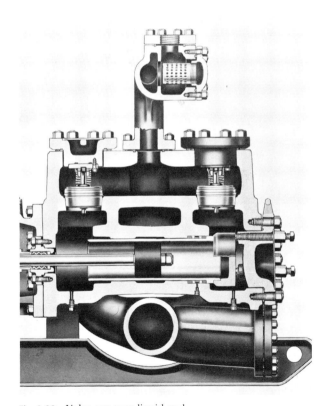

Fig. 3-20   Valve-pot-type liquid end.

Fig. 3-21 Outside-packed plunger pump.

Fig. 3-22 Packing at top of liquid cylinder.

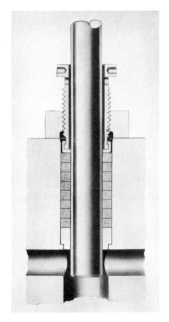

Fig. 3-23 Liquid end of a horizontal triplex pump.

The vertical power pump in Fig. 3-22 has its plunger packing at the top of the liquid cylinder, while the horizontal triplex pump in Fig. 3-23 has the packing in the usual location with the inlet and outlet valves below and above the plungers.   Ball-type suction and discharge valves in the controlled-volume pump in Fig. 3-24 are in a step arrangement. Another liquid-end design for a vertical triplex pump is shown in Fig. 3-25.   Liquid-end design in reciprocating pumps is a function of the pressure developed, liquid handled, pump capacity, etc.

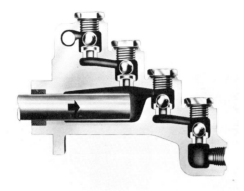

Fig. 3-24 Ball-type suction and discharge valves.

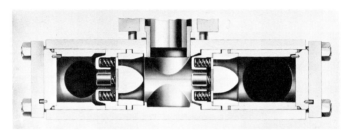

Fig. 3-25 Liquid end of a vertical triplex pump.

In the steam ends of direct-acting pumps, flat or D slide valves find use for steam pressures of about 200 psi, and under. Balanced piston valves are common in large high-pressure pumps. The flat valve in Fig. 3-26 is moved back and forth across its seat by the piston attached to its top.

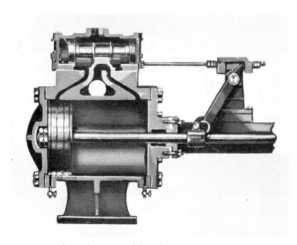

Fig. 3-26 Flat valve moved by piston.

The motion of the valve is regular and positive. In a balanced-piston valve (Fig. 3-27) the unit is arranged to run in a sleeve. It has a minimum of friction and wear. Figure 3-28 shows another balanced-piston valve design. The typical steam-valve linkage (Fig. 3-29) connects

Fig. 3-27 Balanced-piston steam valve.

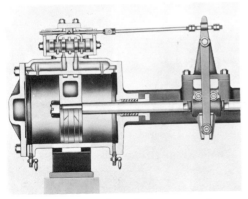

Fig. 3-28 Another balanced-piston-type valve.

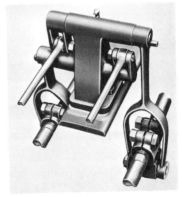

Fig. 3-29 Steam-valve linkage.

the pump piston rod to the steam-valve push rod and transmits the piston-rod movement to the steam valve. The design of steam ends for vertical pumps resembles the horizontal units shown here.

Rod and Piston Packing Any material used to control liquid leakage between a moving and stationary part in a pump is usually called packing. Flexible, and often made of a soft material, it is expendable.

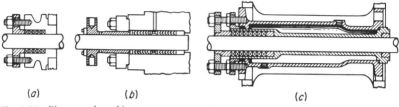

(a)   (b)   (c)

Fig. 3-30 Piston-rod packing.

The simple piston-rod stuffing box in Fig. 3-30*a* has several rings of square packing. On small piston rods, a single nut surrounds the gland, instead of the studs shown. Chevron packing (Fig. 3-30*b*) is often used instead of square packing. The plunger shown is for a high-pressure pump. Jacketed stuffing boxes (Fig. 3-30*c*) are popular for hot-oil and other high-temperature pumps where the liquid temperature exceeds 500 F. Water is circulated through the jacket to cool the portion of the rod contacting the packing. Large plunger pumps for handling high-temperature liquids are also fitted with hollow plungers through which cooling water is continuously circulated.

Packing for liquid pistons takes many forms. The duck-packed piston in Fig. 3-31*A* is for a bronze-fitted general-service pump. Cup-type packing (Fig. 3-31*B*) is standard for oil pumps. Solid-rubber packing rings (Fig. 3-31*C*) are also popular.

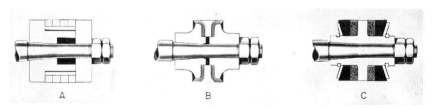

A   B   C

Fig. 3-31 Packing for liquid pistons.

Liquid-end Valves As a general rule, stem-guided disk valves are used for low-pressure pumps, wing-guided (flat- or bevel-faced) for moderate pressures, and bevel-faced wing valves for high pressures. But much depends on the liquid handled, pump design, etc.

The flat-disk valve in Fig. 3-32 has inclined ribs in its seat to direct the liquid so it rotates the disk slightly at each stroke. This equalizes wear on the disk. Ball valves (Fig. 3-33) often find use where a free opening for liquid suction and discharge is desired. The cages guide the balls

Fig. **3-32** Flat-disk valve with inclined ribs.

Fig. **3-33** Ball-type valves.

during their rise and fall. Seats are circular and completely open. The wing-guided valve in Fig. 3-34 for thick gritty liquids can be fitted with renewable rubber inserts for the wings. Another design (Fig. 3-35), for high-pressure clear liquids, has renewable seats.

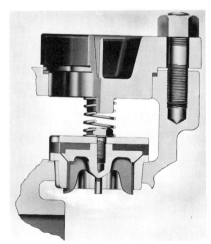

Fig. **3-34** Wing-guided valve for thick liquids.

Fig. **3-35** Wing-guided valve for high-pressure clear liquids.

The low-pressure valve in Fig. 3-36 and the high-pressure valve in Fig. 3-37 for thick liquids are alloy steel with synthetic inserts for all ordinary services. Special materials are used where corrosive liquids are to be handled. Double-ported ring-type valves (Fig. 3-25) are popular in large power pumps.

Fig. 3-36 Low-pressure valve for thick liquids.

Fig. 3-37 High-pressure valve for thick liquids.

Variable-capacity Devices   There seems no end to devices for varying the capacity of small reciprocating pumps. A number of these are described earlier in this chapter. For large power pumps, however, not so many variations exist. Perhaps this is because there are fewer variations in the designs of large pumps.

The suction-valve unloaded in Fig. 3-38a gives a quick but gradual reduction in liquid delivery from full to zero flow in not more than one-half a revolution of the pump. It increases liquid delivery in the same way, and is pneumatically actuated. Output of the stroke transformer in Fig. 3-38b is infinitely variable. It can be arranged to vary

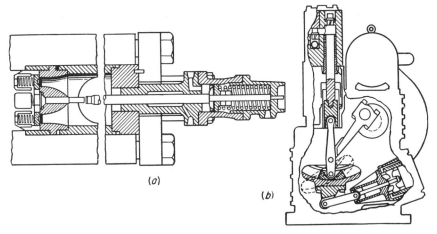

(a)

(b)

Fig. 3-38 Variable-capacity devices. (*Hydraulic Institute.*)

plunger motion manually or automatically from zero to maximum stroke.

Steam-pump Classifications From the standpoint of type, the Hydraulic Institute classifies steam pumps as single, duplex, horizontal, vertical, simple, steam end; compound tandem, crank-and-flywheel, compound cross (crank-and-flywheel only); and trade pumps, single or duplex. A number of these pumps are illustrated in this chapter. Trade pumps are the individual manufacturer's standard-fitted horizontal direct-acting valve-plate-type packed-piston steam pumps of 12-in. stroke or less.

Steam pumps may also be designated by the materials used in their construction. Bronze-fitted pumps (symbol BF) have bronze piston rods (except end-packed plunger pattern), iron liquid pistons or plungers, bronze or rubber-lined valves, bronze liquid-valve seats, guards, and springs, and iron or steel liquid cylinders. BF piston-pattern pumps include bronze-lined liquid cylinders. This type of pump in the plunger pattern has bronze-bushed plunger glands and throats. Fully bronze-fitted pumps (symbol FBF) have bronze piston rods (except end-packed plunger pattern), bronze liquid pistons or plungers, bronze or rubber lined valves, bronze liquid-valve seats, guards, and springs, and iron or steel liquid cylinders. FBF piston-pattern pumps include bronze-lined cylinders. FBF plunger-pattern pumps include bronze-bushed plunger glands and throats.

Acid-resisting pumps (symbol AR) have all parts in direct contact with the liquid handled made of corrosion-resisting materials of suitable properties for the specific application. In all-bronze pumps (symbol AB) all parts of the unit coming in direct contact with the liquid handled are made of bronze. All-iron pumps (symbol AI) have all parts coming in direct contact with the liquid handled made of ferrous metal.

Power-pump Classification The Hydraulic Institute classifies power pumps as single, duplex, triplex, multiplex, horizontal, and vertical units. Four materials designations are used by the Institute for power pumps. Bronze-fitted pumps (symbol BF) consist of bronze piston rods (except end-packed plunger pattern), iron liquid pistons or plungers, bronze or rubber liquid valves, bronze liquid-valve seats, guards, and springs, iron or steel liquid cylinders. BF piston-pattern pumps include bronze-lined liquid cylinders. BF plunger-pattern pumps include bronze-bushed plunger glands and throats. Fully bronze-fitted pumps (symbol FBF) consist of bronze piston rods (except end-packed plunger patterns), bronze liquid pistons or plungers, bronze or rubber liquid valves, bronze liquid-valve seats, guards, and springs, iron or steel liquid cylinders. FBF piston-pattern pumps include bronze-lined liquid cylinders while FBF plunger-pattern pumps include bronze-bushed plunger glands and throats.

Acid-resisting pumps (symbol AR) have all parts in direct contact with the liquid handled made of corrosion-resisting materials of suitable properties for the specific application. All-bronze pumps (symbol AB) have all parts coming in direct contact with the liquid handled made of bronze. All-iron pumps (symbol AI) have all parts coming in direct contact with the liquid handled made of ferrous metal.

PART TWO

Pump Selection

CHAPTER FOUR

Head on a Pump

DURING PUMPING-SYSTEM DESIGN there are a number of elements which must be considered no matter what class or type of pump is ultimately chosen for the installation. These elements include head, capacity, nature of liquid handled, piping, drives, and economics. So, in general, a discussion of any of these factors applies equally well to centrifugal, rotary, or reciprocating pumps. Thus the head on a pump usually will not be altered by the class of unit chosen. The few exceptions which occur are generally limited to a particular type of pump and will be pointed out.

Occasionally neglected during system planning is the important concept of design economies which originate with the project and continue through its useful life. For example, careful study of head conditions and pump location may produce worthwhile power savings over a long period without a major increase in the first cost of the project. Wise choice of pipe sizes, based on predictable or estimated future loads, is another example of how careful design planning can be made to pay off in terms of operating economies. So while this chapter discusses the head on a pump from the standpoint of practical hydraulics, it also considers how the chosen arrangement affects the entire installation.

TERMS AND DEFINITIONS

Precise terms are a necessary part of pump selection because without them the engineer soon becomes lost in a field of confusion. Where pumps are concerned, the current Standards of the Hydraulic Institute are the best guide to terminology. The definitions used in this book are based on today's practice among pump manufacturers and, in the main, follow Institute recommendations.

Pressure   Three pressure terms commonly arise in pumping problems—*absolute, barometric,* and *gage* pressure. A fourth term, *vacuum,* is used with installations operating below atmospheric pressure, but it is not a pressure term in the same sense as the first three.

Absolute pressure (Fig. 4-1) is the pressure above absolute zero. It may be above or below the atmospheric pressure existing at the point under consideration. Barometric pressure is the atmospheric pressure at the locality being studied and varies with altitude and climatic conditions. Gage pressure is the pressure above atmospheric at the locality where it is measured. A vacuum is a negative gage pressure (Fig. 4-1).

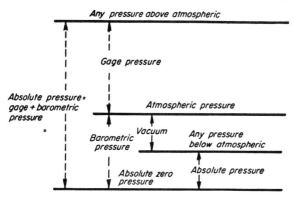

Fig. 4-1   Relation between various pressure terms used in pumping.

While in many pumping problems it is possible to work in terms of gage pressures, there are some instances where use of absolute pressures gives a better concept of the existing conditions and simplifies the required calculations. The decision as to which units are to be used is generally a matter of personal choice and depends on the designer's past experience and his preferences.

Head   A column of water or other liquid in a vertical pipe exerts a certain pressure (force per unit area) on the horizontal surface at the bottom of the pipe. This pressure can be expressed in pounds per square inch (psi) or as the number of feet of liquid which would exert an equal pressure on the same surface.   The height of the column of liquid producing the pressure in question is known as the *head* on the surface. Note that it is the weight of the liquid acting on the surface that produces the pressure.

Consider a vertical column of cold water (32 to 80 F) approximately 2.31 ft high.   A pressure gage connected to the bottom of the column will show a pressure of 1 psi.   But with a column of gasoline, whose specific gravity is 0.75, a height of 3.08 ft is needed to produce the same pressure, 1 psi, at its base.

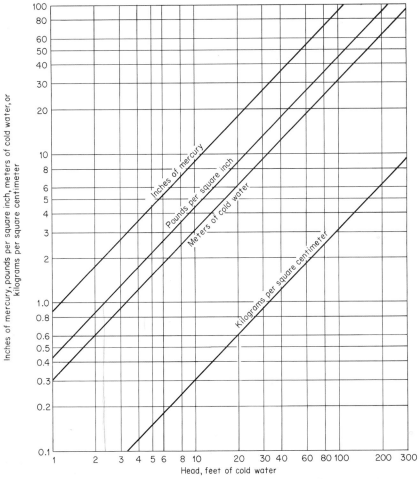

Fig. 4-2 Pressure and head conversion chart. Based on 62-F water, this chart can be used for any temperature between 32 and 80 F. For liquids other than water, divide head in feet of water by the specific gravity of the liquid to get head in feet of liquid. (*Worthington Corp.*)

Thus head and pressure are interchangeable terms, provided that they are expressed in their correct units. To convert from one to the other, use

$$\text{Liquid head, ft} = \frac{2.31 \ (\text{pressure, psi})}{\text{liquid specific gravity}} \qquad (4\text{-}1)$$

Figure 4-2 can be used for this conversion, if desired. It shows a number of other pressure units sometimes used in pump problems.

Static Head   In pump application the height of a column of liquid acting on the pump suction or discharge is often termed the *static head* on the inlet or outlet and is expressed as a certain number of feet of the liquid. Static head is a difference in elevation and can be computed for a variety of conditions surrounding a pump installation.

Vapor Pressure   Every liquid at any temperature above its freezing point exerts a pressure due to formation of vapor at its free surface. This pressure, known as the *vapor pressure* of the liquid, is a function of the temperature of the liquid—the higher the temperature, the higher the vapor pressure. Vapor pressure is an important factor in the suction conditions of pumps handling liquids of all types. In any pumping system the pressure at any point should never be reduced below the vapor pressure corresponding to the temperature of the liquid because the liquid will form vapor which will partially or completely stop liquid flow into the pump.

Static Suction Lift   This (Fig. 4-3a) is the vertical distance, in feet, from the liquid supply level to the pump center line, the pump being above the supply level. Horizontal runs are not considered as part of the static suction lift, so far as lift is concerned.

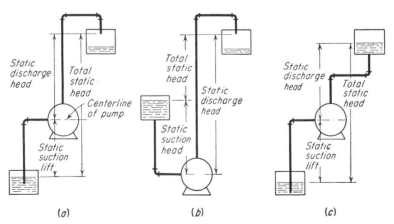

(a)   (b)   (c)

Fig. 4-3   Head terms used in pumping. (*Viking Pump Co.*)

Static Suction Head   Where the pump is below the liquid supply level (Fig. 4-3b) a static suction head exists. Numerically, it is the vertical distance, in feet, between the liquid supply level and the pump center line.

Static Discharge Head   This is the vertical distance, in feet, from the pump center line to the point of *free delivery* of the liquid (Fig. 4-3). Care must be exercised to see that the point of free delivery is used when computing static discharge head. In some layouts the exact point may be difficult to determine. This is discussed in greater detail later in this chapter.

Total Static Head   As shown in Fig. 4-3, total static head on a pump is the vertical distance, in feet, between the supply level and the discharge level of the liquid being handled.

Friction Head   Measured in feet of the liquid handled, this is the equivalent head needed to overcome the resistance of the pipe, valves, and fittings in the pumping system. Friction head exists on both the suction and discharge sides of a pump, and varies with liquid flow rate, pipe size, interior condition of the pipe, type of pipe, and nature of the liquid being handled. Table 4-1 lists the friction loss for typical common sizes of schedule 40 wrought-iron and steel pipes. For an extremely comprehensive set of friction-loss tables, see Standards of the Hydraulic Institute.

The resistance of pipe fittings is usually expressed in terms of the equivalent length of straight pipe of the same size as the fitting. Table 4-2 gives the equivalent pipe length for a number of common fittings used in pumping systems.

Velocity Head   Liquid moving through a pipe at any velocity possesses kinetic energy due to its movement. Velocity head is the distance through which the liquid must fall to acquire a given velocity, and is found from $h_v = v^2/2g$, where h_v = velocity head, ft of liquid; v = liquid velocity, ft per sec; g = acceleration due to gravity = 32.2 ft per sec. Note that the velocity of the liquid at the point being considered must be substituted in this relation for velocity head.

Depending on the nature of the pumping installation, the velocity head may or may not be an important factor in the total head on the pump. This consideration is discussed later in the chapter.

Entrance, Exit Losses   As with liquid flowing through a pipe, there is a frictional loss when a liquid enters a pipe from a free or submerged source of supply, or discharges to a similar region. The loss occurring at the pipe inlet is termed the *entrance loss,* while that at the outlet is the *exit loss.* In both, the loss reduces the velocity head at the point being considered. To alleviate entrance losses, a bellmouthed suction pipe is often used. A long taper ahead of the pipe outlet can be used to reduce exit loss.

TABLE 4-1   Pipe Friction Loss for Water

(Wrought-iron or steel schedule 40 pipe in good condition)

Diam, in.	Flow, gpm	Velocity, ft per sec	Velocity head, ft of water	Friction loss, ft of water per 100 ft pipe
2	50	4.78	0.355	4.67
2	100	9.56	1.42	17.4
2	150	14.3	3.20	38.0
2	200	19.1	5.68	66.3
2	300	28.7	12.8	146
4	200	5.04	0.395	2.27
4	300	7.56	0.888	4.89
4	500	12.6	2.47	13.0
4	1,000	25.2	9.87	50.2
4	2,000	50.4	39.5	196
6	200	2.22	0.0767	0.299
6	500	5.55	0.479	1.66
6	1,000	11.1	1.92	6.17
6	2,000	22.2	7.67	23.8
6	4,000	44.4	30.7	93.1
8	500	3.21	0.160	0.424
8	1,000	6.41	0.639	1.56
8	2,000	12.8	2.56	5.86
8	4,000	25.7	10.2	22.6
8	8,000	51.3	40.9	88.6
10	1,000	3.93	0.240	0.497
10	3,000	11.8	2.16	4.00
10	5,000	19.6	5.99	10.8
10	7,500	29.5	13.5	24.0
10	10,000	39.3	24.0	42.2
12	2,000	5.73	0.511	0.776
12	5,000	14.3	3.19	4.47
12	10,000	28.7	12.8	17.4
12	15,000	43.0	28.7	38.4
12	20,000	57.3	51.1	68.1

A similar loss occurs when liquid flowing through a pipe passes a sudden enlargement of the pipe cross section or meets a sudden reduction of it. Losses at these points and at the entrance and exit of a pipe can be expressed as the product of a coefficient (whose value depends on

TABLE 4-2   Resistance of Fittings and Valves

(Length of straight pipe, ft, giving equivalent resistance)

Pipe size, in.	Standard ell	Medium radius ell	Long- radius ell	45-deg ell	Tee	Gate valve, open	Globe valve, open	Swing check, open
1	2.7	2.3	1.7	1.3	5.8	0.6	27	6.7
2	5.5	4.6	3.5	2.5	11.0	1.2	57	13
3	8.1	6.8	5.1	3.8	17.0	1.7	85	20
4	11.0	9.1	7.0	5.0	22	2.3	110	27
5	14.0	12.0	8.9	6.1	27	2.9	140	33
6	16.0	14.0	11.0	7.7	33	3.5	160	40
8	21	18.0	14.0	10.0	43	4.5	220	53
10	26	22	17.0	13.0	56	5.7	290	67
12	32	26	20.0	15.0	66	6.7	340	80
14	36	31	23	17.0	76	8.0	390	93
16	42	35	27	19.0	87	9.0	430	107
18	46	40	30	21	100	10.2	500	120
20	52	43	34	23	110	12.0	560	134
24	63	53	40	28	140	14.0	680	160
36	94	79	60	43	200	20.0	1,000	240

the fitting and its arrangement) and the velocity head at the fitting. Table 4-3 lists the coefficients for several typical fittings used in pumping systems.

Note that the coefficient from Table 4-3 and the corresponding velocity in the pipe may be used to determine the head loss caused by a fitting, instead of using the equivalent length from Table 4-2.

Total Suction Lift   Numerically this is the sum of the static suction lift, the suction friction head, the entrance loss in the suction pipe, and the suction velocity head. Note that the suction friction head includes the friction in the pipe and all fittings in the suction line.

Total Suction Head   Though a suction lift is a *negative* suction head, the usual practice is to use the term *lift* for a negative suction head when the pump takes its suction from an open tank having the liquid surface exposed to atmospheric pressure. Suction head is the static suction head minus the suction friction head and suction piping entrance loss, plus any pressure in the suction line. Note here that a vacuum in the

TABLE 4-3   Resistance Coefficients for Pipe Fittings*

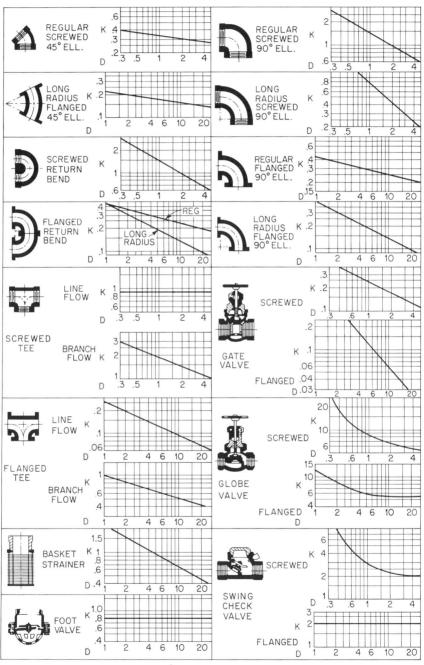

*Hydraulic Institute. $h = k \dfrac{v^2}{2g}$ feet of fluid.

suction line, as in a condenser hotwell, is a negative pressure and is *algebraically* added to the static suction head of the system.

Total Discharge Head   This is the sum of the static discharge head, the discharge friction head, and the discharge velocity head (Fig. 4-4*b*).

Total Head   This is the sum of the suction lift and the discharge head. Where there is a suction head, the total head on the pump is the difference between the discharge and suction heads.

PUMPING-SYSTEM CURVES

Graphical plots of the conditions in a proposed or existing pumping system can be major aids in system analysis. While much has been published on the use of curves in conjunction with centrifugal pumps, it should be remembered that graphical analysis is equally adaptable to rotary and reciprocating pumps.

System Friction Curve   Friction-head loss in a pumping system is a function of pipe size, length, number and type of fittings, liquid flow rate, and nature of the liquid. Figure 4-4*a* shows the piping and fitting

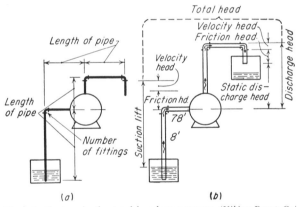

Fig. 4-4   Factors in the total head on a pump. (*Viking Pump Co.*)

elements involved in a typical pumping system. For a turbulent system, the friction-head loss varies roughly as the square of the liquid flow in the system. For laminar systems, the friction-head loss varies directly with liquid flow.

A plot of head vs. capacity (Fig. 4-5) is known as the *system friction curve.* Such a curve always passes through the point 0.0 because, when there is no head developed by the pump, there is no flow through it to the piping system.

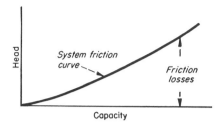

Fig. **4-5** Typical turbulent sys-
tem-friction curve.

System Head Curve   This (Fig. 4.6) is a curve obtained by combining
the system friction curve with a plot of the static head and any pressure
differences in the system.   Superimposing the pump *HQ* curve on the
system-head curve gives the point at which a particular pump will
operate in the system for which the curve is plotted.   Thus, in Fig. 4-6,
point *A* denotes one head-capacity condition for the pump whose *HQ*
curve is plotted.

Changing the resistance of a given piping system by partially closing a
valve or making some other alteration changes the slope of the system-

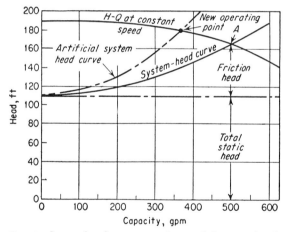

Fig. 4-6   System-head curve superimposed on pump head-capacity
curve.

head curve.   Thus, in Fig. 4-6, partially closing a valve in the discharge
line produces the artificial system-head curve shown, shifting the operating point to a higher head but lower capacity.   Opening a valve wider
has the opposite effect.

Curve Use   To select a pump for a given application properly, at least
one point on the system-head curve must be used.   For some applications, two or more points must be used if the most economical arrangement is to be obtained.

Except for the first two examples to follow, only losses resulting from pipe friction are taken into account. In actual practice, each application should be checked to determine the magnitude of the various hydraulic losses. Once the magnitude is known, a decision can be made as to which losses, if any, can be neglected in the computations of the head in the system.

No Lift   When there is no static lift in a pumping system (Fig. 4-7), the system-head curve starts at zero flow and zero head. For a flow of 900 gpm in this installation, friction loss is computed as follows:

Entrance loss: tank into 10-in. pipe = $0.5V^2/2g$	0.1 ft
Friction loss in 2 ft of suction pipe...	0.02
Loss in 10-in. 90-deg elbow at pump (equivalent to 25 ft of 10-in. pipe)	0.2
Friction loss in 3,000 ft of 8-in. discharge pipe	74.5
Loss in 8-in. gate valve fully open (equivalent to 5 ft of 8-in. pipe)	0.12
Exit loss from 8-in. pipe into tank = $V^2/2g$..............................	0.52
Total friction loss...	75.46 ft

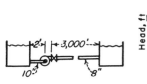
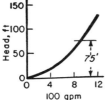

Fig. 4-7   No lift; all friction head. (*Peerless Pump Division, FMC Corp.*)

Friction losses at other flow rates can be computed as shown and plotted to secure the system-head curve shown in Fig. 4-7. If all losses in this system were ignored except friction in the discharge pipe, the total head would not change appreciably. However, this is not true of every pumping system, as will be seen below.

Mostly Static Head, Little Friction   The system-head curve (Fig. 4-8) for the vertical-pump installation shown starts at the total static head, 15 ft, and zero flow. Since the discharge pipe is only 20 ft long and there is no

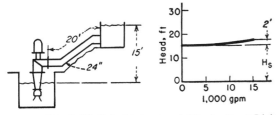

Fig. 4-8   Mostly lift; little friction head. (*Peerless Pump Division, FMC Corp.*)

suction pipe, we can assume the friction loss is small compared with the total static head.

For a flow of 15,000 gpm, losses are:

Friction in 20 ft of 24-in. pipe .... 0.40 ft
Exit loss from 24-in. pipe into tank $= V^2/2g$ .... 1.6 ft
Total friction loss .... 2.0 ft

Hence almost 90 per cent of the total head of 17 ft at 15,000-gpm flow is static. But neglect of the friction and exit losses could result in appreciable error during selection of a pump for the job.

Typical System   While pumping systems vary from one application to the next, a common hookup is shown in Fig. 4-9, where both friction and static head are appreciable. The system-head curve resembles that in Fig. 4-7, with the addition of static head.

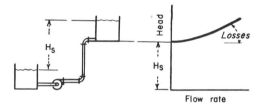

Fig. 4-9   Significant friction loss and lift. (*Peerless Pump Division, FMC Corp.*)

Gravity Head   With the installation in Fig. 4-10, flows up to 7,200 gpm are obtained by gravity head alone. To obtain larger flow rates, a pump is needed to overcome the friction in the piping between the tanks. By

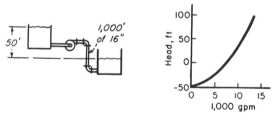

Fig. 4-10   Negative lift (gravity head). (*Peerless Pump Division, FMC Corp.*)

computing the friction loss for three flow rates we can obtain enough points to plot the system-head curve:

At 5,000 gpm, friction loss in 1,000 ft of 16-in. pipe .... 25 ft
At 7,200 gpm, friction loss = available gravity head .... 50 ft
At 13,000 gpm, friction loss .... 150 ft

Different Pipe Sizes   Here (Fig. 4-11) friction loss vs. flow rate is plotted independently for the two pipe sizes. At any given flow rate the total friction loss for the system is the sum of the loss for the two lines. Thus

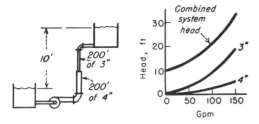

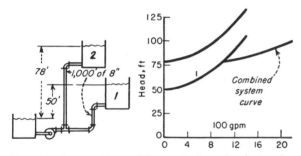

Fig. 4-11 System with two different pipe sizes. (*Peerless Pump Division, FMC Corp.*)

the combined system-head curve represents the sum of static head and the friction losses for all portions of the line.

At 150 gpm, friction loss in 200 ft of 4-in. pipe............. 5 ft
At 150 gpm, friction loss in 200 ft of 3-in. pipe............. 19
Total static head for the 3- and 4-in. pipes 10
                                                          _____
Total head at 150-gpm flow.. 34 ft

Two Discharge Heads   System-head curves (Fig. 4-12) are plotted independently for the two pipes when the discharge heads are different.

Fig. 4-12   System with two different discharge heads. (*Peerless Pump Division, FMC Corp.*)

The combined system-head curve is obtained by adding the flow rates for the two pipes at the same head:

At 550 gpm, friction loss in 1,000 ft of 8-in. pipe.......... 10 ft
At 1,150 gpm, friction loss 38
At 1,150 gpm, friction + lift in line 1 = 38 + 50............ 88
At 550 gpm, friction + lift in line 2 = 10 + 78.............. 88

Therefore, the flow rate for the total or combined system at a head of 88 ft is $550 + 1,150 = 1,700$ gpm. So to produce a flow of 1,700 gpm through this system, a pump capable of producing an 88-ft head is required.

Diverted Flow   In this case (Fig. 4-13) a constant quantity is assumed to be tapped off at the intermediate point. Friction loss vs. flow rate is plotted in the normal manner for line 1. The curve for line 3 is displaced to the right at zero head by an amount equal to Q_2, since this

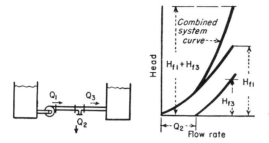

Fig. 4-13   Part of flow diverted from main. (*Peerless Pump Division, FMC Corp.*)

represents the quantity passing through lines 1 and 2 but not through line 3. The combined system-head curve is obtained by adding, at a given flow rate, the head losses for lines 1 and 3. Assume Q_2 is 300 gpm, line 1 has 500 ft of 10-in. pipe, and line 3 has 50 ft of 6-in. pipe:

At 1,500 gpm through line 1, friction loss .... 11 ft
Friction loss for line 3 (1,200-gpm flow)................ 8

Total friction loss at 1,500-gpm delivery............ 19 ft

Operating Point   As was seen earlier (Fig. 4-6) superimposing the pump *HQ* curve on the system-head curve shows at what point the pump operates. The shape or slope of both the system-head curve and the pump *HQ* curve is significant in certain applications, as shown in the following examples.

Pump Wear   When a pump wears there is certain to be a loss in capacity and efficiency. The amount of loss for a given amount of wear, however, depends to a large extent on the shape of the system-head curve. As Fig. 4-14 shows for a centrifugal pump, the capacity loss is

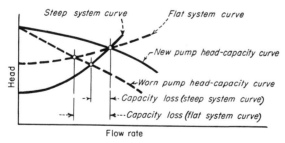

Fig. 4-14   Effect of pump wear on capacity. (*Peerless Pump Division, FMC Corp.*)

greater for a given amount of wear if the system-head curve is flat, as compared with a steep system-head curve.

Shutoff Point   A characteristic of low-specific-speed centrifugal pumps is a relatively flat *HQ* at capacities near shutoff. Occasionally it is necessary to pick such a pump because of a high-head low-flow requirement for the job. While overheating and premature shaft breakage are distinct possibilities, most pumps are physically capable of operating continuously near their shutoff point.   If this is the case, the system curve should certainly be examined before the pump is specified because another class—rotary or reciprocating—might be a better choice from the standpoint of operating reliability.

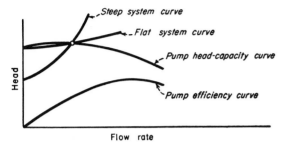

Fig. 4-15 Low-specific-speed pump near shutoff. (*Peerless Pump Division, FMC Corp.*)

Figure 4-15 shows that a shallow or flat system-head curve in conjunction with a flat pump curve can lead to performance trouble if the pump is slightly off in head-capacity or if the system-head curve is calculated slightly too low.   With a steep system-head curve, no hydraulic performance trouble would be expected.

High Specific Speed   A characteristic of some high-specific-speed centrifugal pumps is a dip, or "burble point," in the HQ curve at capacities less than that corresponding to peak efficiency.   Occasionally the need arises to operate a pump at or near the dip.   The system-head curve is a big help in determining if pump performance under these conditions will be hydraulically satisfactory.

Suppose A (Fig. 4-16) is the desired operating point for some transient high-head condition.   With a steep system-head curve, no trouble should be met.   But a shallow system-head curve which intersects the pump curve at three points might result in unsatisfactory operation if the pump hunts, or varies, among the three capacities.

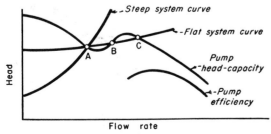

Fig. 4-16 High-specific-speed pump near the dip or "burble" point. (*Peerless Pump Division, FMC Corp.*)

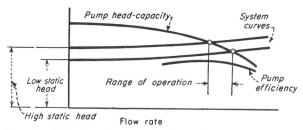

Fig. 4-17   Variations in system head. (*Peerless Pump Division, FMC Corp.*)

Head Changes   Demand of a system can vary as a result of changes in suction or discharge surface levels or greater pipe friction from increased pipe-surface roughness.   The conditions shown in Fig. 4-17, where the static head is varied, represent a favorable operating range so far as the efficiency of this particular pump is concerned.   Figure 4-17 also shows the magnitude of the drop in system flow when a high-head condition occurs.

Parallel or Series Operation   Either type of hookup of any class of pump can introduce problems.   Frequently, where there is a wide range in

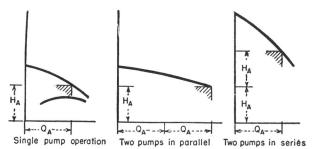

Fig. 4-18   Parallel or series operation. (*Peerless Pump Division, FMC Corp.*)

demand, two or more pumps may be operated in series or in parallel to satisfy the high demand, with just one pump used for the low demands. For proper specification of the pumps and evaluation of their perform-

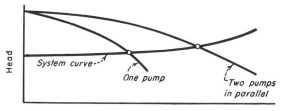

Fig. 4-19   Pump performance in parallel. (*Peerless Pump Division, FMC Corp.*)

ance under various conditions, the system-head curve should be used in conjunction with the composite pump-performance curves.

For pumps in series (Fig. 4-18) performance is obtained by adding the heads at the same capacity. When pumps are operated in parallel (Figs. 4-18 and 4-19) the performance is obtained by adding the capacities at the same head. Superimposing the system-head curve on the pump performance curves clearly indicates what flow can be expected and at what heads each pump will operate. This problem is discussed in greater detail in Chap. 7.

NET POSITIVE SUCTION HEAD

According to field engineers, more pump troubles result from incorrect determination of net positive suction head (npsh) than from any other single cause. Npsh difficulties can reduce pump capacity and efficiency, leading to cavitation damage. They can also cause severe operating problems, reducing plant effectiveness.

Vapor Pressure   Liquids at any temperature above their freezing point have a corresponding vapor pressure which must be taken into account when figuring a pumping system. Figure 4-20 shows the vapor pressure of water at various temperatures. Reducing the pressure in a pump suction pipe below the vapor pressure of the liquid can cause flashing,

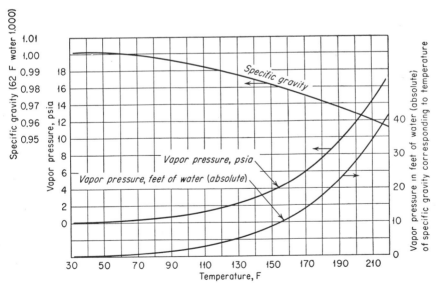

Fig. 4-20   Vapor pressure and specific gravity of water at various temperatures. (*Worthington Corp.*)

i.e., formation of vapor from the liquid.   Since no liquid pump of ordinary design can pump only vapor, liquid flow to the pump falls off and the unit is said to be steam- or vapor-bound.   The most common method used to avoid this condition is to provide enough head on the pump suction so that the pressure in the suction pipe is always greater than the vapor pressure of the liquid handled.

Available NPSH   This is a function of the system—its suction head or lift, friction head, and the vapor pressure of the liquid being handled. Depending on job conditions, the available npsh can be altered to suit that required by the pump for satisfactory operation, if changes can be made in the piping, level of the liquid supply, etc.   Thus, by altering the physical arrangement of an installation, it is possible to control one phase of available npsh.   But the vapor pressure of the liquid cannot be changed without increasing or decreasing the temperature of the liquid —and this is not always feasible.   Hence it may be a deterrent to easy alteration of the available npsh.

Required NPSH   This is a function of pump design and varies from one make of pump to the next, between different models of one manufacturer, and with the capacity and speed of any given pump.   Thus, while the available npsh is easily calculated for a given set of conditions, the required npsh for a particular pump must be obtained from the builder.

The manufacturer can plot the required npsh characteristics for a given pump on a performance curve.   Figure 4-21 shows one such curve for a turbine pump, while Fig. 4-22 shows how npsh characteristics can be plotted on a typical rating curve for a centrifugal pump.

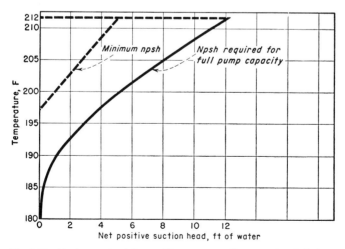

Fig. 4-21   Npsh required for turbine pumps of conventional design. (*Aurora Pump, A Unit of General Signal Corp.*)

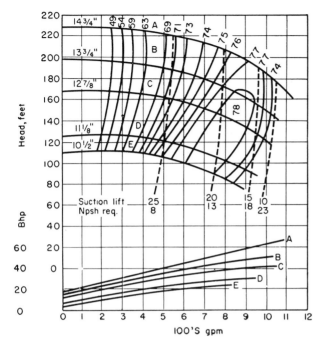

Fig. 4-22 Characteristics curves for a 5- by 4-in. 1,750-rpm centrifugal pump for cold-water service. (*Allis-Chalmers.*)

Note that in both cases, npsh is the pressure available or required to force a given flow, stated in gallons per minute, through the suction piping into the impeller eye, cylinder, or casing of a pump. For uniformity, npsh is stated as feet of the liquid handled equivalent to the pressure in pounds per square inch required to force the liquid into the pump. Values given by the pump manufacturer are based on tests and are regularly corrected to the center line of the pump.

With the liquid supply above the pump center line, and the surface of the liquid exposed to the atmosphere, npsh is the sum of the barometric pressure + the static suction head minus friction-head loss in the suction piping and the vapor pressure of the liquid, all expressed in feet of the liquid handled. When the suction supply is from a closed tank or vessel, the tank pressure is substituted for the barometric pressure (a vacuum being expressed as a negative pressure). Tank pressure must be converted to feet of the liquid being handled before it is entered in the equation for npsh.

When the liquid supply is below the pump in a tank open to the atmosphere, npsh is the difference between the barometric pressure and the sum of the static suction lift + friction-head loss in the suction piping + the vapor pressure of the liquid. All are expressed in feet of the liquid pumped. Where the liquid supply comes from a closed tank or vessel below the pump, the tank pressure is used instead of the barometric pressure. It must be converted to equivalent feet of liquid, however, before being substituted in the equation for the npsh.

Using NPSH   Assume a pump in a plant 1,500 ft above sea level requires 10 ft npsh when handling 150-F water at rated capacity. What is the allowable suction lift of this unit if entrance and suction friction losses are 5 ft of liquid?

Figure 4-23*a* shows the conditions in this problem. At sea level and normal atmospheric pressure, 14.7 psia, the maximum theoretical static suction lift possible is 33.9 ft when the pump handles cold water. (For convenience, this is usually rounded off to 34 ft.) As the altitude above sea level increases, the atmospheric pressure decreases, reducing the maximum theoretical static suction lift. But as the temperature of water is raised above 39.2 F, its point of maximum density, the density decreases. So at altitudes above sea level when warm or hot water is being handled, and at sea level too, a density correction is necessary to determine the maximum theoretical static suction lift.

From Table 4-4 the maximum theoretical suction lift possible with cold water at 1,500 ft altitude is 32.1 ft. However, 150-F water has a specific gravity of 0.981 (Fig. 4-20). So the maximum theoretical lift with 150-F water at 1,500 ft altitude is 32.1/0.981 = 32.7 ft of water.

Plot this to scale (Fig. 4-23*a*). Allowable suction lift, if one exists, is a

Table 4-4   Properties of Water at Various Altitudes

Altitude, ft	Barometric pressure, in. Hg	Atmospheric pressure		Boiling point, F
		Psia	Ft water	
−1,000	31.0	15.2	35.2	213.8
−500	30.5	15.0	34.6	212.9
0 (sea level)	29.9	14.7	33.9 *	212.0
1,000	28.9	14.2	32.8	210.2
1,500	28.3	13.9	32.1	209.3
2,000	27.8	13.7	31.5	208.4
4,000	25.8	12.7	29.2	204.7
6,000	24.0	11.8	27.2	201.0
8,000	22.2	10.9	25.2	197.4
10,000	20.6	10.1	23.4	193.7
15,000	16.9	8.3	19.2	184.0

* Commonly taken as 34.0 ft.

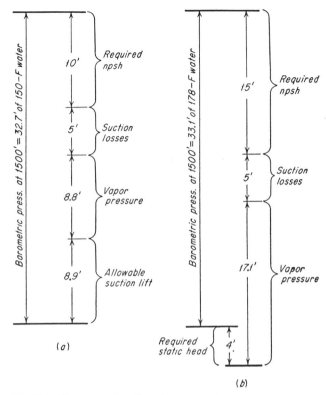

Fig. 4-23 Two examples of npsh computation.

function of the maximum theoretical lift, required npsh, suction losses, and liquid vapor pressure at the temperature in the pump inlet.

Lay off to scale the required npsh, 10 ft (Fig. 4-23*a*). Next, lay off the suction losses, 5 ft. The vapor pressure of 150-F water is 8.8 ft (Fig. 4-20). Lay this off as shown. The remaining distance, 8.9 ft, is the maximum allowable suction lift under the existing conditions. Using a greater lift would lead to partial or complete vapor binding of the pump, depending on a number of factors.

If a similar pump at the same elevation above sea level has the same suction and entrance losses but handles 178-F water and has a required npsh of 15 ft, Fig. 4-23*b* shows that a static suction head of 4 ft of water is required. In other words, this pump will not give its guaranteed performance unless a static head is maintained in its suction pipe.

These two examples illustrate the importance of careful computations when dealing with npsh. While approximations may be made for other heads in a pumping system, npsh must be computed using the data relating to the specific conditions encountered. Though water was chosen for these examples, it must be remembered that npsh is a factor no matter what liquid is handled. This is discussed later, in Chap. 6.

Capacity Reduction   As the available npsh for a given pump decreases, its capacity falls off. In Fig. 4-22, operation at any point along the 20-ft suction-lift 13-ft npsh line gives a flow ranging between somewhat more than 700 gpm and a little less than 800 gpm. But where the available npsh is only 8 ft and the suction lift 25 ft (Fig. 4-22), pump capacity falls to the 500-gpm range.

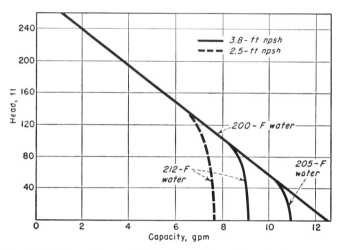

Fig. 4-24   Effect of low suction head on delivery of a typical regenerative-turbine pump. (*Aurora Pump, A Unit of General Signal Corp.*)

Figure 4-24 shows a similar effect for a regenerative turbine pump. As can be seen, reducing the available npsh on either a small or large pump can have a major effect on its capacity in the system.

Cavitation   With abnormally high suction lift or insufficient npsh, cavitation may occur in the pumping installation. Steps in this phenomenon are illustrated in Figs. 4-25 through 4-28. The exact process is usually far more complex, but these illustrations serve to convey a picture of what is believed to occur during cavitation in a centrifugal pump.

With the pump operating at excessive lift, a low suction pressure is developed at the pump inlet (Fig. 4-25). The pressure decreases until a vacuum may be created and the liquid flashes to vapor if the pressure in the pipe is lower than the liquid vapor pressure. Liquid flow into the pump ceases (Fig. 4-26). This is known as the *break-off point* because the capacity limit of the pump at this inlet pressure has been reached. The pump is approaching an operating condition which can cause damage.

When the inlet pressure has almost reached the flash point, vapor

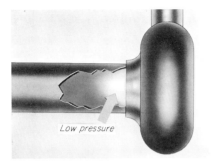

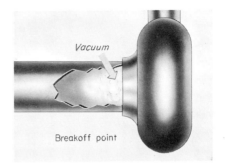

Fig. 4-25 Low pressure preceding cavitation (*Allis-Chalmers.*)

Fig. 4-26 Vacuum exists at the break-off point. (*Allis-Chalmers.*)

pockets form bubbles on the underside of the impeller vane, near its base (Fig. 4-27). As a bubble moves from the low-pressure area at the inlet to the high-pressure area near the tip of the vane, the bubble

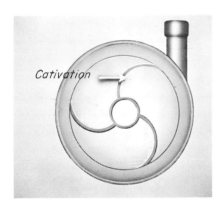

Fig. 4-27 Cavitation in a centrifugal-pump impeller. (*Allis-Chalmers.*)

collapses (Fig. 4-28). It breaks so quickly that the liquid hits the vane with extreme force, often hard enough to gouge out small pieces of the impeller. The damage is generally termed pitting and the noise heard outside the pump during the cavitation is caused by the collapse of the vapor bubbles.

While this description applies specifically to a centrifugal pump, a similar condition can occur in rotary and reciprocating pumps. Excessive suction lift, insufficient npsh, or operation at too high a speed are common causes of cavitation. Pitting, vibration, and noise are common troubles stemming from cavitation. While severe cavitation is usually accompanied by excessive noise and damage to the pump, mild cavitation may produce nothing more than a small reduction in pump efficiency and moderate wear of pump parts.

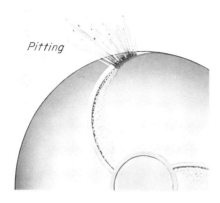

Pitting

Fig. 4-28 Pitting accompanying cavitation in a centrifugal pump. (*Allis-Chalmers.*)

Since any pump can be made to cavitate, care must be taken in unit selection and planning of the installation. For centrifugal pumps, Fairbanks, Morse & Co. recommends that the following five conditions be avoided as much as possible: (1) heads much lower than head at peak efficiency of the pump, (2) capacity much higher than capacity at peak efficiency of the pump, (3) suction lift higher or positive head lower than that recommended by the manufacturer, (4) liquid temperatures higher than that for which the system was originally designed, and (5) speeds higher than the manufacturer's recommendation.

For propeller-type pumps, this firm recommends avoiding: (1) heads much higher than at peak efficiency of the pump, (2) capacity much lower than capacity at peak efficiency of the pump, (3) suction lift higher or positive head lower than that recommended by the manufacturer, (4) liquid temperatures higher than that for which the system was originally designed, and (5) speeds higher than the manufacturer's recommendation.

Specific Speed   In centrifugal pumps, suction lift is related to the specific speed of the unit. Tests show that, for a given specific-speed range, there is a relation between head, capacity, and safe suction lift for various types of pumps. Figure 4-29 shows the Hydraulic Institute recommendations for the upper limits of specific speed for certain types of centrifugal pumps.

These curves are useful in checking the suitability of a pump chosen from rating tables or charts in a manufacturer's catalog and for verifying the recommendations in a pump bid or proposal.

   example: What is the upper limit of the specific speed of a single-stage double-suction centrifugal pump having a shaft which passes through the impeller eye if it handles clear water at 85 F at sea level and total head is 280 ft while suction lift is 10 ft?

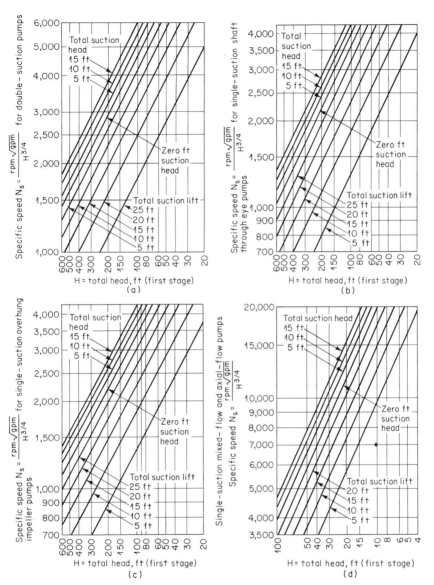

Fig. 4-29 Upper specific-speed limits for (*a*) double-suction pumps (shaft through impeller eye) handling clear water at 85 F at sea level, (*b*) single-suction pumps (shaft through impeller eye) handling clear water at 85 F at sea level, (*c*) single-suction pumps (overhung-impeller type) handling clear water at 85 F at sea level, (*d*) single-suction mixed- and axial-flow pumps (overhung-impeller type) handling clear water at 85 F at sea level. (*Hydraulic Institute.*)

solution: Enter the bottom of Fig. 4-29*a* at 280-ft total head and project vertically upward until the 10-ft suction-lift curve is intersected. At the left read the specific speed as 1,850. This is the maximum safe specific speed for this type of pump under the state conditions. Any proposed unit having a higher specific speed would be rejected as unsatisfactory for this particular job.

SIPHONS

One way of reducing the total head on a pump during operation is by using a siphon in the discharge line. Figure 4-30 shows a typical siphon arrangement for condenser circulating-water piping. With a suitable hookup, a siphon can reduce the required power input to handle a specified flow.

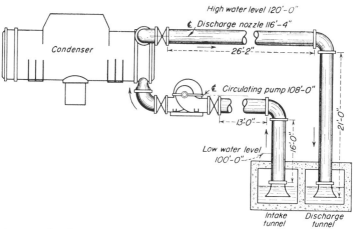

Fig. 4-30 Condenser with siphon setting. (*Allis-Chalmers.*)

Requirements. To secure satisfactory results from a siphon installation, the pump and piping arrangement must meet several requirements that are not always too obvious. These relate to the pump capacity, liquid velocity in the pipe, pipe construction, and submerged depth of the discharge pipe.

To start flow through a siphon, the high point or peak must be filled with water to the full diameter of the pipe so air in the discharge leg is pushed out to the atmosphere. Experience shows that liquid velocity past the peak must be at least 5 ft per sec. This velocity, combined with sufficient flow to fill the pipe cross section at the peak, ensures suitable priming conditions. Besides delivering this flow, the pump must be able to develop a head equal to that at the elevation of the peak, plus any intervening losses in the piping system.

To prevent vaporization of the liquid at the peak of the siphon, the pressure at this point must exceed the vapor pressure of the flowing liquid, once the siphon has been primed. In other words, the liquid must not flow from the discharge pipe so rapidly that an excessively low pressure occurs at the peak. If this happens, and the pressure in the pipe is less than the vapor pressure of the liquid, vaporization will begin. Such a condition is sometimes termed cavitation because it resembles and has effects similar to suction-pipe vapor binding of a pump.

Even where the pressure at the peak exceeds the liquid vapor pressure, air or vapor may collect in the peak if the siphon is not correctly designed. An automatic air-relief valve at the high point in the pipe is the best safeguard against air locking.

Since the pressure at the siphon peak may fall below atmospheric, the pipe must be airtight throughout. Otherwise, air may leak into the pipe and collect at the peak. Also, the pipe must be able to withstand the external pressure load of the atmosphere. With the internal pressure less than atmospheric, the external pressure acts to crush the pipe walls inward.

To ensure best performance, the end of the discharge pipe should be submerged to a depth of at least one pipe diameter in the liquid. This prevents entrance of air. A bellmouthed discharge (Fig. 4-30) will reduce the pipe exit loss and permit recovery of a major portion of the liquid velocity head.

Siphonic Assistance At sea level the maximum siphonic assistance theoretically possible is about 34 ft of water. Exact theoretical value depends on the water temperature and existing atmospheric pressure. For other liquids, the maximum assistance is influenced by the specific gravity of the liquid, with respect to water, in addition to temperature and atmospheric pressure.

TABLE 4-5   Siphonic Assistance

Static head when pump starts	H
Pumping assistance	0.75–0.8S
Pumping static head	$H - (0.75–0.8S)$

From a practical standpoint, the theoretical value of maximum assistance must be revised as in Table 4-5, using the values shown in Fig. 4-31.

Siphon Calculation Figure 4-30 shows a siphon setting for a power-plant steam condenser. However, the general arrangement shown could be used for any of a number of industrial applications where a siphon arrangement was suitable. The only difference would be a change in the head loss through the apparatus served by the siphon.

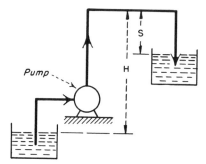

Fig. 4-31 Head relations in a siphon.

example: Compute the total head in the siphon system shown in Fig. 4-30 for a low of 9,000 gpm and a condenser friction loss of 13.7 ft of water when 24-in.-diameter suction and discharge piping is used.

solution: As Fig. 4-30 shows, the low-water level of the supply river is below the center line of the pump, and the high-water level is above the pump center line. First, consider the heads for the low-water condition:

Item	*Ft of Water*
Friction head:	
Condenser	13.7
Piping: 76 ft of 24-in. pipe, four 90-deg ells, three 24-in. gate valves	2.3
Velocity head	0.63
Static suction lift	8.0
Static discharge head (pump center line to center line of condenser discharge nozzle)	8.33
Static-head gain from siphon effect (this is measured from discharge nozzle to water level in the discharge tunnel)	−16.33
Siphon loss (assumed)	1.63
Entering and leaving losses (assumed)	1.0
Tunnel loss (assumed)	1.0
Total head	20.26

Note in this example how the static head between the water level in the intake tunnel and the center line of the condenser discharge nozzle is balanced by the siphon regain in the discharge line from the nozzle to the discharge tunnel, with the exception of the small siphon loss.

The siphon loss varies with water velocity, water temperature, physical structure of the siphon, amount of entrained gas in the water, and the barometric pressure. With a correctly designed siphon, an efficiency of 75 to 80 per cent should be obtained.

Entering and leaving losses in this type of installation vary, depending upon how the circulating-water piping is installed, upon the shape of the inlet and outlet sections of the pipe, and upon the water velocity in the pipe. They generally do not exceed 1 ft of water in well-designed systems.

The tunnel loss consists of friction in the intake screen or rack, if used, and in the tunnel itself. In an average condenser setting with half the rack clogged by debris, the tunnel loss should not exceed 1 ft of water.

With the river at high-water level, the head losses in the pipe and fittings remain the same, as do the velocity, condenser, tunnel, entering, and leaving losses. However, the total static head changes, as shown below.

Item	Ft of Water
Friction head:	
Condenser	13.7
Piping	2.3
Velocity head	0.63
Entering and leaving losses	1.0
Tunnel loss	1.0
Static suction head (pump center line to high-water level)	−12.00
Static discharge head (pump center line to center line of condenser discharge nozzle)	8.33
Static discharge head (center line condenser discharge nozzle to high-water level)	3.66
Total head	18.62

By comparing the total head obtained in both these cases it will be seen that the only individual head that varies is the head due to siphon loss. When choosing a pump for this type of service, both head conditions should be computed and the pump selected for the maximum total head expected, plus any safety factors which are necessary. For the installation considered here, the pump would be chosen for the low-water-level condition because it requires the maximum total head.

SPECIFYING PUMP HEAD

Though the total head on a pump may be computed with great accuracy, careful specification of the exact conditions existing at the installation is necessary to prevent costly errors in pump selection. Perhaps the most common cause of such errors is an accumulation of safety factors applied before the final operating point of the pump is chosen. Too many or too liberal safety factors can produce overrating of the pump, leading to excessive power consumption and, possibly, higher maintenance costs.

Safety Factors Head computations, when accurately made, give the total head that must be developed by the pump for a given installation. It is usual practice to make the capacity estimate or computation at the same time the head is being determined. Once these two values are known, the engineer usually applies a safety factor to each, before referring to a rating table or chart. The exact value of these factors varies from one individual to another, and from one set of conditions to another. Usual practice places them anywhere from 10 to 50 per cent of the required head and capacity of the pump.

When the pump manufacturer receives the request for a quotation on the unit, another safety factor may be added by use of a larger-diameter impeller, or a higher speed for belt-driven pumps. If the pump is driven by a squirrel-cage induction motor, its actual speed may be higher than the guaranteed rated speed of the motor, unless the motor is overloaded. This results from the safety factor the motor manufacturer introduces.

When all these safety factors are used the result may be a pump greatly overrated for its job. Figure 4-32 shows the effect of safety factors on the performance of a centrifugal pump.

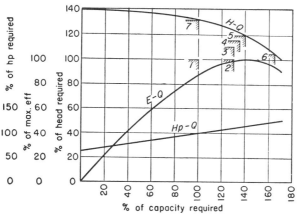

Fig. 4-32   Effect of safety factors on pump operation. (*Allis-Chalmers.*)

Point 1 in Fig. 4-32 indicates 100 per cent head and 100 per cent capacity. This is the actual head and capacity requirement for the pump under a given set of conditions. To assure adequate gallonage, a 30 per cent capacity safety factor, point 2, is applied by the plant engineer. A head safety factor of 10 per cent may also be used, giving point 3. This is the required operating point supplied to the manufacturer as a basis for his quotation. Of course, the point given will vary from one individual to another, depending on the factors used.

To meet the requested guaranteed conditions, the manufacturer may use an impeller of larger diameter, giving point 4. If the motor speed exceeds that guaranteed for full-load conditions, the point of operation of this pump will move to 5. Point 6 indicates the actual operating point which can be determined by using the pump speed, rpm, discharge and suction pressure, and the manufacturer's characteristic curve for the unit.

Should the discharge valve on this pump be partially closed to throttle flow so 100 per cent capacity is delivered, the operating point moves to 7. The difference between points 1 and 7 represents the power unnecessarily lost during throttled flow through the valve.

While Fig. 4-32 has been set up to show the effect of safety factors on one particular pump, the general results obtained are true for almost any class or type of unit. Of course, much depends on the actual shape of the head-capacity curve of the pump being considered. Safety factors applied to Fig. 4-32 , while typical of standard practice in certain fields, should not be used without a complete study of the hydraulic needs of a given system.

Cost, Noise   An oversize pump has a higher first cost than a correctly sized unit. For example, a 7.5-hp circulating pump installed in a hot-water heating system was found unsuitable. A check of flow conditions showed that a 1.5-hp unit would be satisfactory. Not only did the bigger unit cost more when first installed, it did not operate satisfactorily.

Noise, a factor of increasing importance in industrial pumping systems of all types, is often traceable to an oversize pump. A larger pump than actually needed discharges more liquid into the piping system than required. Since the pipe size cannot be changed, the liquid must flow through the pipe at a greater velocity than anticipated. This can lead to excessive noise and complaints from personnel in areas through which the piping passes. In fact, these complaints are often the first clue to the existence of an oversize pump in a plant.

Efficiency   Note that the pump in Fig. 4-32, when operating at point 7, is out of the best efficiency range corresponding to 100 per cent required head. Any point between 130 and 150 per cent rated capacity lies in a good efficiency range. At 7 the efficiency is several points below that in this range.

Should this pump be handling liquid containing abrasive solids which prevent use of a throttling valve in the discharge line, it will operate at point 6, discharging 164 per cent of the required capacity and using about 123 per cent of the required horsepower. There may be extreme abrasive action because flow into the impeller and through its passages does not follow the internal metal contours. This condition can lead to excessive maintenance needs.

Other Factors   Besides the application of arbitrary safety factors, there are other considerations in specifying pump head. These include friction-loss calculations, new or future demands on the pump, variation in suction or discharge level, parallel operation, and liquid handled.

Friction-loss Calculations Over the years, a number of tables and charts of pipe friction loss have been published. Today some of these are outdated and can cause serious errors in head computations. Since no more labor or time is required to compute heads by means of modern data, it is far safer to use only the latest and most accurate available.

Perhaps the most reliable friction-loss data available today are found in the "Pipe Friction Manual," published by the Hydraulic Institute. This manual contains a wealth of information on pipe friction and is recommended to all engineers dealing with pump selection.

New or Future Demands If an existing pump is to be used for a new service or a pump being purchased today will have a different demand in the future, the specified head must be considered from several different aspects.

For example, changing the width of the impeller of a centrifugal pump will usually alter the head-capacity curve (Fig. 4-33). A wider impeller discharges a greater volume of water than a narrow one and generally has a flat HQ curve (Fig. 4-33). The pump with the narrower impeller has less capacity and a steeper HQ curve.

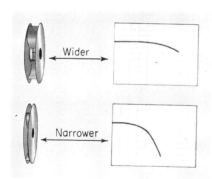

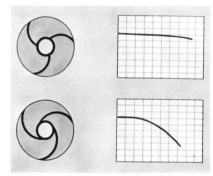

Fig. 4-33 Wider impeller has flatter HQ curve. (*Allis-Chalmers.*)

Fig. 4-34 Changing impeller-vane pitch alters HQ curve. (*Allis-Chalmers.*)

Changing the pitch of the impeller vanes also alters performance. A radial or spokelike vane usually has a flat HQ curve (Fig. 4-34). Swept-back vanes give a steeper HQ curve. Increasing the number of vanes in an impeller also produces a flatter HQ curve (Fig. 4-35). Reducing the number of vanes gives the steeper curve shown in Fig. 4-35.

Ordinarily none of these changes can be made in a pump after it has been installed. So if new or future conditions will vary much from the

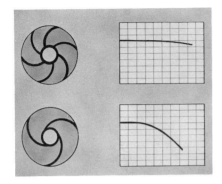

Fig. 4-35 Increasing number of impeller vanes flattens *HQ* curve. (*Allis-Chalmers.*)

existing ones, the head should be based on the ultimate operating conditions or another pump which is suitable for them should be chosen.

In many cases sizing a pump for today's needs and purchasing a new one in the future to meet increased demands is better from an economic standpoint than buying an oversize unit today. At best, the oversize unit is a compromise for today's demands and it probably will operate at reduced efficiency. Should the expected increase in future demand fail to occur, the pump may operate for years at a lower efficiency, wasting power and requiring a higher write-off. Studies of many plants show this condition is rather common. Purchase of the correct size pump to handle new demands, when they occur, ensures higher efficiency, longer life, and less maintenance.

Liquid Level   When either the suction or discharge liquid level varies during pump operation, there is likely to be an accompanying change in the static head on the pump. Depending on the class and type of pump being used, there may be a variation in the quantity of liquid delivered at different static heads. The change in capacity leads to a change in friction loss and, as a result, the range in total head on the pump may be smaller than the static-head range.

To determine the effect of such a variation on a particular pump, the total head must be computed for each set of operating conditions. While this procedure may seem tedious, it is the only safe way of evaluating exactly what occurs in the system.

Parallel Operation   Two or more pumps operating in parallel on a common system introduce additional head considerations because the friction head varies with capacity. Depending on the number of pumps in operation, the head will vary from a certain minimum to a maximum for the given system. These factors are further discussed in Chap. 7.

Liquid Handled   Viscous and volatile liquids, and those carrying suspended solids, affect the performance of any pump. As a result, a complete statement of liquid characteristics is necessary before a pump manufacturer can prepare a bid for a given set of conditions. The various elements in this problem are discussed in Chap. 6.

Pipe Aging   In general, the resistance of a given pipe increases with age, the magnitude of the increase depending on the chemical characteristics of the liquid flowing and the properties of the material of which the pipe is made. Chemical deposits which reduce the area of the pipe, or a greater interior roughness caused by corrosion or corrosion products, are the common symptoms leading to increased flow resistance.

A general statement relating resistance increase with age is almost impossible to make because liquids vary so much from one area to another and from one application to the next. Where economics of a project do not warrant the expense of a detailed investigation of the effect of aging, or there are no records of the effect of local or similar waters on aging, the factors in Table 4-6 can be applied. These values, however, must be used with caution and discretion and are to be applied with friction-loss data given in the "Pipe Friction Manual." No aging allowance was used when the friction-loss tables in the manual were set up.

Factors given in Table 4-6 are composites of many tests. To ensure satisfactory results when using age factors, the entire pumping system must be carefully studied and factors applied in light of the particular

TABLE 4-6   Increase in Friction Loss Due to Aging of Pipe

(*Multipliers to be used with Table 4-1 and similar values*)

Pipe age, years	Small pipes, 4–10 in.	Large pipes, 12–60 in.
New	1.00	1.00
5	1.40	1.30
10	2.20	1.60
15	3.60	1.80
20	5.00	2.00
25	6.30	2.10
30	7.25	2.20
35	8.10	2.30
40	8.75	2.40
45	9.25	2.60
50	9.60	2.86
55	9.80	3.27
60	10.00	3.70
65	10.05	4.25
70	10.10	4.70

conditions existing.   Some test data may vary as much as 50 per cent from those given in Table 4-6.

Specifying Head   Chapter 9 gives the essential data required for correct selection of various classes of pumps.   So far as head is concerned, a pump specification should include the following, depending on the exact layout involved: (1) suction lift, (2) suction head, (3) length and diameter of suction pipe, (4) number and type of fittings in suction line, (5) static head, (6) friction head, (7) maximum discharge pressure against which pump must deliver, (8) discharge line length and diameter, (9) number of type of fittings in the discharge line, and (10) any special considerations influencing head.

CHAPTER FIVE

Pump Capacity

In terms of importance in pump application, head and capacity deserve about equal rank. While other factors like liquid handled, piping arrangement, and type of driver are also important, the primary requirement of a pump is that it deliver the correct amount of liquid against the head existing in the system. Chapter 4 covers the major factors deserving consideration in determining the head a pump must develop. The present chapter discusses the various items which must be studied before the capacity of a proposed pump can be specified.

Units Because the quantity of liquid handled by a pump varies so much from one application to another, a number of different units are used to express pump capacity. Probably the most common unit in the United States is U.S. gallons per minute. In Great Britain and its territories, imperial gallons per minute is the common unit of capacity. For countries using the metric system, pump capacity is usually expressed in cubic meters per hour.

Other units used in the United States include million gallons per day, cubic feet per second, gallons per hour, barrels per day, pounds per hour, acre-feet per day, and milliliters per hour. For any given pump, the capacity units chosen should be in keeping with those commonly used in the industry served. But regardless of the capacity units used, they should specifically state the volume and time requirements. Do not

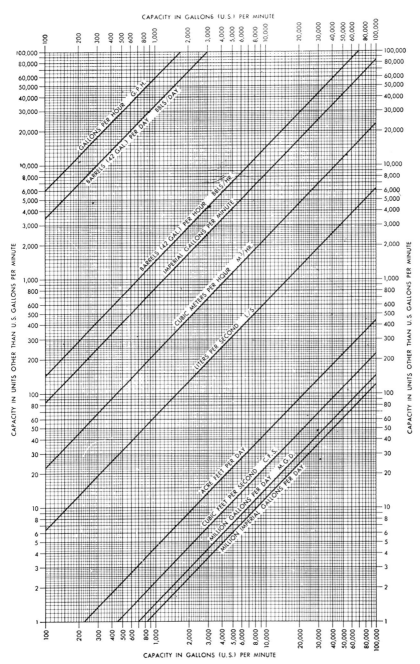

Fig. 5-1 Capacity conversion chart. (*Worthington Corp.*)

specify that a pump deliver 45,000 gal per day without stating the time duration of the day—8 or 24 hr.

For the majority of small and medium-sized centrifugal pumps, capacity is generally stated in U.S. gallons per minute, while for the largest units, million gallons per day or acre-feet per day may be used. Small reciprocating pumps for chemical-feed and similar services are commonly rated in gallons per hour or milliliters per hour. Larger units of this type may be rated in gallons per minute. General-service, slush, condensate-return, and other types of reciprocating pumps are usually rated in gallons per minute. Rotary pumps of various types are also rated in gallons per minute, except in the very smallest sizes where gallons per hour is used. Figure 5-1 permits rapid conversion from one capacity unit to another and is satisfactory for all routine design calculations.

Temperature   Since liquid density changes with temperature, it is important to state the liquid temperature at pumping conditions when specifying the capacity required. Cold water between 32 and 80 F is generally assumed to have a constant density, so far as pumping calculations are concerned. Above 80 F, density change becomes a factor and must be considered in all usual calculations.

For liquids other than water the change in density at reduced or elevated temperatures can be extremely important in its effect on pumping conditions. Therefore, liquid temperature should be made an integral part of all capacity specifications.

Determining Liquid Flow   Before the required capacity of a pump can be specified, the flow requirements of the system served must be known. In some installations, like boiler feed, condenser cooling, etc., flow requirements are fairly easy to determine because there is a certain minimum demand and a predictable maximum demand. But in other installations, such as general industrial processing, it is usually more difficult to predict liquid demand accurately.

Table 5-1 lists water and steam requirements for a number of industrial operations. Tabulated water consumption given can be used as a rough indicator of the pump capacity needed for a particular process. However, it must be remembered that reuse of water is common in many industries today and that a single pump, or group of pumps, may be sufficient to supply all needs except certain specialized ones. Where condensate is to be returned to the plant boiler or return tank, the steam consumption given in Table 5-1 is helpful in indicating how much condensate must be handled. Allowances must be made for those processes in which the steam becomes part of the product, leaving the plant in a container or other vessel, and for exhaust of all or some of the steam to the atmosphere.

TABLE 5-1 Industrial Water and Steam Requirements*

	Water	Steam
Acetic acid from carbide		7300 lb per ton HAc (3)
Acetic acid from pyroligneous acid.	100,000 gal per ton HAc (3)ᵃ	15,700 lb per ton HAc (3)
Acetic acid from pyroligneous liquor.	240 M gal per ton HAc (3)	64,000 to 74,000 lb per ton HAc (3)
Acetic acid, direct (Othmer process).		54,200 lb per ton HAc (3)
Alcohol, industrial	120 gal per gal 100 proof alcohol (5)	
	52 gal per gal 190 proof alcohol (3)	50 lb per gal 190 proof alcohol (3)
	100 gal per gal alcohol (2)	
	20,000 gal per ton grain (1)	
	600,000 gal per 1000 bu grain mashed (5)	
Alumina (Bayer process)	6300 gal per ton Al₂O₃·3H₂O (3)	15,000 lb per ton Al₂O₃·3H₂O (3)
Ammonia, synthetic	31,000 gal per ton liquid NH₃ (1, 3)	
Ammoniated superphosphate	27 to 30 gal per ton ammoniated superphosphate (3)	
Ammonium sulfate	200,000 gal per ton salt (1)	
Buna S	173,000,000 gal per day for 100,000 tons Buna S per year (3)	
Butadiene	320,000 gal per ton butadiene (2)	
Calcium metaphosphate	4000 gal per ton Ca(PO₃)₂	
Carbon dioxide	23,000 gal per ton CO₂ (1)	
	20,000 gal per ton solid CO₂ from 18 per cent flue gas (3)	20,000 lb per ton solid CO₂ from 18 per cent flue gas (3)
Casein (grain-curd process)		2400 lb per ton casein
Caustic soda (lime-soda process)	18,000 lb per ton NaOH in 11 per cent solution (3)	2700 lb per ton NaOH in 11 per cent solution (3)
	21,000 gal per ton NaOH in 11 per cent solution (1)	
Caustic soda (electrolytic)		20,000 lb per ton 76 per cent NaOH (3)
Cellulose nitrate	50 gal per lb cellulose nitrate (3)	
	10,000 gal per ton cellulose nitrate (1)	
Charcoal and wood chemicals	65,000 gal per ton crude CaAc₂ (3)	64,000 lb per ton crude CaAc₂ (3)
Cottonseed oil	20 gal per gal oil (3)	15 lb per gal oil (3)
	0.6 gal per gal hardened oil (3)	0.5 lb per gal hardened oil (3)
Coumarin (synthetic)		3000 lb per ton coumarin or 0.75 ton salicylaldehyde (3)
Cuprammonium rayon	90,000 to 160,000 gal per ton 11 per cent moisture rayon (3)	
Fatty acid refining, continuous		1390 lb per ton stock charged (3)
Gelatin		400 lb per ton gelatin (3)
Glycerine	1100 gal per ton glycerine (1)	8000 lb per ton glycerine (3)
Gunpowder	200,000 gal per ton gunpowder (1) or explosives (2)	
Hydrochloric acid (salt process)	2900 gal per ton 20 Bé HCl (3)	
Hydrochloric acid (synthetic process)	500 to 1000 gal per ton 20 Bé HCl (3)	
Hydrogen	660,000 gal per ton H₂ (1)	
Lactose (milk sugar)	200,000 to 220,000 gal per ton lactose (3)	80,000 lb per ton lactose (3)
Magnesium carbonate, basic	4320 gal per ton basic MgCO₃ (3)	18,000 lb per ton basic MgCO₃ (3)
	39,000 gal per ton MgCO₃ (1)	
Magnesium hydroxide from sea water and dolomite	Sea water 58,000 gal and fresh water 500 gal per ton Mg(OH)₂ (3)	800 lb per ton Mg(OH)₂ (3)
Oxygen, liquid	2000 gal per 1000 cu ft O₂ (3)	
Phenol, synthetic		4000 lb per ton phenol (3)
Phosphoric acid (blast furnace)	75,000 gal per ton 100 per cent H₃PO₄ (3)	
Phosphoric acid (Dorr strong-acid process)	7500 gal per ton 35 per cent P₂O₅ acid (1, 3)	780 lb per ton 35 per cent P₂O₅ acid (3)
Potassium chloride from Sylvinite	40,000 to 50,000 gal per ton KCl (3)	2500 lb per ton KCl (3)
Soap, laundry	230 gal per ton soap (3)	4000 lb per ton soap (3)
	500 gal per ton soap (2)	
Soda ash (ammonia-soda process)	15,000 to 18,000 gal per ton 58 per cent soda ash (1, 3)	
Sodium bichromate		6000 lb per ton sodium bichromate (3)
Sodium chlorate	60,000 gal per ton sodium chlorate (3)	11,000 lb per ton sodium chlorate (3)
Sodium silicate	160 gal per ton 40 Bé water glass (3)	1040 lb per ton 40 Bé water glass (3)
Sodium sulfate, natural		3650 lb per ton anhydrous Na₂SO₄ (95 + per cent) (3)
Stearic acid and red oil		18,000 lb per ton stearic acid (3)
Sulfur dioxide, liquid	18,000 gal per ton liquid SO₂ (3)	6800 lb per ton liquid SO₂ (3)
Sulfuric acid (chamber process)	2500 gal per ton 100 per cent H₂SO₄ (3)	

* American Society for Testing Materials.

ᵃ The boldface numbers in parentheses refer to the list of references appended to this table.

TABLE 5-1   Industrial Water and Steam Requirements (*Continued*)

	Water	Steam
Sulfuric acid (contact process).....	4000 gal per ton 100 per cent H_2SO_4 (3) 5000 gal per ton H_2SO_4 (2)	
Trisodium orthophosphate.........		150 lb per ton $Na_3PO_4 \cdot 12H_2O$ (3) 30,800 lb per ton vanillin (3)
Vanillin (synthetic)...............		
Viscose rayon....................	180,000 to 200,000 gal per ton viscose yarn (3)	140,000 lb per ton viscose yarn (3)
	FOOD INDUSTRY	
Bread...........................	500 to 1000 gal per ton bread (4)	600 to 1000 lb per ton bread (4)
Brewing		
Beer..........................	470 gal per bbl beer (5)	
Whiskey.......................	80 gal per gal whiskey (5)	
Canning		
Apricots.......................	8000 gal per 100 cases No. 2 cans (5)	
Asparagus.....................	7000 gal per 100 cases No. 2 cans (1, 5)	
Beans		
Green........................	3500 gal per 100 cases No. 2 cans (1, 5)	
Lima........................	25,000 gal per 100 cases No. 2 cans (1, 5)	
Pork and beans...............	3500 gal per 100 cases No. 2 cans (1)	
Beets.........................	2500 gal per 100 cases No. 2 cans (5)	
Corn.........................	2500 gal per 100 cases No. 2 cans (5)	
Cream or Whole...............	4000 gal per 100 cases No. 2 cans (1)	
Peas.........................	3000 gal per 100 cases No. 2 cans (1)	
Sauerkraut....................	300 gal per 100 cases No. 2 cans (1, 5)	
Spinach.......................	16,000 gal per 100 cases No. 2 cans (1, 5)	
Succotash.....................	12,500 gal per 100 cases No. 2 cans (5)	
Tomatoes		
Products......................	7000 gal per 100 cases No. 2 cans (1)	
Whole.......................	750 gal per 100 cases No. 2 cans (1)	
Corn Refining...................	333 gal per ton corn (1)	
Edible Gelatin	13,200 to 20,000 gal per ton gelatin (4)	
Edible Oil......................	22 gal per gal oil (3)	
Meat packing		
Packing house..................	55,000 gal per 100 hog units (1, 5)	
Poultry.......................	4400 gal per ton live weight (1)	
Slaughter house................	16,000 gal per 100 hog units (1, 5)	
Stockyards....................	160 gal per acre (5)	
Milk and milk products		
Butter........................	5000 gal per ton butter (1)	
Cheese.......................	4000 gal per ton cheese (1, 5)	
Dairies.......................	3 gal per qt milk (2)	
Receiving and bottling..........	450 gal per 100 gal milk (1, 5)	
Creamery.....................	220 gal per ton raw (5)	
Restaurants....................	0.5 to 4.0 gal per meal (2, 5)	
Sugar		
Beet..........................	2160 gal per ton refined sugar (3) 20,000 to 25,000 gal per ton sugar (1) 2600 to 3200 gal per ton beets (1)	
Refined cane..................	1000 gal per ton sugar (2) Condensing 4800 to 8400 gal per ton (3) Pure water 1400 gal per ton refined sugar	3500 lb per ton refined sugar (3)
Vegetable dehydration		
Beets.........................	37,400 gal per ton product (1)	
Cabbage......................	15,000 gal per ton product (1)	
Carrots.......................	31,600 gal per ton product (1)	
Potatoes......................	11,200 to 25,000 gal per ton product (1)	
Rutabagas.....................	30,400 gal per ton product (1)	
Sweet potatoes................	18,000 gal per ton product (1)	
	TEXTILE INDUSTRY	
Cotton		
Bleaching......................	25 to 38 gal per yd (2)	
Dyeing........................	1000 to 2000 gal per 100 lb goods (1)	
Finishing......................	10 to 15 gal per yd (2)	
Processing.....................	3800 gal per 100 lb goods (1)	
Knit goods, bleaching...........	16,000 gal per ton goods (2)	
Linen.........................	200,000 gal per ton goods (1)	
Rayon		
Cuprammonium yarn...........	160,000 gal per ton yarn (1)	
Dissolving pulp................	190,000 gal per ton pulp (1)	
Viscose yarn..................	200,000 gal per ton yarn (1)	
Silk, hosiery dyeing.............	6000 to 8000 gal per ton goods (2)	
Wool		
Scouring......................	2000 to 15,000 gal per 100 lb raw wool (1)	
Scouring and bleaching.........	40,000 gal per ton goods (2)	
	MISCELLANEOUS INDUSTRIES	
Air conditioning.................	6000 to 15,000 gal per person per season (1)	
Aluminum......................	1,920,000 gal per ton aluminum (2)	
Buildings, office................	27 to 45 gal per day per capita (2, 5)	
Cement, portland...............	750 gal per ton cement (2, 3)	
Cement rock, beneficiation........	720 gal per ton raw rock (3)	

TABLE 5-1   Industrial Water and Steam Requirements (*Continued*)

	Water	Steam
Coal		
By-product coke	1430 to 2860 gal per ton coke (3)	570 to 860 lb per ton coke (3)
Carbonizing	3500 gal per ton coal carbonized (1)	
Washing	125 gal per ton coal (1)	
Electricity	80 gal per kw electricity (2, 5)	
	120,000 gal per ton coal burned (1)	
Hospitals	135 to 350 gal per day per bed (2, 5)	
Hotels	300 to 525 gal per day per guest room (2, 5)	
Laundries		
Commercial	8600 to 11,400 gal per ton "work" (2, 5)	
Institutional	6000 gal per ton "work" (2, 5)	
Leather tannery	375 gal per ton vegetable tan (3)	
	600 gal per ton chrome tan (3)	
	6000 to 16,000 gal per ton leather (2)	
	16,000 gal per ton hides (1)	
Petroleum		
Airplane engine (to test)	125,000 gal per airplane engine (2)	
Gasoline	7 to 10 gal per gal gasoline (2)	
Gasoline, aviation	25 gal per gal aviation gasoline (2)	
Gasoline, natural	20 gal per gal gasoline (3) and 2000 cu ft stripped gas at 150 lb pressure	6 lb per gal gasoline (3) and 2000 cu ft stripped gas at 150 lb pressure
Gasoline, polymerization	34 gal per gal polymer gasoline (3)	2.7 lb per gal polymer gasoline (3)
Oil, Fischer-Tropsch synthesis	150,000 gal per 100 bbl oil (7)	
Oil fields	18,000 gal per 100 bbl crude oil (1)	
Oil refinery	77,000 gal per 100 bbl crude oil (1)	
Pulp and paper mills	50,000 to 150,000 gal per ton pulp (2)	
De-inking paper	38,000 gal per ton paper (1)	
Paper board	14,000 gal per ton paper board (1)	
Soda pulp		13,000 lb per ton dried soda pulp (3)
Strawboard	26,000 gal per ton strawboard (1)	
Sulfate pulp (Kraft)		10,000 lb per ton dried sulfate pulp (3)
Sulfate pulp bleaching	60,224 gal to bleach 1 ton (3) dry pulp of 80 to 85 G.E. brightness	3120 lb to bleach 1 ton (3) dry pulp of 80 to 85 G.E. brightness
Sulfate pulp		5000 to 7000 lb per ton dried pulp (3)
Rock wool	4000 to 5000 gal per ton rock wool (1, 3)	3000 lb per ton rock wool (3)
Rubber (auto tire)		120 lb per auto tire (3)
Steel plant	20,000 to 35,000 gal per ton steel (1)	
Fabricated steel	42,000 gal per ton steel (2)	
Ingot steel	18,000 gal per ton steel (2)	
Pig iron	4000 gal per ton pig iron (1)	
Sulfur mining	3000 gal per ton sulfur (1)	

REFERENCES

(1) G. E. Symons, "Treatment of Industrial Wastes," *Water and Sewage*, Vol. 82, No. 11, November, 1944, p. 44.
(2) *Journal*, Am. Water Works Assn., Vol. 37, No. 9, September, 1945, p. 4.
(3) Chemical and Metallurgical Engineering Flow Sheets, 4th Ed. (1944).
(4) Food Industries Flow Sheets of the Food Producing Industry, 2nd Ed. (1947).
(5) H. E. Jordan, "Industrial Requirements for Water," *The Johnson National Drillers' Journal*, July–August, 1948, p. 7.
(6) W. L. Faith, "Plant Location in Agricultural Process Industries," *Chemical Engineering Progress*, Vol. 45, May, 1949, p. 313.
(7) W. C. Schroeder, "Comparison of Major Processes for Synthetic Liquid Fuels," *Chemical Industries*, Vol. 62, No. 4, p. 577 (1948).
(8) S. T. Powell and L. G. von Lossberg, "Relation of Water Supply to Chemical Plant Location," *Chemical Engineering Progress*, Vol. 45, May, 1949, pp. 289–300.

Estimating Water Needs Since water is probably industry's most commonly used liquid, many problems arising in practice relate to it. And since pumping capacity cannot be chosen until water requirements are known, a systematic water-consumption survey is a necessary first step in pump selection. Table 5-2 shows one way in which water needs can be summarized for a projected or existing plant. Where only a portion of the plant is being considered, the number of entries can be

TABLE 5-2 Form for Estimating Plant Water Consumption

	Gal per day, max	Temp, F	Total hardness, ppm	Suspended solids, ppm	Dissolved solids, ppm	pH	Organic matter	Microorganisms	Color	Turbidity	Silica	Iron and manganese	Fluorides	Carbon dioxide	Hydrogen sulfide	Dissolved O_2	Chlorine	Radioactive wastes	Oil
Drinking water																			
Other sanitary																			
Boiler make-up																			
Air conditioning																			
Refrigeration																			
Compressed air																			
Diesel cooling																			
Steam condensing																			
Fire protection																			
Process A																			
Process B																			
Process C																			
Total																			

reduced as desired. Table 5-2 helps avoid one of the commonest pitfalls in industrial pump selection-wasting money and power to bring all water up to the strict requirements of some minor application—drinking water, for example. It also aids in planning for water recirculation and multiple use, and can indicate possibilities for segregating special needs from others.

Other Flow Data Besides the flow data given in Table 5-1, later chapters on pump application in various industries give details of flow rates used for a variety of services. Liquids other than water are considered there because they can be better discussed in terms of specific applications than in a general chapter like the present one.

Capacity Control The discharge of a centrifugal pump can be varied by a number of different methods, depending on whether the unit runs at constant or varied speed. With constant-speed units the capacity can be changed by (1) throttling the discharge of the pump, or bypassing all or some of the discharged liquid; (2) using more than one pump, proportioning the capacity of each so that one or more can be shut down or started to provide the needed flow; (3) using a storage tank or reservoir and operating the pump intermittently to maintain a certain minimum level; and (4) using an adjustable-capacity pump—such as the unit described in Chap. 15, or a low-head adjustable-vane axial-flow pump.

Speed variation, either manual or automatic, is another means of varying the output of a centrifugal pump. This requires either a variable-speed driver or coupling, and can give extremely economical operation because there are no throttling or bypass losses. Pump drives are discussed in greater detail in Chap. 8.

Figure 5-2 shows a constant-speed centrifugal pump arranged for throttling of the discharge. This unit handles 800 gpm at 100-ft head. Figure 5-3 shows one of six 116-in. horizontal axial-flow propeller pumps in a Florida flood-control station. Each pump is driven by a diesel engine through double-reduction chain drive. The rated speed of these pumps is 124 rpm but they may be driven at other speeds to meet changing load conditions.

Positive-displacement reciprocating and rotary pumps may be controlled by varying the pump speed, stroke, amount of liquid bypassed, or one of several other methods. A number of these are discussed in Chaps. 2 and 3, while others are covered in later chapters related to specific applications.

Demand Variations Extreme care is necessary in specifying pump capacity when plant load, production rate, or other factors will cause a change in liquid demand. Variations in the amount of liquid delivered generally influence the efficiency of a pump. Usual practice is to choose

Fig. 5-2 Centrifugal pump fitted with globe valve for throttling the discharge. (*Goulds Pumps, Inc.*)

Fig. 5-3 Axial-flow propeller pump in a Florida flood-control station. (*Fairbanks, Morse Pump Division, Colt Industries.*)

a pump so that when operating at normal capacity its efficiency is a maximum, or nearly so.

Where a wide range in capacity is required, the duration of the operating periods at various capacities, up to and including rated capacity, should be carefully studied. With the usual centrifugal pump, an increase in capacity is generally accompanied by a decrease in the head developed. And since pipe friction increases with the quantity of liquid flowing, the pump may not be able to deliver the desired capacity because it cannot develop sufficient head. Excessive operation of the pump at capacities greater than normal may lead to premature maintenance difficulties. Where a pump operates a good part of the time at rated capacity, with only occasional overloads, the duration of the overload period is not quite so important. But when overloads are excessive, it is wise to consider use of two pumps instead of one.

Using two pumps instead of one permits each to operate in its best efficiency range most of the time. While initial costs may be higher, the lower operating cost and greater flexibility help pay off the added investment. Later chapters discuss specific examples of the use of two pumps instead of one in various industrial operations.

Capacity Choice Assume an industrial plant has an average water demand which is 70 per cent of the maximum demand. Two pumps, each rated at 75 per cent of the maximum demand, might be chosen. This would allow either pump to handle the average demand and it would probably operate alone most of the time. When the demand exceeds 75 per cent of maximum, the second pump starts and operates in parallel with the first. Both operate at or near the maximum efficiency.

In some instances an arrangement of this type is not suitable. When total demand is so low that it is impossible to choose two pumps which will operate in their most efficient ranges, a single unit may have to be substituted because two pumps would probably have higher initial and operating costs than a single pump under these circumstances.

With extremely large capacities it may be impossible to find a unit which operates at its best efficiency or the most economical speed when meeting the average demand. Two half-size or other capacity units may then have to be chosen, regardless of how the liquid demand may vary. When demand fluctuates markedly from a low to a high value, this arrangement may permit a better over-all efficiency because capacity can be split as desired—say 25 and 75 per cent of the total in the two pumps. Both will have to operate to meet the maximum, or 100 per cent demand. The exact hookup used depends on how liquid demand varies in different areas of the system served. For example, where one area of an industrial plant uses about 25 per cent of the total flow, and another

75 per cent, the pumps can be piped independently of each other. But where a single area is served, or the liquid is pumped to a tank before distribution, the units would probably be operated in parallel.

Where operating efficiency is not too important a factor, the number of units chosen to deliver the required capacity is generally a function of first cost. If a single unit has lowest first cost, it is used. The same is true if more than one has lowest initial cost for the capacity desired. While it is not too often that a condition exists where operating efficiency can be neglected, there are situations where energy which might otherwise be wasted can be used in the prime mover driving the pump. Typical examples include utilization of exhaust steam, waste refinery gas, and water or other liquids supplied under pressure when the pressure energy is not needed to overcome flow resistances.

In installations where failure of liquid supply might seriously endanger life, equipment, or safety, two pumps may be installed to operate in parallel at all times. Each pump is rated for the maximum demand and can carry the entire load if the other unit fails. There are disadvantages to this scheme: (1) Since both pumps operate at 50 per cent or less of rated capacity, the head developed exceeds that required by the system, and a throttling device must be fitted to the discharge if both pumps run at constant speed. (2) The useful life of the pumps may be shortened. (3) Overheating may occur when flow drops to extremely low values. This factor is discussed later in the present chapter. Despite these disadvantages, there are installations where the added safety of two duplicate pumps makes their use worthwhile.

Parallel Operation   As indicated earlier, parallel operation of pumps is often the best solution to a problem in capacity variation.

In general, reciprocating and rotary pumps present fewer problems when operated in parallel than do centrifugal pumps. One reason for this difference is that most reciprocating and rotary pumps are positive-displacement units, whereas the capacity of a centrifugal pump varies with the head it develops. And, in turn, the head a centrifugal pump develops is a function of impeller design, speed, and other factors. So the following discussion deals specifically with centrifugal pumps even though some of the principles apply equally well to reciprocating and rotary units.

THROTTLED SYSTEMS

Industrial pumping systems served by centrifugal pumps are generally either the throttled or unthrottled type. In the throttled type, flow (or capacity) is controlled by use of one or more valves which dissipate the excess head developed by the pumps. In some systems, like those for boiler feed, a throttle valve placed directly in the pump discharge line controls flow. In others, like industrial water-supply systems without a standpipe or reservoir floating on the distribution mains, the water users control pump discharge as they open or close valves to suit their needs.

In unthrottled systems, where the pumps often discharge into a standpipe or reservoir, flow depends on the head developed and the normal characteristics of the system.

Throttled System Figure 5-4 shows one example of a typical industrial throttled system with flow controlled by valve A. This valve can be posi-

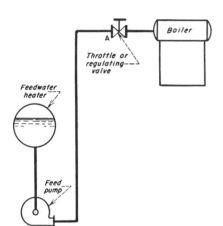

Fig. 5-4 Simplified diagram of a boiler-feed-water system. (*Worthington Corp.*)

tioned manually or automatically. Figure 5-5 shows the system-head curves superimposed on the pump HQ curve. Line CB represents boiler pressure plus the static portion of the pumping head. This pressure is practically constant.

While a boiler is shown as the vessel under pressure, any other similar device might be substituted without changing the relations in the system. Therefore the following analysis can be assumed to apply to any industrial system resembling that in Fig. 5-4.

System Operation When water is supplied the boiler in Fig. 5-4 by the pump, the pressure against which the pump operates increases because of the friction-head losses in the piping, valves, fittings, and other devices in the line. With valve A open, the pressure due to friction losses can be assumed to increase along curve CD (Fig. 5-5). Partially closing valve A produces a series of system-head curves like CE, CF, and CG, depending on how much the valve is closed. Complete closure of the valve causes the discharge pressure to rise to the shutoff valve J at zero capacity. By varying the degree of opening of valve A a family of system-head curves can be produced between points J (valve fully closed) and D (valve wide open).

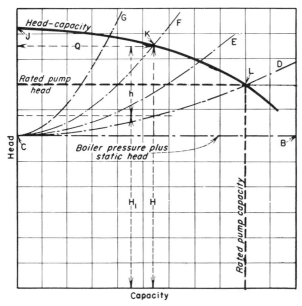

Fig. 5-5 System-head curves plotted on pump HQ curve show capacity at various heads. (*Worthington Corp.*)

To deliver a required capacity, say Q, the throttle valve must be adjusted until the system-head curve is CF. This crosses the pump HQ curve at K and the head against which the pump operates is represented by the vertical distance H. The actual head required to deliver capacity Q on the normal system-head curve CD is represented by H_1. Since the pump develops head H at capacity Q, valve A must dissipate a head equal to $H - H_1$, or h.

The wasted head h represents a loss of power input because the driver must deliver enough power to the pump to produce head H. This loss can be evaluated as discussed in Chap. 8.

Rising Head Curves When two like pumps with steadily rising HQ curves operate in parallel, system load should divide equally between them. Figure 5-6 shows the HQ curve of either of two like pumps. To draw the HQ curve for the two pumps operating in parallel, it is only necessary to project capacity points at various heads on the curve to the right to double their value. For example, point A, 1,000 gpm and 244-ft head, is extended to the right to 2,000 gpm and 244-ft head, point B. The 2,000-gpm intersection, C, is projected to 4,000 gpm, point D, and in like manner 3,000 gpm is projected to 6,000 and 4,000 to 8,000 gpm, as the horizontal dotted lines indicate. The curve drawn through these points is the HQ characteristic for the two pumps. Each pump is rated

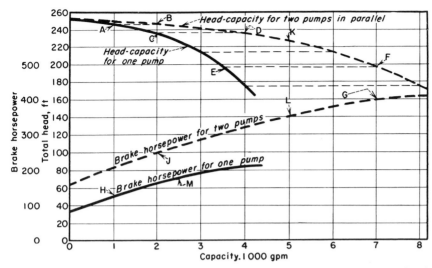

Fig. 5-6 *HQ* and brake-horsepower curves for two like pumps. (*Worthington Corp.*)

3,500 gpm at 197-ft head, point *E*, so two pumps will handle 7,000 gpm at the same head, point *F*, on the two-pump curve shown in Fig. 5-6.

Power input to the two pumps is about 400 hp, point *G* on the horse-power curve for the two pumps (Fig. 5-6). This curve was plotted by doubling the horsepower for one pump and doubling the capacity. For example, at 1,000 gpm, the power for one pump is 125 hp, point *H*. So for two pumps it is 250 hp at 2,000 gpm, point *J*. Repeating this procedure at different capacity values, a series of points is obtained, through which the two-pump power curve is drawn.

If these two pumps operate on a throttled system and demand decreases, they must deliver less capacity. As a result, the head developed increases and more throttling is necessary. For example, if demand decreases to 5,000 gpm, head developed increases to 227 ft, point *K*, and power required is 356 hp, point *L*. Each pump delivers 2,500 gpm against 227 ft and requires 178 hp, point *M*.

Unlike Characteristics Two unlike pumps can operate in parallel, but they present different problems from two that are alike. Figure 5-7 shows the characteristics of two pumps, either of which can be used alone on a throttled system. If power costs are high, the pump with the characteristics shown by the dotted lines would be preferred because its power requirement at part capacities is lower. In general, however, a unit with a steep *HQ* curve (full line, Fig. 5-7) has the advantage in a single-pump throttled system requiring a greater travel of the throttling

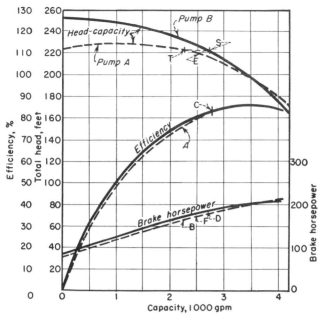

Fig. 5-7 Two pumps with unlike *HQ* curves may operate satisfactorily in parallel. (*Worthington Corp.*)

valve, because more head has to be lost across the valve.   Pump operation is more stable.

Drooping Head Curve   In spite of having a drooping *HQ* curve (dotted line, Fig. 5-7) a pump of this design can usually be used on a single-pump throttled system because the operating point is determined by system head.   On some throttled systems with a single pump having a drooping *HQ* curve, flow surging may occur.   This has also developed to a lesser degree with pumps having a stable, constantly rising *HQ* curve (full line, Fig. 5-7) but such cases are rare.

Piping Hookup   When two or more pumps operate in parallel with piping like that in Fig. 5-8, each has its own piping and fitting losses, as

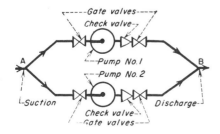

Fig. 5-8 Simplified piping hookup for two centrifugal pumps operated in parallel. (*Worthington Corp.*)

well as losses common to both pumps.   Instead of using the true HQ curve, one measured from A to B (Fig. 5-8) is used in studies of the system.   This is the true characteristic of the pumps, minus piping and fitting head losses from A to B.   Usually in throttled systems, particularly in high-head applications like boiler feed, losses in the piping of each pump are such a small part of the total head that they may be neglected.

Combined Head Curves   Figure 5-9 shows characteristic curves of two pumps, each designed for 3,500 gpm, 197-ft total head, point N, but with different-shaped HQ curves. A and B are HQ curves of the two pumps; C is the combined HQ curve.   It includes values from D to E on curve B, to which is added curve A, and extends from E to F.   Point G on curve A is at 0 gpm, while E on B is at 2,650 gpm, with 223-ft head on both curves.   The combined capacity of the two pumps at 223-ft head is $0 + 2,650 = 2,650$ gpm.   Point H on curve A is at 228 ft and 1,000 gpm. To get a corresponding point for curve C move to $2,450 + 1,000 = 3,450$ gpm and 228-ft total head.

At 2,000 gpm on curve A, point K, the head is 224 ft.   Again, to obtain a corresponding point for curve C, project to $2,600 + 4,600$ gpm and 224-ft head, point L (Fig. 5-9).   Repeating this operation, a series of points is obtained through which curve C can be extended to represent the combined characteristics of the two pumps.   Construction of this curve is the first step in finding how two pumps with unlike HQ curves perform on the same system.

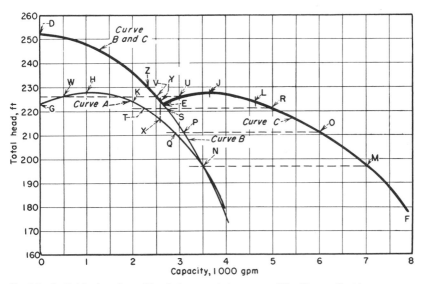

Fig. 5-9   Individual and combined characteristic curves. (*Worthington Corp.*)

Waterworks System   Assume these pumps are in a water-distribution system, either industrial or municipal, and that they discharge into a distribution system at a normal total head of 197 ft. At this head the two bumps discharge 7,000 gpm, point M, where both have the same capacity, point N. If flow in the system falls to 6,000 gpm, point 0, the two pumps develop 211-ft total head. Under this condition, pressure in the main is higher than desired by an amount found by taking the difference between the head at 6,000 gpm and that desired in the system, or $211 - 197 = 14$ ft, or 6.06 psi. This must be accepted unless a speed adjustment of some type is used to control pump output. Of the 6,000 gpm discharge, 3,100 gpm is supplied by pump B, point P, and 2,900 gpm by A, point Q.

To find the load division between the two pumps, after the points of combined discharge and total head are determined on curve C, project horizontally to the left from C to intersect curves A and B, as indicated by the dotted line from O to Q (Fig. 5-9). For example, if flow from the system decreases to 5,000 gpm, the total head increases to 221 ft, point R. Projecting to the left from R, curve B is intersected at 2,730 gpm, point T. On reduction of pump capacity to 3,000 gpm, point U, total head rises to 226 ft. The discharge of pump A drops to 500 gpm, point W, while that of pump B falls to 2,500 gpm, point V.

Throttling the discharge until system head rises to 230 ft causes the discharge of pump B to drop to 2,350 gpm, point Z. Pump A will not discharge because the system head has increased above the maximum of 227 ft, point H, for this unit. Pump A is therefore backed off the line by B, and operates at its shutoff head. This could be a dangerous condition, even if it lasted for only a short time. It is discussed in greater detail later in this chapter. Two pumps should not be operated in parallel at a capacity below the safe minimum for the particular units in question.

Load Division   Assume pump A is operating alone and discharging 2,600 gpm at 216-ft head, point X, when pump B starts. The latter has an HQ curve higher than A and picks up the load to back A off the line. If flow in the system stays constant at 2,600 gpm, the head rises to 224 ft, point Y. But if B had been discharging 2,600 gpm alone and A started, A could not supply any water to the system.

Suppose these pumps operate on an industrial system where the demand changes slowly and one unit is taken out of service when the load falls below 3,500 gpm. Under careful supervision these two pumps could operate safely in parallel. At 5,000 gpm combined flow, A discharges 2,270 gpm at 221-ft head, point T (Figs. 5-7 and 5-9). Efficiency is 76.8 per cent, point A; power input is 165 hp, point B.

Pump *B* discharges 2,730 gpm, point *S*, while operating at an efficiency of 82.3 per cent, point *C*, and an input of 185 hp, point *D*. Total power input to the two pumps is $165 + 185 = 350$ hp.

If both pumps were like *A*, a flow of 5,000 gpm would take 2,500 gpm from each unit, point *E* (Fig. 5-7), and 173 hp per pump, or 346 hp for the two. With both pumps like *B*, the power for each is 178 hp, point *F*, or 356 hp for the two pumps. In one case, with the two units alike, the power input is less than for two unlike pumps, while in the other it is greater. This shows that having both pumps alike so the load divides equally between them may not result in an input-power reduction.

Drooping Head Curve   Two similar pumps with drooping *HQ* character-istics (dotted curve, Fig. 5-7) can cause operating difficulties when run on a throttled system. Curve *AB* (Fig. 5-10) shows the *HQ* characteristic of either of two like pumps operated in parallel. Curve *AD* is the theoreti-cal combined *HQ* relation of the two pumps. This curve was developed as explained for Fig. 5-6, with the assumption that the friction losses are so low in the piping of each pump that they can be neglected.

Assume a water demand of 1,750 gpm at 226-ft head, point *C*, with pump No. 1 operating alone, when No. 2 starts. Pump No. 1, operating at 226-ft head, applies this pressure to the check valve of pump No. 2 to hold the valve closed. This head is greater than the shutoff head, 223 ft, point *A*, of pump No. 2. So when this pump comes up to speed on shutoff it can develop only a 223-ft head, which is not enough to open the check valve against 226 ft, and the pump cannot discharge into the system.

There are several means to overcome this difficulty. Consider the two pumps in Fig. 5-8, with No. 1 operating at 1,750 gpm. By throttling the discharge valve of this pump, the pressure at point *B* in the piping system and on the check valve of pump No. 2 can be reduced below the shutoff head of No. 2. When pump No. 2 is started it can open the check valve and take part of the load. This, and other methods to get a second pump on the system, generally require careful timing and han-dling of the valves.

Division of Flow   Two pumps having drooping *HQ* curves (Fig. 5-10) may not divide the flow equally between themselves, even if they are hydraulic duplicates and operate at the same speed. For example, to supply a demand of 2,250 gpm, point *E*, the load could be divided equally with the two operating at 1,125 gpm and 228-ft head, point *F*. Or under this condition, one pump may operate at point *C*, 1,750 gpm, and the other at 500 gpm, point *G*. Though the heads at these points are equal, this does not ensure stable operation.

It is not good practice to operate in parallel two pumps with drooping

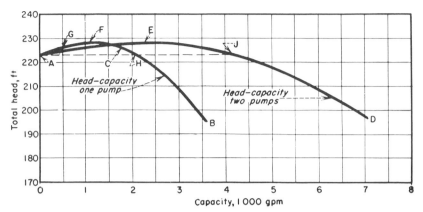

Fig. 5-10 Two similar centrifugal pumps with drooping *HQ* curves can give trouble when paralleled on a throttled system. (*Worthington Corp.*)

HQ curves (Fig. 5-10) at capacities at which the system head exceeds the shutoff head of the pumps. In Fig. 5-10 this is 2,100 gpm and 223-ft head for one pump, point *H*, or 4,200 gpm and the same head, point *J*, for two. When the two operate at this or a greater capacity, they are on the stable part of their *HQ* curves and divide flow about equally. Also, one pump cannot take all the load and shut off the other.

Operating Requirements For two pumps to run in parallel on a throttled system, they should (1) have *HQ* curves that rise steadily to shutoff and (2) have the same percentage reduction in capacity over their probable operating-head range, or at least deliver some liquid over this entire range. Usually, two or more units with stable *HQ* curves and about equal shutoff heads share the load nearly equally at their lower capacities.

UNTHROTTLED SYSTEMS

On an unthrottled system, pumps of similar or dissimilar characteristics can operate in parallel in most applications without trouble. To do so, the system head must not exceed the shutoff head of any pump at any capacity that can be produced by the combination of other pumps delivering into the same system.

Three Pumps in Parallel Take, as an example, the three pumps with *HQ* curves HQ_1, HQ_2, and HQ_3 (Fig. 5-11). Assume they operate on the system in Fig. 5-12, each pump having separate piping and valves to point *B*. As the pumps differ in size, the head losses from *A* and *B* will probably be different for each size. Also, since they have different characteristics, operating two or three in parallel causes their capacity to change as the system head changes. So when combining pump charac- teristics, it is necessary to use the net characteristic of each pump measured from A to B. Also, the combined pump characteristic must be compared with the net system characteristics, including friction losses from B to the reservoir.

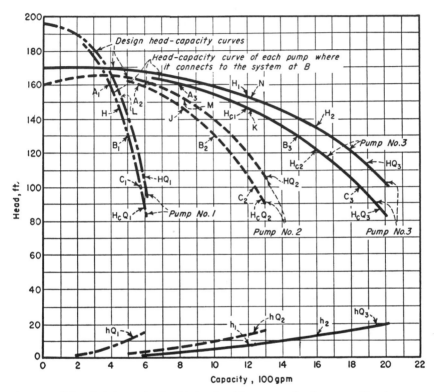

Fig. 5-11 HQ and piping and valve friction-head-loss curves for three centrifugal pumps with dissimilar characteristics. (*Worthington Corp.*)

Shown at the bottom of Fig. 5-11 are friction-head-loss curves for each pump from A to B. These curves are labeled hQ_1, hQ_2, and hQ_3. The head produced by each pump, measured from A to B for any capacity, is the head shown on its HQ curve less that on the hQ curve at the same ca- pacity. For example, pump No. 3 at 1,200 gpm delivers 153-ft head, point H_1, as measured from its suction to its discharge nozzle. Friction loss on curve hQ_3 at 1,200 gpm is 7 ft, point h_1. So the head available to overcome the system head common to the three pumps is $153 - 7 = 146$ ft, point H_{c1}.

At 1,600 gpm, pump No. 3 develops 135-ft head, point H_2, and the friction loss on curve hQ_3 is 12 ft, point h_2. Then the head available to discharge water into the reservoir against the system head is $134 - 12 = 122$ ft, point H_{c2}. In this way, the other points can be located to draw curve H_cQ_3. Plot H_cQ_2 the same way.

Since friction losses increase with system capacity, the H_cQ curves are steeper than the HQ curves. With lower head pumps, for which friction loss in the piping of each pump is a larger percentage of the total head, the H_cQ curve may rise steadily to shutoff, even though the HQ curve is slightly drooping.

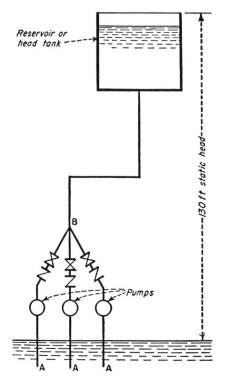

Fig. 5-12 Three pumps connected in parallel on an unthrottled system with a reservoir. (*Worthington Corp.*)

System Characteristics   Figure 5-13 shows the system characteristic SS_1, with losses from A to B, Fig. 5-12 excluded. This is the 130-ft static head plus friction losses from B to the reservoir, which increase with flow, starting at 130 ft, point S, for zero flow. Also plotted on Fig. 5-13 are the H_cQ curves of pumps 1, 2, 3, identified as in Fig. 5-11 as H_cQ_1, H_cQ_2, and H_cQ_3, and their combined H_cQ characteristics. Like any combined head-capacity curves, these are found by adding the capacities of the pumps at the same head. This gives the capacity of the combina-

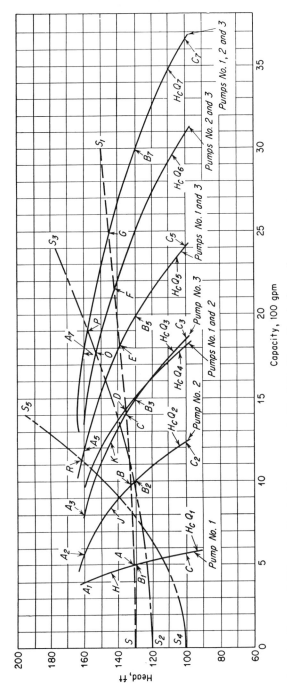

Fig. 5-13 Corrected *HQ* curves for the three pumps in Fig. 5-11. Operation is shown for the pumps working as single units and in combinations of two and three. (*Worthington Corp.*)

tion at several heads, and a curve is drawn through the plotted points. For example, for heads measured from A to B, pump No. 1 will deliver 387 gpm at 160-ft head, 497 at 130-ft head, and 572 gpm at 100-ft head, points A_1, B_1, and C_1, respectively (Fig. 5-13). No. 2 handles 565 gpm at 160 ft, point A_2; 1,000 gpm at 130 ft, B_2; and 1,238 gpm at 100 ft, C_2. No. 3 discharges 800 gpm at 160 ft, point A_3; 1,480 gpm at 130 ft, B_3; and 1,845 gpm at 100 ft, C_3. These values are also identified by the same symbols on curves H_cQ_1, H_cQ_2, and H_cQ_3 (Fig. 5-11).

If pumps No. 1 and 3 operated in parallel, the combined capacity would be $387 + 800 = 1,187$ gpm at 160-ft head, point A_5; $497 + 1,480 = 1,977$ gpm at 130-ft head, point B_5; $572 + 1,845 = 2,417$ gpm at 100 ft, point C_5, curve H_cQ_5 (Fig. 5-13). Similarly, if the three pumps operated in parallel, their combined capacity would be $387 + 565 + 800 = 1,752$ at 160-ft head, point A_7; $497 + 1,000 + 1,480 = 2,977$ gpm at 130-ft head, point B_7; and $572 + 1,238 + 1,845 = 3,655$ gpm at 100-ft head, point C_7 on curve H_cQ_7 (Fig. 5-13). Head-capacity curves H_cQ_4 and H_cQ_6 for No. 1 and 2 and No. 2 and 3 operating in parallel can be plotted in a similar manner.

Points at which the different H_cQ curves cross the system-head curve SS_1 indicate the capacity that each pump or combinations of pumps will deliver and also the net system head against which they will be operating. Thus No. 1 operating alone will deliver 495 gpm at 131-ft head, point A. No. 2 will discharge 975 gpm at 133-ft head, and No. 3, 1,390 gpm at 135-ft head, points B and C, respectively. When they operate in parallel, No. 1 and 2 will discharge 1,420 gpm against 136-ft head, point D; No. 1 and 3 at 138-ft head, 1,810 gpm, point E; No. 2 and 3 at 142-ft head, 2,160 gpm, point F; No. 1, 2, and 3 at 145-ft head, point G, 2,500 gpm.

Pump Capacity   To find what capacity each pump delivers, refer to its H_cQ curve at the operating head. For example, No. 1, 2, and 3 together discharge 2,500 gpm against a head of 145 ft as measured from A to B (Fig. 5-12). From the H_cQ curve of Fig. 5-13, we find that No. 1 supplies 450 gpm, point H; No. 2 supplies 835 gpm, point J; No. 3 supplies 1,215 gpm—all at 145-ft head, point K. These values are also identified on the H_cQ curves in Fig. 5-11. It also shows on the HQ curves that the head the pumps develop from suction to discharge nozzle is 153 ft for No. 1, 151 ft for No. 2, and 152 ft for No. 3, points L, M, and N, respectively.

Different Systems   The head of most unthrottled systems consists of a large static component with a relatively small friction component at maximum pumping capacity such as curve SS_1 (Fig. 5-13) . On this system the total operating head range is roughly 140 to 153 ft. Curve S_2S_3 in Fig. 5-13 shows a system with 10 ft less static head and greater friction head than SS_1. The steeper system head results in less increase in capacity when another pump is added. For example, going from a condition with No. 2 and 3 in service to all three running increases delivered capacity from 2,170 to 2,500 gpm, or 330 gpm, points F to G on system curve $SS1$. On system S_2S_3, capacity increases from 1,765 gpm, point O, to 1,915, point P, or 150 gpm compared with 330 gpm on system SS_1.

It was mentioned earlier that pumps of dissimilar characteristics can operate in parallel if the system head does not exceed the shutoff head of any pump at any capacity produced by a combination of other pumps on the system. If the three pumps in Fig. 5-11 are on a system with the head curve S_4S_5 (Fig. 5-13), No. 1 and 3 together can deliver 1,120 gpm, 162-ft head, point R. No. 2, if started, cannot discharge into the system as its shutoff head is only 160 ft (Fig. 5-11). Thus for system S_4S_5, parallel operation will be limited to two pumps, never all three. Also, at 162-ft head, No. 2 and 3, operating in parallel on system head S_4S_5, have practically the same capacity as No. 1 and 3.

Friction Systems The head on a condenser circulating-water system is nearly all friction. Such a system generally uses two equal-capacity pumps. Both are in service in summer when water temperatures are high, and one in winter when they are low. Figure 5-14 shows, without going into details of how the system-head curves were developed, the results obtained with two duplicate pumps on a circulating-water system. Single-pump operation gives 2,140 gpm at about 18-ft total head, point

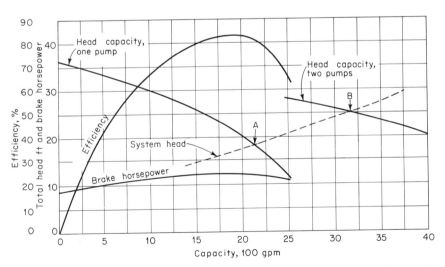

Fig. 5-14 Characteristic curves of two duplicate pumps run singly and in parallel. (*Worthington Corp.*)

A, while two-pump operating capacity is 3,200 gpm at 25-ft head, point *B*. With these pumps and system characteristics, capacity increase is only 3,200 − 2,150 = 1,050 gpm, or less than half of one pump operating alone.

For convenience, when quoting on several duplicate pumps in parallel, manufacturers show, on their proposal curve of the individual pump characteristic, system-head curves plotted against pump capacity (Fig. 5-15) instead of giving curves as in Fig. 5-14 where single- and two-pump characteristics are plotted against system capacity. Either method shows the results that can be obtained, but Fig. 5-15 indicates

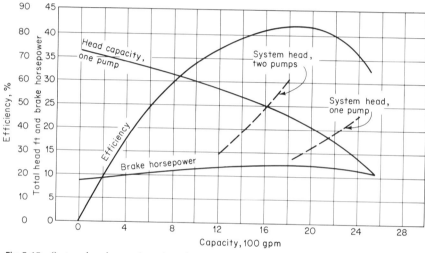

Fig. 5-15 System heads are plotted against the capacity of one pump when a manufacturer quotes on duplicate pumps to run in parallel. (*Worthington Corp.*)

the efficiency and horsepower for single- and two-pump operation in simpler form.

Pump Selection   When a number of pumps operate on an unthrottled system, they are often purchased for the capacities and total head corresponding to that at maximum demand. For example, in the system in Fig. 5-12, whose characteristics less losses from *A* to *B* are curve SS_1 (Fig. 5-13), 3,000-gpm delivery with all pumps running might be desired. This would require pumps giving a total of 3,000 gpm against 150-ft head plus losses from *A* to *B*. This is approximately 10 ft; so pumps for 160-ft total head would be purchased.

In such a system, demand might be constant or vary over a narrow or wide range. As a result, selection of a number of pumps and their individual ratings would depend on these factors. In selecting the pumps watch several points if the head changes when fewer units are in service:

1. Select driver sizes for the power required over the head range on which the pumps will operate.

2. If considerable operation will be at reduced capacity with resulting reduced total head, select the pumps so maximum efficiency is at the average operating heads. As an example, take a pump for 1,200 gpm at 143-ft head to be used with others on a system where operating head ranges from 120 to 143, with 135 ft as average. The pump in Fig. 5-16 would be an ideal selection with good efficiency at high- and low-head points *A* and *B* and maximum efficiency roughly at 135 ft.

3. With reduced total head, a centrifugal pump delivers more capacity. This means less available net positive suction head (npsh) because of increased friction in the suction line, causing reduced suction head or increased suction lift. However, with increased capacity, the pump requires more npsh. Take care that available npsh at reduced-head operating point, like *B*, Fig. 5-16, equals or exceeds what the pump requires.

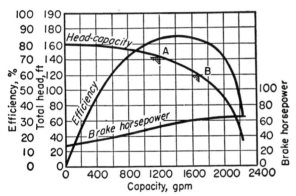

Fig. 5-16 Select pumps so that best efficiency occurs at a head between the maximum and minimum operating points when one of several units in parallel is frequently idle. (*Worthington Corp.*)

Existing Pumps, New Ones Successful operation of two or more centrifugal pumps in parallel depends on the characteristics of the system and pumps. To determine if existing pumps are suitable to operate in parallel, the characteristics of the pumps and system must be known.

For details of the effect of a change in pump speed on capacity, and the characteristic curves obtained with various drives and drive arrangements, see Chap. 8.

Safe Flow for Pump Two other factors deserving consideration where flow demands vary are power input and pump overheating. In a unit having a rising power-input curve with either increase or reduction in flow, a large change in flow needs may overload the pump driver, leading to damage. This can cause motor burn-out or overheating of a gasoline or diesel engine.

At reduced capacities, centrifugal pumps are likely to overheat because units of this class have what is known as a *minimum safe flow*. If the unit runs at a lower capacity, the temperature of the liquid pumped rises until flashing occurs. The unit becomes vapor-bound and stops pumping.

The brake horsepower (bhp) needed to drive a centrifugal pump is that supplied to the coupling by the driver. Water horsepower (whp) developed is the useful work the pump does. Efficiency of the pump is whp/bhp. Bhp includes all losses, mechanical and hydraulic. Mechanical losses are bearing, packing, and disk friction. Hydraulic losses occur in the impeller and volute passages, including leakage. Leakage losses occur at wearing rings, interstage or leakoff bushings, and balancing devices, when used. Heat generated by hydraulic and disk friction causes the fluid temperature to rise during passage through the pump.

Bearing and packing losses are 1 to 2 per cent of total; they are usually neglected when figuring fluid-temperature rise. The rise is caused by the difference between bhp and whp and is called heat horsepower.

Temperature-rise Formula   Figure 5-17 shows the characteristics of a centrifugal pump. Its head, efficiency, bhp, and temperature rise are plotted against capacity in gpm. The temperature-rise curve is calculated from

$$t = \frac{(1-E)H}{778E} \tag{5-1}$$

where t = temperature rise, F; E = efficiency, expressed as a decimal; H = total dynamic head, ft.

Minimum safe flow through a centrifugal pump depends on the suction conditions, especially in boiler-feed service. There, net positive suction head (npsh) determines the minimum safe flow because it fixes the temperature rise of the water before it flashes to steam. This is called the allowable temperature rise.

Minimum Safe Flow   Assume the pump in Fig. 5-17 handles 220-F water and npsh is 18.8 ft. At 220 F, vapor pressure is 17.19 psia and liquid specific gravity is 0.955. The npsh of 18.8 ft = 18.8 × 0.433 × 0.955 = 7.78 psia at 220 F. The vapor pressure to which water may rise before it flashes is 17.19 + 7.78 = 24.97 psia. This pressure corresponds to 240 F. The allowable temperature rise of water is then 240 − 220 = 20 F. A rise of 20 F on the temperature curve corresponds to 47 gpm,

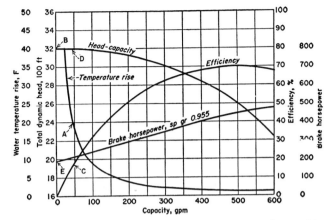

Fig. 5-17 Characteristic curves for a centrifugal pump discussed in text.

point A (Fig. 5-17). This is the minimum safe flow for this pump when handling 220-F water with 18.8-ft npsh.

Trial and Error Plotting the temperature rise of a pump and then finding the capacity at which this equals the allowable rise is an accurate method of calculating the minimum safe flow. This value is often approximated by trial and error, as follows: The temperature rise for a selected minimum capacity is found by trial. If this is about the same as the allowable rise, the minimum safe flow is chosen accordingly. If there is too much difference between the trial and calculated temperature rise, use other trial capacities until a satisfactory one is found. With practice, a minimum safe flow is easily found.

Other Methods A value of the minimum safe flow, accurate enough for all practical purposes without using trial and error, may be found from

$$N = \frac{100h}{778t + h} \tag{5-2}$$

where $N =$ per cent pump efficiency at safe minimum flow; $h =$ shutoff head of pump, ft; $t =$ allowable temperature rise, F.

Assume the pump in Fig. 5-17 handles 220-F water, npsh is 18.8 ft. As figured before, the allowable temperature rise is 20 F. The shutoff head h is 3,200 ft, point B. Then the efficiency for minimum safe flow is

$$N = \frac{100 \times 3{,}200}{(778 \times 20) + 3{,}200} = 17 \text{ per cent}$$

At this efficiency, the capacity is about 47 gpm, point C. This is the same

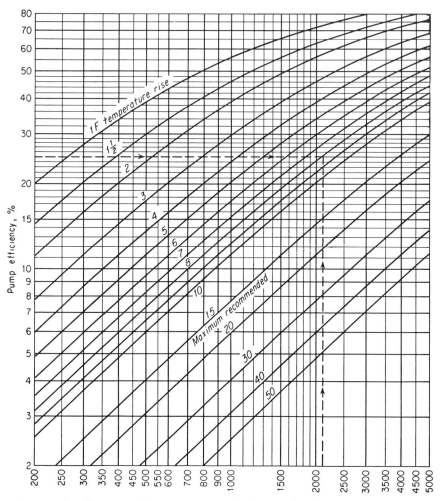

Fig. 5-18 Chart for determining temperature rise in a centrifugal pump.

as obtained by the accurate method of plotting the temperature-rise curve.

The total head at 47 gpm is 3,195 ft, point *D*. The temperature rise for these conditions is

$$t = \frac{(1 - 0.17) \times 3,195}{778 \times 0.17} = 19.99 \text{ F}$$

This shows that the rise will be slightly less than allowable because the head at minimum safe flow is less than shutoff. Thus the error is on the safe side. A satisfactory answer is obtained the first time by Eq. (5-2).

If desired, Fig. 5-18 can be used instead of Eq. (5-2). It gives essentially the same results, as can be verified by solving the above problem by means of the chart.

Other Pump Classes In general, rotary and reciprocating pumps do not have any minimum-flow restrictions. However, it is wise to choose these units as near their normal capacity as possible because better

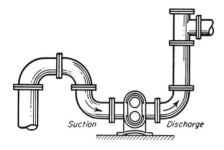

Suction Discharge

Fig. 5-19 Loop in suction piping of rotary pump prevents dry starts. (*Worthington Corp.*)

efficiency is obtained. While rotary pumps may be started dry, a U-shaped loop (Fig. 5-19) or a check valve in the suction line is recommended. Then the moving parts will be coated with liquid when the pump is started. This prevents metal-to-metal contact. Reciprocating pumps should not be started dry.

CHAPTER SIX

Liquid Handled

THE LIQUID HANDLED BY A PUMP affects (1) the head and capacity at which the unit can operate, (2) the required power input to the pump, and (3) the materials of construction that must be used to ensure a satisfactory life for the pump. Four classes of liquids, other than water, are met in pumping problems: (1) viscous, (2) volatile, (3) chemical, and (4) liquids carrying solids in suspension. All require careful study because their handling may present difficult problems in pump selection, construction, and use.

VISCOUS LIQUIDS

Viscosity This is one of the two properties requiring consideration when pumping viscous liquids. The other is liquid specific gravity. Texts on fluid mechanics discuss viscosity units, their derivation and use. For usual pumping problems, viscosity can be considered a measure of the internal friction of a liquid which produces a resistance to flow through a pipe, valve, pump, etc. Though a large number of different viscosity units are used, most engineers commonly work with about three units—Saybolt seconds universal (SSU), centistokes (kinematic viscosity), or centipoises (absolute viscosity).

Figure 6-1 shows the relation between viscosity and temperature for a

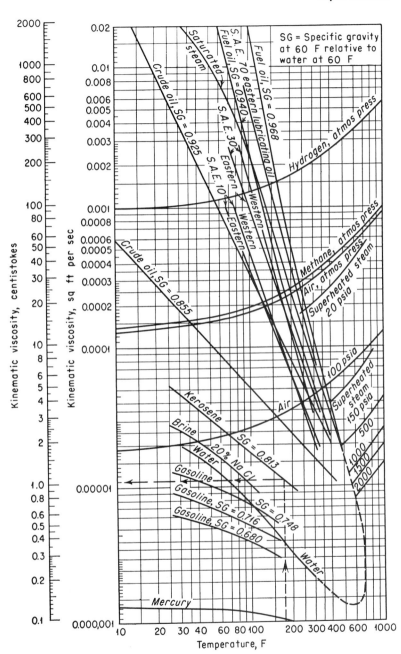

Fig. 6-1 Viscosities of common liquids.

number of common liquids of various specific gravities. It is useful in determining the viscosity of many liquids commonly met in pumping problems. Viscosity data are important in computing the pressure loss in piping systems conveying viscous liquids.

Liquid Flow During passage through a pipe, the flow of a liquid is said to be *laminar* or *turbulent*, depending on the liquid velocity, pipe diameter, liquid viscosity, and density. For any given liquid and pipe these four factors are expressed in terms of a dimensionless number, called the Reynolds number R. For values of R less than about 2,000 the flow is laminar. Particles of the liquid follow separate nonintersecting paths with little or no eddying or turbulence.

When R is higher than about 4,000, turbulent flow is said to exist. Here the paths of the liquid particles are extremely irregular, with much crossing and recrossing.

In the *critical zone*, between R of 2,000 to 4,000, flow is generally assumed to be turbulent, when pressure-loss problems are being solved. This gives safe results because the friction loss is higher for turbulent flow than for laminar flow. Should the flow be laminar, the pipe will be oversized, but not enough so to be uneconomical.

Pressure Loss—Laminar Flow To determine the pressure loss in pipes in which the flow is laminar: (1) Compute the Reynolds number from Eq. (6-1). (2) Express the liquid viscosity in centipoises, using Fig. 6-2, if necessary. (3) Enter Fig. 6-3 with the viscosity, flow rate, and pipe diameter to determine the pressure loss, psi per 100 ft. (4) Convert this loss to that for the actual pipe length.

example: Determine the pressure loss in 500 ft of 4-in. schedule 40 clean steel pipe when the flow rate is 200 gpm of 1,000-SSU crude oil having a specific gravity of 0.9250.

solution: (1) Determine the Reynolds number from

$$R = 3,162s(\text{gpm})/dc \tag{6-1}$$

where R = Reynolds number; s = liquid specific gravity, with respect to water; gpm = flow rate, gallons per minute; d = inside diameter of the pipe, in.; c = liquid absolute viscosity, centipoises. Using Fig. 6-2, enter on the left at 1,000 SSU and project through the specific gravity, 0.925, on the right. At the intersection with the central scale read the absolute viscosity as 200 centipoises. Then, $R = 3,162$ (0.925) (200)/(4)(200) = 732. Flow is laminar because R is less than 2,000. (2) This step was performed as part of (1), above. (3) Enter Fig. 6-3 on the left at 200 centipoises and project through 200 gpm. From the intersection of this trace with the index stem, draw a line through the 4-in. schedule 40 pipe size. Extend this line to the right and read the pressure loss as 4 psi per 100 ft. (4) Pressure loss in 500 ft of pipe is therefore (500/100)(4) = 20 psi.

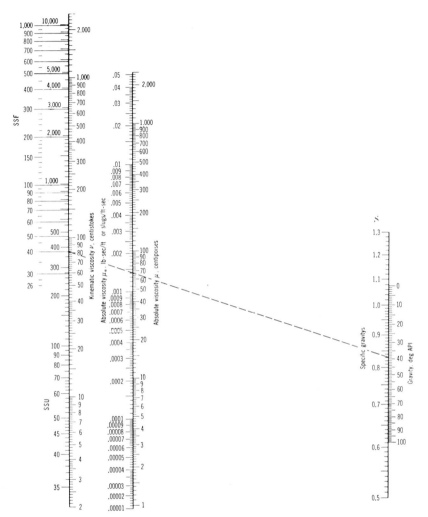

Fig. 6-2 Viscosity conversion chart. (*Crane Co.*)

Pressure Loss—Turbulent Flow   Where the Reynolds number is between 2,000 and 4,000, the safest procedure is to assume turbulent flow exists. For values greater than 4,000, turbulent flow exists for all liquids. To solve these problems: (1) Determine the Reynolds number from Eq. (6-1). (2) Find the corresponding friction factor from Fig. 6-4. (3) Using this factor, the flow rate, specific gravity, and pipe size, find the pressure loss, psi per 100 ft, from Fig. 6-5. (4) Convert this loss to that corresponding to the actual pipe length.

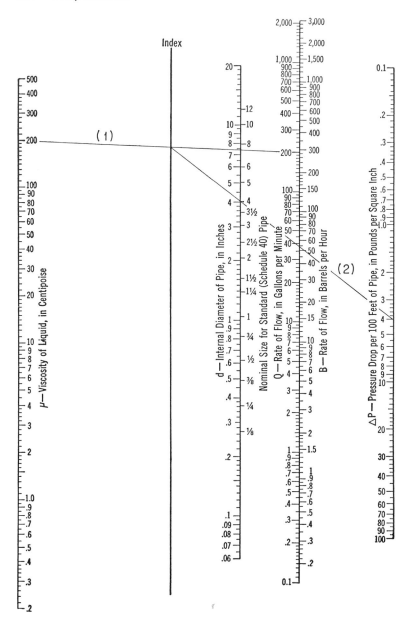

Fig. 6-3 Pressure loss with laminar flow. (*Crane Co.*)

Nominal pipe size

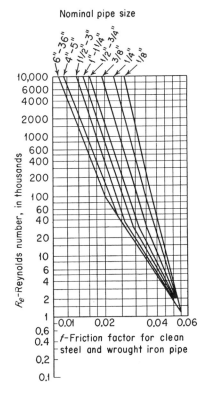

Fig. 6-4 Friction factors for clean steel, wrought-iron, galvanized, cast-iron, and cement pipe.

example: Determine the pressure loss in 5,000 ft of 12-in. ID commercial steel pipe conveying 2,500 gpm of crude oil having a specific gravity of 0.90 and a viscosity of 40 centipoises.

solution: For this pipeline, $R = 3,162(0.90)(2,500)/(12)(40) = 14,850.$ (2) Enter Fig. 6-4 at this value of R and project to the 12-in. pipe curve. Read the friction factor as 0.031. (3) Using this value enter Fig. 6-5 and project as shown to read the head loss as 0.97 psi per 100 ft. (4) Hence, pressure loss in 5,000 ft of pipe is $(5,000/100)(0.97) = 48.5$ psi.

Note several facts about Figs. 6-2 through 6-5. Any liquid-flow problem can be solved by using these charts as directed above because all pressure-loss computations are for either laminar or turbulent flow. These charts may be used for any liquid, provided the Reynolds number is determined prior to entering the viscous- or turbulent-flow chart. If desired, the following equation may be used instead of Fig. 6-5:

$$h = fLV^2/2gD \qquad (6-2)$$

where $h =$ friction-head loss, ft of liquid pumped; $f =$ friction factor from Fig. 6-4; $V =$ average liquid velocity in pipe, ft per sec; $D =$ inter-

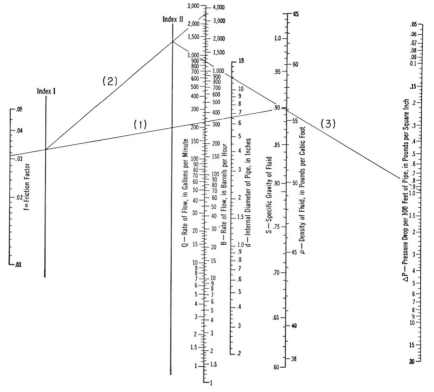

Fig. 6-5   Pressure loss with turbulent flow. (*Crane Co.*)

nal diameter of pipe, ft; $g = 32.17$ ft per sec^2. Liquid velocity may be found from $V = 0.408$ gpm/d^2, where $V =$ liquid velocity, ft per sec; other symbols as before. Pipe diameter can be found from $d = (gpm/2.45V)^{1/2}$.

Use Fig. 6-3 only for laminar flow; Fig. 6-5 is suitable only for turbulent flow. These charts have sufficient range and accuracy for all usual pressure-loss problems met in pumping systems. For other methods of determining pressure loss with laminar or turbulent flow, see the Hydraulic Institute "Pipe Friction Manual."

Viscosity Blending Chart   When two liquids are blended before being pumped through a pipe, the final viscosity after blending will usually be different from the initial viscosity of either liquid. To determine the viscosity to be used for pressure-loss calculations of such liquids, Fig. 6-6 may be used.

To find the final viscosity of a blend, plot the initial lower viscosity on

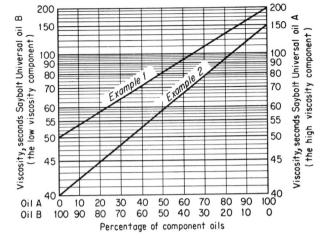

Fig. 6-6   Viscosity blending chart.

the left side of the chart and the initial higher viscosity on the right side. Draw a straight line between these two points. Determine the final viscosity of the blend by entering the bottom of the chart at the percentage composition, projecting vertically upward until the proper line is intersected, and reading the blend viscosity on the right or left scale. For instance, in Example 1 in Fig. 6-6, a 20-80 blend has a final viscosity of 56 SSU, while in Example 2 a 55-45 blend has a final viscosity of 62 SSU. Note that the final viscosity may be read on either side of the chart because both vertical scales contain the same subdivisions.

Fitting Losses   As for other liquids, the losses in fittings are found by using the equivalent length of the particular fitting and determining the loss for a straight pipe of this length, if the flow is turbulent. For laminar flow, it is somewhat more difficult to find fitting losses because few data are available giving exact values needed for computations.

When liquid viscosity is low and flow is near the turbulent region, the equivalent fitting lengths given in Table 4-2 should be used. With high viscosities, above 500,000 SSU, the effect of the fitting is small when the flow is laminar. Usual practice under these conditions is to use only the actual length of the fitting, instead of its equivalent length. Where the pipeline is extremely long compared with the fitting length, or its equivalent length—for example, 10 ft in a 5,000-ft-long pipe—the fitting length or its equivalent length can be neglected in all usual pressure-loss calculations. For intermediate viscosities, 0 to 500,000 SSU, and flows to 1,000 gpm, the fitting equivalent length obtained from Table 4-2 should be multiplied by the applicable factor from Table 6-1.

TABLE 6-1   Fitting Factors for Laminar Flow

(*Flow rate of 0 to 1,000 gpm*)

Liquid Viscosity, SSU	Multiply Equivalent Length of Fitting by
To 500	1.0
501–5,000	0.75
5,001–50,000	0.50
50,001–500,000	0.25

Centrifugal Pumps   Viscous liquids affect the performance of centrifugal pumps in three ways: (1) the pump develops a lower head than when handling water, (2) pump capacity is reduced when moderate- or high-viscosity liquids are handled, and (3) the horsepower input required is higher. In each case, the performance of a pump handling a viscous liquid is compared with the same pump handling water at normal temperature—about 32 to 80 F. The larger power input required results from increased impeller friction, and is constant over a rather wide capacity range.

Performance Corrections   Figure 6-7 provides factors for finding the performance of a conventional centrifugal pump handling a viscous liquid, when its water performance is known or can be found from curves or tables. Use this chart only within its scale limits; do not *extrapolate*. Do not use for mixed-flow or axial-flow pumps, or for pumps of special design. Use only for uniform liquids—slurries, gels, paper stock, etc., may cause incorrect results. Available npsh is assumed adequate for the pump.

example: Select a pump to deliver 750 gpm of 1,000-SSU oil at a total head of 100 ft. The oil has a specific gravity of 0.90 at the pumping temperature.

solution: Enter the bottom of Fig. 6-7 at 750 gpm and project vertically to intersect the 100-ft head curve. From here project horizontally to the correct viscosity curve and then upward to the correction-factor curves. For this example, C_Q = 0.95, C_H = 0.92 (for $1.0Q_{NW}$), and C_E = 0.635. The subscripts Q, H, and E refer to correction factors for capacity, head, and efficiency, respectively, while $_{NW}$ refers to the water capacity at a particular efficiency. At maximum efficiency, water capacity is given as $1.0Q_{NW}$; other capacities are expressed by numbers greater than or less than unity.

Water capacity required = Q_v/C_Q, where Q_v = viscous capacity, gpm. For this pump, Q_w = 750/0.95 = 790 gpm. Likewise water head = H_w = H_v/C_H, where H_v = viscous head. Or H_w = 100/0.92 = 108.8, say 109 ft of water. Choose a pump to deliver 790 gpm of water at 109-ft head and the required viscous head and capacity will be obtained. The pump chosen should be operating at or near its maximum efficiency on water. If this is 81 per cent at 790 gpm, the efficiency when handling the viscous liquid is $E_v = E_w(C_E)$, where E_w = water efficiency. For this unit E_v = 0.81(0.635) = 0.515, or 51.5 per cent. Power input to the pump

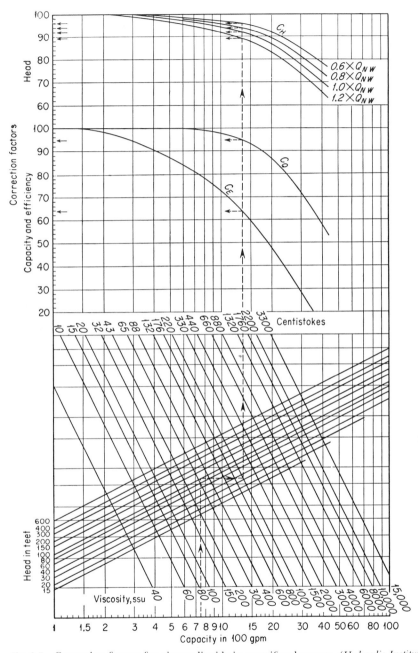

Fig. 6-7 Correction factors for viscous liquids in centrifugal pumps. (*Hydraulic Institute.*)

when handling viscous liquids is given by $P_v = Q_v(H_v)(sg)/3,960(E_v)$, where P_v = horsepower input on viscous liquids; sg = specific gravity of viscous liquid; other symbols as before. Hence, for this unit $P_v = 750(100)$ $(0.90)/3,960(0.515) = 33.1$ hp.

Plotting Characteristic Curves   Figure 6-7 can also be used to plot the complete characteristic curves of a centrifugal pump handling a viscous liquid when the water characteristics are known.   To do this: (1) Secure a characteristic curve of the pump to be used.   (2) Locate the point of maximum efficiency when handling water.   (3) Read the capacity at this point.   (4) Compute the values of 0.6, 0.8, and 1.2 times the capacity at maximum efficiency.   (5) Using Fig. 6-7, determine the correction factors at the capacities given in (3) and (4), above.   Where a multistage pump is being considered, use the head per stage instead of the total head developed, when entering Fig. 6-7.   (6) Correct the head, capacity, and efficiency for each of the flow rates in (3) and (4).   (7) Plot the corrected head and efficiency against the corrected capacity, as in Fig. 6-8.   (8) Compute the power input at each flow rate and plot.   Draw smooth curves through the points obtained.

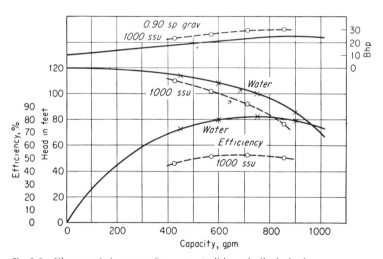

Fig. 6-8   Characteristic curves for water (solid) and oil (dashed)

Figure 6-8 shows the corrected curves for a pump handling 1,000-SSU oil having a specific gravity of 0.90.   The capacity of this pump at its best efficiency point on water is 750 gpm.   The dashed-line curves are the corrected characteristics when handling oil.   Note that the total head of single-stage pumps and the head per stage of multistage pumps are the only suitable values for use in Fig. 6-7.

Entrained Air or Gas These severely reduce the capacity of horizontal centrifugal pumps if the air or gas occupies more than 5 per cent of the liquid volume. In vertical pumps the reduction becomes severe when the amount of air or gas exceeds 10 per cent, except in the propeller and mixed-flow types, where the reduction is only moderate.

Regenerative Pumps Turbine* or regenerative-type pumps handling viscous liquids show the same effects as discussed above—reduced head and capacity, increased power input. When liquid viscosity exceeds 500 SSU, capacity is considerably reduced and there is a large increase in the power input. The resulting efficiency decrease is so large, compared with a positive-displacement pump, that the regenerative pump is seldom used for high-viscosity liquids. Figure 6-9 shows the typical characteristics of a turbine pump handling liquids of various viscosities.

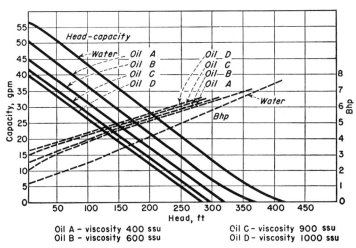

Oil A – viscosity 400 ssu
Oil B – viscosity 600 ssu
Oil C – viscosity 900 ssu
Oil D – viscosity 1000 ssu

Fig. 6-9 Regenerative-turbine pump performance on water and oil (*Aurora Pump, A Unit of General Signal Corp.*)

Rotary Pumps Close-clearance rotary pumps of various types are particularly well suited for handling clear viscous liquids of widely varying viscosities. Some rotary pumps are steam-jacketed to reduce liquid viscosity and permit pumping at lower power inputs. Liquids with viscosities to 250,000 SSU are handled by standard-model rotary pumps of several types—for higher viscosities special designs are usually required. Rotary pumps handle a variety of viscous and nonviscous liquids, including water, gasoline, kerosene, ammonia, freon, etc.

* Throughout this book the word *turbine* is accompanied by another definitive word, like *regenerative, deepwell,* or *close-coupled,* so that the type of pump being referred to can be easily identified. See Chap. 1.

When the temperature of the liquid handled by a rotary pump exceeds 350 F, a water-jacketed stuffing box should be used. At temperatures over 450 F, a forged or cast-steel pump casing is often recommended. When liquid viscosity exceeds 600 SSU, many manufacturers recommend that the speed of a rotary pump be reduced to permit operation without excessive noise or vibration. Table 6-2 shows the typical speed reductions recommended by one builder. Thus a pump handling 100,000-SSU oil and having a rated speed of 800 rpm should run at $800 - (800)(0.50) = 400$ rpm, if the values in the table are followed. The capacity of the pump will vary directly with speed, and in this case will be half that at the rated speed.

TABLE 6-2   Rotary-pump Speed Reduction

Liquid Viscosity, SSU	Speed Reduction, % of Rated Pump Speed
600	2
800	6
1,000	10
1,500	12
2,000	14
4,000	20
6,000	30
8,000	40
10,000	50
20,000	55
30,000	57
40,000	60

Entrained Air or Gas   Many of the liquids handled by rotary pumps contain entrained or dissolved gases. For example, lubricating oils at atmospheric pressure and temperature may contain up to 10 per cent dissolved air by volume. Gasoline, under similar conditions, may contain as much as 20 per cent. When the inlet pressure of a rotary pump is below atmospheric, entrained or dissolved gas in the liquid expands, taking up part of the pump displacement volume and reducing the liquid capacity.

Table 6-3 shows the effect of entrained or dissolved gas on the liquid capacity of a rotary pump having an inlet pressure below atmospheric. The liquid capacity of a rotary pump is theoretically reduced at least in direct proportion to the amount of air or gas presented at the inlet flange of the pump. The actual reduction is greater, depending on the amount of pump clearance volume and the reduction in pressure inside the pump.

TABLE 6-3 Effect of Entrained or Dissolved Gas on the Liquid Displacement of Rotary Pumps

(Liquid displacement: % of displacement)

Vacuum at pump inlet, in. Hg	Gas entrainment					Gas solubility					Gas entrainment and gas solubility combined				
	1%	2%	3%	4%	5%	2%	4%	6%	8%	10%	1% 2%	2% 4%	3% 6%	4% 8%	5% 10%
5	99	97½	96½	95	93½	99½	99	98½	97	97½	98½	96½	96	92	91
10	98½	97¼	95½	94	92	99	97½	97	95	95	97½	95	93	90	88¼
15	98	96½	94½	92½	90½	97	96	94	92	90½	96	93	89½	86½	83¼
20	97½	94½	92	89	86½	96	92	89	86	83	94	88	83	78	74
25	94	89	84	79	75½	90	83	76½	71	66	85½	75½	68	61	55

For example: with 5% gas entrainment at 15 in. Hg vacuum, the liquid displacement will be 90½% of the pump displacement neglecting slip or with 10% dissolved gas liquid displacement will be 90½% of pump displacement, and with 5% entrained gas combined with 10% dissolved gas, the liquid displacement will be 83¼% of pump displacement.

Reciprocating Pumps For viscous liquids, reduce the piston speed of direct-acting and power-type reciprocating pumps. Since pump capacity is a function of piston speed, these pumps deliver less capacity with a viscous liquid than with water because the pump speed is lower. Table 3-3 shows the effect of viscosity on pump speed. Thus, a pump handling 100 gpm of water at 50 revolutions or strokes per minute should run at only $0.80(50) = 40$ rpm and will deliver 20 per cent less capacity, or 80 gpm, when handling 2,000-SSU crude oil, if the values from Table 3-3 are applicable. See Chap. 7 for design pointers on piping for pumps handling viscous liquids. Entrained air or gas has only a mild effect on the capacity of reciprocating pumps.

VOLATILE LIQUIDS

Gasoline, kerosene, naphtha, refrigerants, and similar liquids are generally classed as volatile because they vaporize readily at normal atmospheric temperatures and pressures. However, any liquid at or near its boiling temperature is not a volatile state and may be classed as volatile, so far as its effect on a pump is concerned.

The principal problem encountered in pumping volatile liquids is one of net positive suction head (npsh). As discussed earlier (Chap. 4), the system npsh must equal or be greater than the required npsh of the pump to prevent vaporization of the liquid in the suction pipe. Vaporization on the discharge side of the pump is rarely a problem because the pressure is usually sufficiently high to produce a much higher boiling temperature.

Vapor Pressure   Before npsh calculations can be made, the vapor pressure of the volatile liquid at suction conditions of the pump must be obtained.   Figure 6-10 gives the approximate vapor pressure of a number of different volatile liquids at various temperatures.   Values read from this chart are satisfactory for many problems, but wherever possible the actual vapor pressure of the liquid being handled should be obtained by test.   See Chap. 9 for a number of pointers on suction piping for pumps handling volatile liquids.

When gasoline, gas oil, kerosene, or similar hydrocarbons are handled

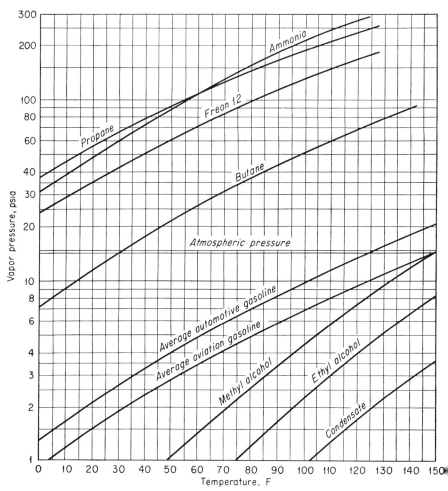

Fig. 6-10   Approximate vapor pressure of various liquids. (*Johnston Pump Co.*)

with only nominal suction pressure, a stick-grease-type lubricator can be used to inject grease into the packing-box seal cage, if grease is not objectionable from a contamination standpoint. Figures 1-32 and 1-33 show a number of seal rings and stuffing boxes. Where the pump suction pressure is so high that the packing will not function properly, a bleed-off bushing can be fitted at the inlet end of the box. This bushing resembles a seal cage and the liquid bled through it passes to a bleed-off connection which is run to a region of lower pressure.

Wherever volatile liquids are handled, mechanical seals should be considered. They give excellent service in all types of pumps in refineries, petrochemical plants, tankers, barges, etc.

CHEMICALS

Many acidic and basic liquids are handled by all classes of pumps. Compared with water, these liquids are often more difficult to pump because they corrode or attack various parts of the pump, reducing its life unless special precautions are taken in the construction of the unit. Generally, eight factors affect the choice of materials for pumps: liquid temperature, head per stage, discharge pressure, pump load factor, suspended-solids abrasiveness, corrosion resistance of the metal, electrochemical action, and structural considerations.

Tables 6-4 and 6-5 summarize the broad experience of pump manufacturers and users with various materials and liquids, as designated by the Hydraulic Institute. The materials listed are applicable to all classes of pumps so far as corrosion-resistance characteristics are concerned. But in some pumps, design considerations impose differing physical requirements, which may eliminate some materials listed as suitable for a given liquid. Table 6-5 is a major condensation of a much longer tabulation of materials for handling various liquids given in the Standards of the Hydraulic Institute. For liquids not listed here, the reader is referred to this standard.

Obviously, tabulations cannot recognize all possible construction variations but must concern themselves primarily with the general problem of corrosion in pumps. This must be remembered when evaluating the merits of the various materials listed in Tables 6-4 and 6-5.

The materials listed are those commonly used in the principal parts of the pump, such as the casing, impeller, and cylinders. Where possible, wrought materials, such as for shafts, should be of similar composition to the castings used. Trim items like sleeves, glands, rings, and valves may be made of different materials to meet some mechanical requirement

TABLE 6-4   Hydraulic Institute Materials Selections and Designations

Material selection No.	Corresponding natural society* standards designation			Remarks
	ASTM	ACI	AISI	
1	A48, classes 20, 25, 30, 35, 40, and 50	. . . .	. . .	Gray iron—six grades
1(a)	A339, A395, and A396	. . . .	. . .	Nodular cast iron—five grades
2	B143, 1B, and 2A; B144, 3A; B145, 4A	. . . .	. . .	Tin bronze—six grades [includes two grades not covered by ASTM specifications as explained in Hydraulic Institute *Standards*]
3	A216, WCB	. . . .	1030	Carbon steel
4	A217, C5	. . . .	501	5% chromium steel
5	N296, CA15	CA15	410	13% chromium steel
6	A296, CB30	CB30	. . .	20% chromium steel
7	A296, CC50	CC50	446	28% chromium steel
8	A296, CF8	CF8	304	18-8 austenitic steel
9	A296, CF8M	CF8M	316	18-8 molybdenum austenitic steel
10	A296, 60T	Cn-7M	. . .	A series of highly alloyed steels normally used where the corrosive conditions are severe
11	. . . .	. . . .	. . .	A series of nickel-base alloys
12	. . . .	. . . .	. . .	High-silicon cast iron
13	. . . .	. . . .	. . .	Austenitic cast iron
13(a)	A439	. . . .	. . .	Nodular austenitic cast iron
14	. . . .	. . . .	. . .	Nickel-copper alloy
15	. . . .	. . . .	. . .	Nickel

*ASTM, American Society for Testing Materials; ACI, Alloy Casting Institute; AISI, American Iron and Steel Institute.

but the material selected should be suitable for the environment. Because of the many variables, it is impossible to say with certainty that some one material will best withstand the corrosive attack of a given liquid. So more than one material is usually listed for a given liquid, as shown in Table 6-5. The order of the listing does not indicate relative superiority.

The symbols used in Table 6-5 are: A, an all-bronze pump; B, bronze-fitted pump; C, all-iron pumps. Other materials, including a large number of special corrosion-resistant types, are listed in Table 6-4. To aid the pump manufacturer in choosing the correct materials for a given application, it is wise to use the Hydraulic Institute form shown in its

TABLE 6-5   Hydraulic Institute Standard Materials for Pumping
Various Liquids

Liquid	Condition	Material selection
Acid, acetic...........	Conc. cold	8, 9, 10, 11, 12
Acid, hydrochloric......	Commercial conc.	11, 12
Acid, nitric...........	Conc. boiling	6, 7, 10, 12
Acid, sulfuric..........	>77%, cold	C, 10, 11, 12
Alcohols..............		A, B
Asphalt..............	Hot	C, 5
Beer.................	......	A, 8
Beet juice............	.....	A, 8
Brine, CaCl..........	pH >8	C
Brine, CaCl..........	pH <8	A, 10, 11, 13, 14
Brine, NaCl..........	<3%, cold	A, 6, 13
Brine, NaCl..........	>3%, cold	A, 8, 9, 10, 11, 13, 14
Calcium hypochlorite...	......	C, 10, 11, 12
Carbon tet...........	Anhydrous	B, C
Formaldehyde........	......	A, 8, 9, 10, 11
Gasoline.............	......	B, C
Glue.................	Hot	B, C
Lead................	Molten	C, 3
Milk................	......	8
Molasses.............	......	A, B
Mustard.............	......	A, 8, 9, 10, 11, 12
Oil, crude............	Hot	3
Oil, fuel.............	......	B, C
Oil, kerosene.........	......	B, C
Oil, lube.............	......	B, C
Sewage..............	......	A, B, C
Soda ash.............	Hot	8, 9, 10, 11
Sodium hydroxide......	Aqueous sol.	C, 5, 8, 9, 10, 11, 13, 14
Sugar................	Aqueous sol.	A, 8, 9, 10, 11, 13
Tar.................	Hot	C, 3
Water, feed..........	Not evaporated	
	pH >8.5	C
	pH <8.5	B
	Evaporated, any pH	4, 5, 8, 14

Standards when supplying data for the order. This form is also useful for reviewing the factors which are important in selecting materials to handle any chemical liquid. Table 6-6 gives several general rules useful in preliminary selection of centrifugal pumps for a number of different liquids varying in pH from 0 to 14. Table 6-7 lists the pH of a number of liquids commonly handled by pumps.

TABLE 6-6   Pump Materials Chosen for Various Liquids

Liquid pH	Type of Pump Often Chosen
0–4	Stainless steel
4–6	All-bronze
6–9	Standard-fitted
9–14	All-iron

TABLE 6-7   Typical pH Values of Various Liquids

Acids		Foods	
Acetic acid, N	2.4	Beer	4.0–5.0
Arsenious acid (saturated)	5.0	Cider	2.9–3.3
Citric acid, 0.1N	2.2	Drinks, soft	2.0–4.0
Formic acid, 0.1N	2.3	Eggs, fresh white	7.6–8.0
Hydrochloric acid, N	0.1	Grapefruit	3.0–3.3
Hydrocyanic acid, 0.1N	5.1	Grapes	3.5–4.5
Lactic acid, 0.1N	2.4	Lemons	2.2–2.4
Sulfuric acid, N	0.3	Maple sirup	6.5–7.0
Bases		Milk, cow's	6.3–6.6
		Oranges	3.0–4.0
Ammonia, N	11.6	Shrimps	6.8–7.0
Blood plasma, human	7.3–7.5	Tomatoes	4.0–4.4
Calcium carbonate	9.4	Turnips	5.2–5.6
Lime (saturated)	12.4	Vinegar	2.4–3.4
Potassium hydroxide, N	14.0	Water, drinking	6.5–8.0
Sodium carbonate, 0.1N	11.6	Wines, grape	2.8–3.8
Sodium hydroxide, N	14.0		

Special Considerations In certain applications—for example, food processing—porcelain, glass, and certain synthetic materials may be used in pump construction because they are better suited to the pumping requirements. Where a pump must handle liquids at extremely low temperatures (like brine), the casing or cylinder is often made of alloy cast iron or cast steel because ordinary cast iron becomes brittle at low temperatures. For handling strong electrolytes, an all-iron pump is generally recommended. Stainless-steel alloys, now available at lower cost, are replacing common steel in many pumps. One typical example is stainless-steel shaft sleeves in boiler-feed pumps. These replace the older bronze sleeves. With flammable liquids, bronze-bushed iron or steel glands should be used to prevent the possiblity of sparks.

While it might seem an obvious course, many engineers overlook the help manufacturers can give in materials choice. The wide experience of most manufacturers provides them with an excellent background for solving a variety of materials problems with little or no difficulty. So every engineer who must select and apply pumps is wise to use the services of one or more manufacturers. They can save the engineer time and help ensure more satisfactory results.

SOLIDS IN SUSPENSION

Perhaps the most difficult liquids to handle satisfactorily in pumps of any class are those which contain solids in suspension. These liquids may contain sewage, paper stock, slurries, sand, and foods of various types. However, the many successful installations of various classes of pumps for these services show that careful selection and application will provide the desired head and capacity, together with long pump life. Today, with such a wide variety of special pump designs available, it is relatively easy to choose a unit for a specific application and know that it will perform for long periods with a minimum of trouble. Many of these specialized designs are discussed in later chapters dealing with various industries.

Liquid Velocity In every pump installation, liquid velocity is important from an economic standpoint, as discussed in Chap. 10. An additional consideration enters in installations where a pump is handling a liquid containing solids in suspension. This is the *fall velocity* or hydraulic subsiding values necessary to prevent the solids from depositing out of the liquid and collecting on the bottom of the pipe. In selecting the velocity for the pump suction and discharge pipes, effort should be made to see that the flow will be such that the solids travel in or near the center of the pipe. This type of flow condition keeps friction in the pipe to a minimum, protects the solids from disintegration on the pipe wall, and reduces wear caused by abrasion of the pipe walls by the solids. The most economical velocity is the lowest producing flow in which these three conditions prevail. In general, reciprocating and rotary pumps are suitable for handling liquids with little or no abrasiveness, while centrifugal pumps will handle the most abrasive types of suspensions. There are, however, exceptions—for example, single-screw rotary

pumps with rubber-coated impellers handle abrasive suspensions exceptionally well.

Fluid Flow   It is difficult to describe or visualize the fluidity of a
mixture of water and fibrous material. But paper stock resembles
cooked cereal—say oatmeal or cornmeal—and has similar flow characteristics. The same is true of many other similar mixtures—foods,
sewage, slurries, etc. If a jar is filled with cooked cereal or paper stock
(Fig. 6-11) it will have to be tilted at an angle to make the mixture flow
from the jar. The heavier or thicker the material, the greater the angle
through which the jar must be tilted to make the mix flow. A few
examples will show how far the jar must be tilted for various mix consistencies.

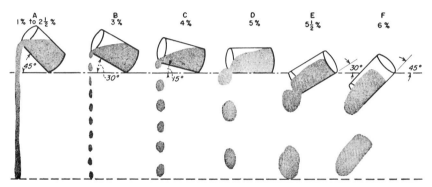

Fig. 6-11   Flow characteristics of thick liquids may be visualized as shown here. (*Goulds
Pumps, Inc.*)

Between 1 and 2.5 per cent consistency (Fig. 6-11*A*) the mix flows in a
continuous stream, much like water, when the jar is tilted at an angle of
about 45 deg. The surface of the mixture in the jar remains nearly flat.
At 3 per cent consistency (Fig. 6-11*B*) the jar must be tilted to 30 deg.
Flow changes from a solid stream to small continuous droplets. The
surface of the mix in the jar begins to show a convex shape. With a mix
consistency of 4 per cent (Fig. 6-11*C*) the angle of tilt is 15 deg. The mix
emerges in larger drops and less continuously. The surface is more
convex. At 5 per cent (Fig. 6-11*D*) the jar side is horizontal; the mix
drops in quite large gobs. The surface of the mix in the jar is quite
rounded, and V-shaped cracks appear as the mix begins to drop from
the jar.

With a 5.5 per cent mix, the jar must be tilted so its side is about 30 deg
below the horizontal. The surface of the mix shows large V-shaped
cracks as it leaves the jar. The entire contents may leave in two or three
large gobs. Note how the mix removes itself from the bottom of the jar

(Fig. 6-11*E*). When the mix is 6 per cent, the jar must be tilted 45 deg below the horizontal and the whole mass leaves as one large single body or slug (Fig. 6-11*F*).

Accuracy   While Fig. 6-11 may not be exactly correct as to the angle of tilt in relation to the consistency or behavior of all thick liquids, it can serve as a rough guide in finding the fluidity of many thick mixtures.   If their flow is observed they can be compared with the thick liquid about which the greatest amount of data have probably been accumulated— paper stock.   A comparison like this can be a big help in choosing the proper type of pump and impeller for a given job.   Figures 6-12 and

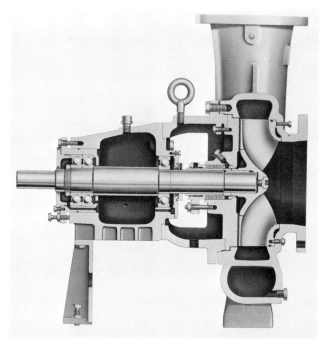

Fig. 6-12 Paper-stock pump with open end-suction impeller. Back pull-out design permits disassembly without disturbing piping connections. (*Goulds Pumps, Inc.*)

6-13 show typical centrifugal pumps designed to handle paper stock, fruit pulp, mashes, and similar thick liquids.

Pointers   When pumping thick liquids made up of a mixture of a liquid and a solid, the most important consideration is to provide suitable suction head and piping so the material reaches the impeller entrance freely and easily.   Here the special warped vanes of the impeller or the special piston or plunger can pick it up without unwater-

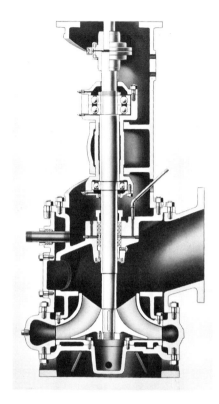

Fig. 6-13 Vertical centrifugal stock pump has open-type impeller. (*Goulds Pumps, Inc.*)

ing or clogging. When proper suction flow to the pump is secured, the remainder of the problem is simple. Almost any centrifugal or reciprocating pump will force the mix into the discharge line. Discharge head is figured in the usual manner, using the resistance found for the consistency being handled.

Friction Losses   The curves in Figs. 6-14 through 6-21 give the friction loss of paper-stock suspensions of various consistencies in schedule 40 steel pipe. They are based on work done by the University of Maine on the data of Brecht and Heller of the Technical College, Darmstadt, Germany. The charts may be used for all pipe materials. But friction in cement-asbestos pipe may be lower than the values shown.

With some types of stock the friction loss read from the charts must be corrected by a suitable factor. For soda, sulfate, bleached sulfite, and reclaimed paper stocks, multiply by 0.90; for ground wood, by 1.40.

For stock consistencies below 1.5 per cent, use water-friction values. Below a consistency of 3 per cent, the velocity of flow should not exceed

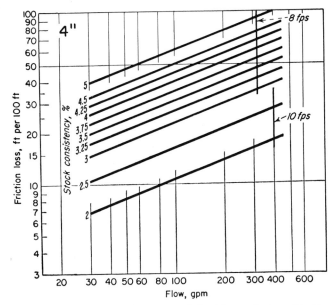

Fig. 6-14 Friction loss of paper stock in 4-in. steel pipe. (*Goulds Pumps, Inc.*)

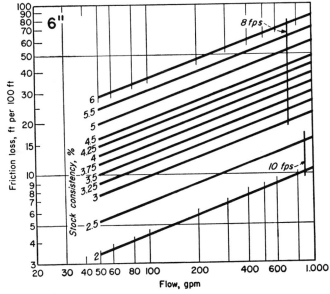

Fig. 6-15 Friction loss of paper stock in 6-in. steel pipe. (*Goulds Pumps, Inc.*)

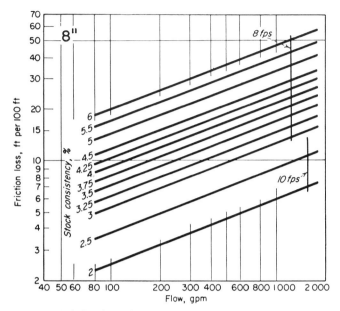

Fig. 6-16 Friction loss of paper stock in 8-in. steel pipe. (*Goulds Pumps, Inc.*)

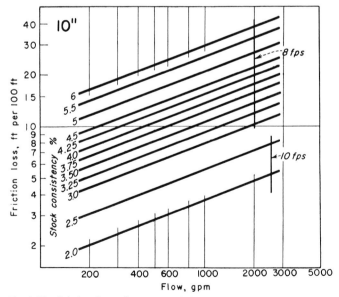

Fig. 6-17 Friction loss of paper stock in 10-in. steel pipe. (*Goulds Pumps, Inc.*)

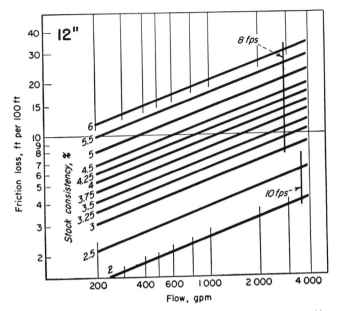

Fig. 6-18 Friction loss of paper stock in 12-in. steel pipe. (*Goulds Pumps, Inc.*)

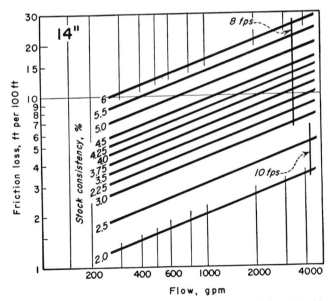

Fig. 6-19 Friction loss of paper stock in 14-in. steel pipe. (*Goulds Pumps, Inc.*)

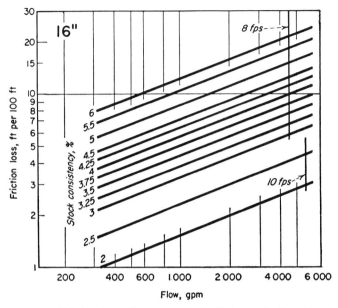

Fig. 6-20 Friction loss of paper stock in 16-in. steel pipe. (*Goulds Pumps, Inc.*)

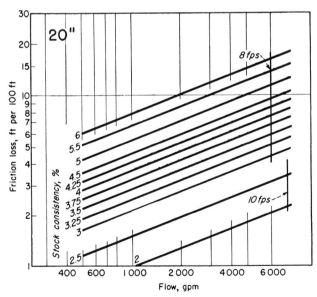

Fig. 6-21 Friction loss of paper stock in 20-in. steel pipe. (*Goulds Pumps, Inc.*)

10 ft per sec. For suspensions of 3 per cent and above, the maximum velocity in the pipe should be limited to 8 ft per sec.

example: Find the friction loss in 100 ft of 12-in. schedule 40 pipe when 1,000 gpm of sulfate stock is flowing. The consistency of the stock is 4.5 per cent.

solution: Enter the chart for 12-in. schedule 40 pipe (Fig. 6-18) at a flow of 1,000 gpm and project upward until the 4.5 per cent consistency curve is intersected. At the left read the loss as 12 ft per 100 ft of pipe. Correcting for sulfate stock, using the value from the data above, loss per 100 ft = (0.90) (12) = 10.8 ft. Note that these charts may be used for both suction and discharge piping.

Pump Selection Centrifugal pumps, perhaps the most widely used class for handling paper stock and similar liquids, usually develop less head and deliver less capacity with thick liquids than when handling pure water. For this reason it is often necessary to correct the water performance of a pump before its suitability for paper-stock use can be determined. Figures 6-22 and 6-23 give head and capacity correction factors for centrifugal pumps handling paper stock. Here is an example of their use in pump application.

example: Approximately what capacity and head must a pump develop to deliver 1,000 gpm of 5 per cent air-dry kraft stock at 75 ft total head? What do these values mean?

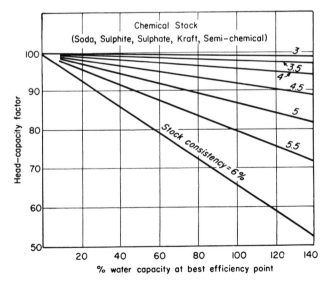

Fig. 6-22 Head-capacity correction factors for chemical stock. (*Goulds Pumps, Inc.*)

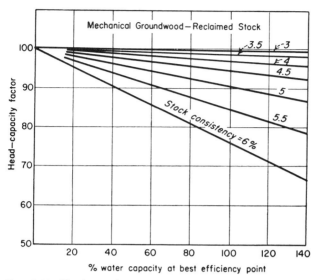

Fig. 6-23 Head-capacity correction factors for mechanical groundwood and reclaimed stock. (*Goulds Pumps, Inc.*)

solution: Enter Fig. 6-22, the correction chart for chemical stock, at 100 per cent capacity and project upward to the 5 per cent consistency curve. At the left read the water-rating correction factor as 0.87. Divide the stock head and capacity by the correction factor, or 1,000/0.87 = 1,150 gpm, and 75/0.87 = 86.2 ft total head.

These values mean that a pump rated at 1,150 gpm of water at a total head of 86.2 ft will deliver at least 1,000 gpm of 5 per cent kraft stock at 75 ft total head. Water ratings are used by manufacturers for almost all centrifugal pumps because it is extremely difficult to test pumps for the large number of other liquids which they may handle in actual service. Use of a correction factor as shown here is easier and saves time and money.

The correction factors in Figs. 6-22 and 6-23 are approximate for normally refined stock of 400 to 600 SR sec freeness, with short direct suction lines of ample size. Faster or freer stocks, and those with entrained air, may require greater conditions. Slow stocks and those containing fillers and additives may require less correction because the stock stays more readily in suspension. But overdosing with additives like alum may cause gas formation on the stock fibers, interrupting pumping.

Actual Pump Figure 6-24 shows the characteristics of a pump which might be suitable for the conditions given in the above example. To use this curve, the exact impeller diameter required for the water conditions must be found if a sizable power saving is desired when the unit handles paper stock.   Here is how to do this.

1. Locate the water-rating point *A* (Fig. 6-24) at 1,150 gpm and 86 ft total head.   To deliver this head and capacity a 15-in.-diameter impeller would be used because point *A* just about falls on the curve for this diameter.

2. The capacity of this pump at its best efficiency point (bep) *B* (Fig. 6-24) is 1,600 gpm for an impeller of slightly smaller diameter than that for the water rating.   Multiply this capacity by the correction factor obtained above to find the stock capacity at the bep, or 1,600 (0.87) = 1,392 gpm.

3. Next, find the ratio of the stock capacity at the required rating (1,000 gpm) and the stock capacity found in step 2 (1,392 gpm).   Or, (1,000/1,392) = 0.72.   This can be taken as equivalent to 72 per cent efficiency of the pump.

4. Where the ratio obtained in step 3 is not close to the decimal expression of the best efficiency of the pump when handling water, 0.82 for this unit, and the stock consistency is greater than 3.5 per cent, subtract the per cent consistency of the stock from the ratio to secure a closer approximation to the per cent water capacity at the bep.   Or 0.72 − 0.05 = 0.67, for this pump.

5. Find the correction factor for 67 per cent capacity by entering Fig. 6-22 at 67 per cent and projecting upward to a consistency of 5 per cent.

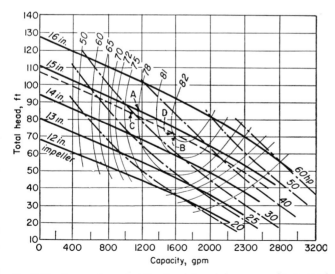

Fig. 6-24   Pump characteristics for determining approximate stock performance of unit discussed in text. (*Goulds Pumps, Inc.*)

At the left read the correction factor as 0.91. Divide the stock head and capacity by this correction factor to determine the final water rating of the pump. This gives 1,000/0.91 = 1,100 gpm, and 75/0.91 = 82.4 ft total head. Locate this point in Fig. 6-24. It occurs at an impeller diameter of 14.75 in., point C. To check the validity of this impeller diameter find the water capacity at the bep. This value must approximately equal the capacity used in step 2. Figure 6-24 shows a capacity of 1,650 gpm. This compares favorably with the 1,600 gpm used in step 2. If the capacities vary appreciably, repeat steps 2, 3, 4, and 5, using another value in step 2.

6. To plot approximate head-capacity curves for this pump handling water and stock, tabulate the water head, efficiency, and bhp from Fig. 6-24 at 40, 60, 80, 100, and 120 per cent of rated capacity at the bep for the impeller diameter found in step 5, or 14.75 in.

7. Tabulate the correction factors, using the above per cent capacities and Fig. 6-22. Using these, compute the stock gpm and total head, as above. Horsepower input will remain approximately the same. Plot the computed values (Fig. 6-25). Pump efficiency is computed from $E =$

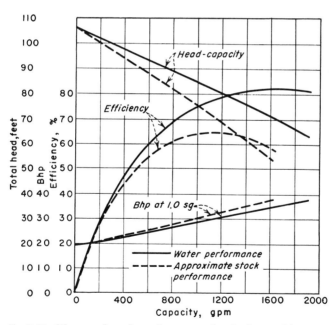

Fig. 6-25 Water and stock performance of unit discussed in text. (*Goulds Pumps, Inc.*)

(gpm) (head, ft)/(3,960)(bhp)(stock specific gravity). Note that the horsepower on stock is almost the same as on water.

STUFFING-BOX SEAL LIQUID

A flow of from 40 to 60 drops per minute out of a normal packed-type stuffing box is required to provide lubrication and to dissipate the heat generated. An external sealing liquid is needed if the internal pump pressure acting against the inside of the stuffing box is negative (suction lift), the liquid being pumped has solids in suspension, is highly volatile, or has poor lubricating qualities.

If the internal pump pressure acting against the inside of the stuffing box is negative (suction lift), air in the atmosphere will be at a higher pressure. A liquid barrier is required to prevent air from entering the pump through the stuffing box and causing loss of prime and possible seizing of rotating parts. The sealing liquid is piped externally to the stuffing box and distributed between rows of packing through a seal cage. Some of the sealing liquid will be "sucked" into the pump, and the remainder will trickle out of the stuffing box. In this case the sealing liquid acts as both a lubricant and a barrier against entrance of air. If the liquid pumped is clean and has lubricating qualities, seal lines may be connected between the stuffing box and the discharge-pressure area of the pump casing. If the liquid pumped has solids in suspension, the stuffing box should be sealed with clean water from the plant water-supply lines or some other clean liquid available under pressure. The use of plant water assures a positive seal on the stuffing box during pump priming.

If the internal pump pressure acting against the inside of the stuffing box is positive, from either suction or discharge head, the natural flow of liquid trying to escape to the atmosphere will seal and lubricate the packing, and *the seal lines from the pump casing should not be connected.* Adding pumped liquid under discharge pressure will not improve sealing. It will greatly increase stuffing-box pressure, maintenance, and frequency of repacking.

If the pumped liquid contains solids in suspension, leakage through the stuffing box must be minimized. The solids will impinge on the packing and score the shaft sleeve. External sealing with clean plant water is recommended. The pressure in the seal lines should be at least 10 psi above the pressure acting on the stuffing box, to ensure a flow of liquid into the pump and prevent the entrance of solids.

Figure 6-26 shows typical impellers used to pump paper stock.   These impellers have back pump-out vanes to prevent paper-stock buildup on the impeller.

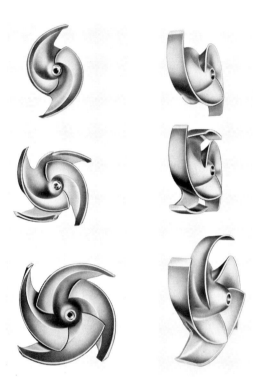

Fig. 6-26   Two views of paper-stock impellers show special warped vanes. These open impellers have back pump-out vanes to prevent build-up of material that could bind and prevent free rotation. (*Goulds Pumps, Inc.*)

CHAPTER SEVEN

Piping Systems

Piping for pumps can be conveniently classified into three broad categories—suction, discharge, and auxiliary lines. Since, in many installations, the head to be developed is principally a function of the piping resistance, extreme care is necessary in choosing pipe size and arrangement. Friction losses were studied in earlier chapters; here the discussion is confined to piping size and arrangement. The success or failure of any pumping system is usually a direct function of the degree of suitability of its piping.

SUCTION PIPING

From the standpoint of importance, suction piping probably rates somewhat higher than discharge piping because fewer serious difficulties can occur from wrongly sized discharge lines than from suction pipes. Insufficient npsh, hydraulic instability with formation of a strong vortex leading to vibration, noise, cavitation, and excessive bearing wear are but a few of the troubles experienced with poorly designed suction piping. Others include reduced capacity, water hammer, overheating of the pump, and reduced life of the operating parts.

Suction-pipe Inlet Figure 7-1 shows three common forms of inlets

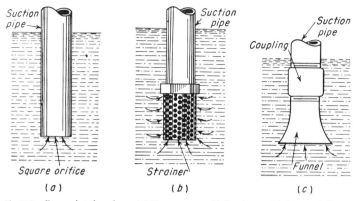

Fig. 7-1 Pump intake pipes. (*a*) Temporary. (*b*) Strainer with integral foot valve. (*c*) Belled or funnel-type intake.

used for industrial pumps. A simple pipe (Fig. 7-1*a*) is suitable only for temporary installations because its entrance loss is likely to be excessive. A strainer with an integral foot valve (Figs. 7-1*b* and 7-2) is preferred because there is less danger of foreign matter entering the intake piping. Also, the water held in the suction pipe by the foot valve will eliminate the need for priming the pump after a shutdown. Suction bells (Fig. 7-1*c*) can be built with or without foot valves and are useful where suction losses must be kept to a minimum. Good practice dictates use of a belled intake with a foot valve and strainer wherever possible in permanent installations of most reciprocating and centrifugal pumps.

Several common troubles are often met when a vertical intake pipe is

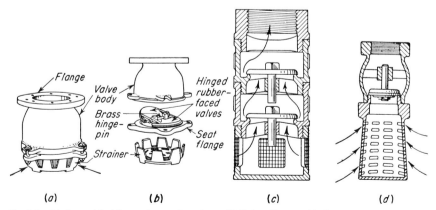

Fig. 7-2   (*a*) Assembled foot valve and strainer. (*b*) Valve disassembled. (*c*) Strainer with two foot valves. (*d*) Strainer with single poppet valve.

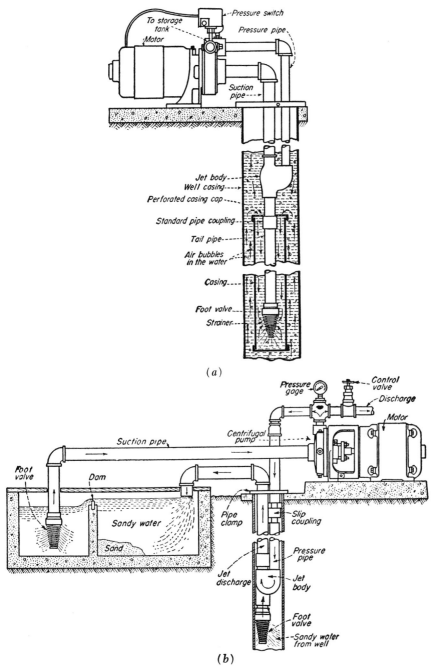

Fig. 7-3 (a) Enclosing foot valve helps prevent entrance of air bubbles. (b) Settling chamber removes sand from water.

used for a pump. With a low liquid level in the pump intake chamber, or insufficient submergence of the intake pipe, a vortex may be set up, entraining air in the liquid. Best remedy is use of a belled inlet sized to keep the liquid velocity lower than 3 ft per sec. When the liquid supply line enters the intake chamber close to the intake pipe, formation of air bubbles can be prevented by submerging the outlet end of the supply pipe. To prevent swirl at the intake when the supply pipe enters at one side of a cylindrical chamber, a baffle can be used at the outlet of the supply pipe. Figure 7-3 shows schemes helpful in planning intake piping for jet pumps.

Intake Design Many useful data, especially for vertical pumps, have been obtained by model tests of various intake designs. Channels guiding water to vertical pumps measurably affect unit performance and dependability. Though intakes for single-pump installations are relatively simple, multipump intakes require much ingenuity during design, particularly where the capacity of existing facilities is increased by insertion of more pumps into a given space. Figure 7-4 shows a

Fig. 7-4 Intake model fitted with flow-indicated flags. (*Ingersoll-Rand Co.*)

typical intake model fitted with flow-indicating flags. In a similar model (Fig. 7-5) the effect of reducing intake-bell clearance is shown in the lower photo. Note, in the upper photo, how flow is either across the bottom of the intake bell or in a vortex pattern, as indicated by the positions of the flow flags. Lowering the bells gives uniform radial flow into each suction pipe, as the flags in the lower photograph show.

Channel flow is the most important aspect in hydraulic-stability

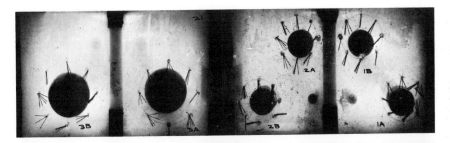

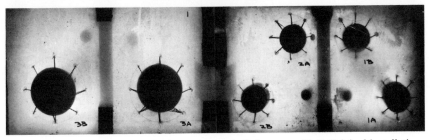

Fig. 7-5 Test models before and after reducing bell clearance. (*Top*) Original installation. (*Bottom*) Revised installation. (*Ingersoll-Rand Co.*)

studies. Sharp corners, abrupt turns, and nonsymmetry cause channel-flow disturbances. With vertical pumps, hydraulic losses are seldom a consideration because the velocities in the intake are low. Prevention of eddies and vortices that disturb flow is the principal problem to solve. This may be done by changing a channel design to eliminate the disturbances or by isolating them in areas where they will do no harm.

Typical Designs Figure 7-6 shows an installation proposed for a new

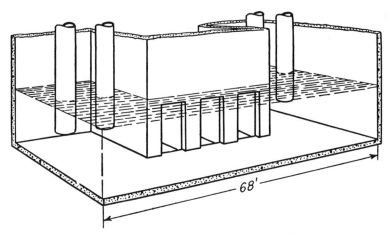

Fig. 7-6 Proposed intake for four vertical pumps. (*Ingersoll-Rand Co.*)

building.   Four large vertical centrifugal pumps were to be used—two in each bay of the intake.   Model tests immediately showed that the large quantity of water emerging from the sluice gates impinged against the facing wall, creating a strong flow distortion.   The abrupt 90-deg turn ahead of the pumps aggravated this condition.   The average velocities around the pumps appeared low, but the lack of guides allowed disturbances to reach the pumps and set up vortices.   Figure 7-7 shows the

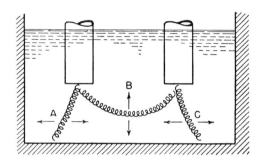

Fig. 7-7   Vortices set up in intake in Fig. 7-6. (*Ingersoll-Rand Co.*)

vortices observed with various combinations of pumps and sluice gates. Because of high bell clearances and the large area behind them, rotating bodies of water under the pumps with vortices *A* and *C* (Fig. 7-7) moved about whenever disturbing influences from the sluice gate reached them.   The water surface gave no sign of the conditions below it.

At increased capacities, the vortices became more positive, showing that the flow pattern was unsound and needed correction.   Figure 7-8

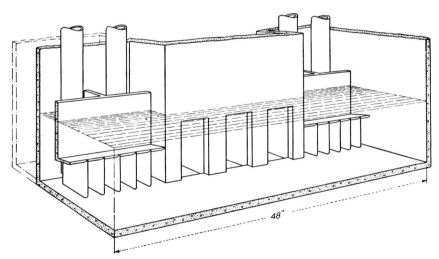

Fig. 7-8   Intake in Fig. 7-6 rearranged after model tests. (*Ingersoll-Rand Co.*)

shows how the intake was rearranged. The magnitude of the vortices *A* and *C* was reduced while holding them firmly in position under the bells. Vortex *B* was eliminated entirely. The front wall prevents disturbances from reaching the pumps, while the tunnels with their sloping ceilings have enough length to straighten the flow into the pumps.

Figure 7-9 shows a plan view of an existing pump installation in which

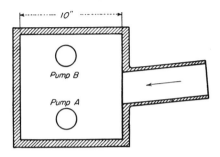

Fig. 7-9 Plan view of pump intake causing excessive vibration. (*Ingersoll-Rand Co.*)

there was excessive vibration. Jetting of the inlet flow between the pumps created a number of vortices on each side. Figure 7-10 shows the corrected intake adopted. It contains a dam of suitable proportions between the pumps and the inlet, 0.75 of a bell diameter from the center of the pumps. The suction bells were lowered to a clearance of 0.4 of a bell diameter. Lastly, another wall was inserted between the dam and the back wall, separating the pumps into symmetrical cells.

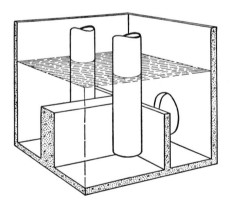

Fig. 7-10 Intake of Fig. 7-9 after correction by model tests. (*Ingersoll-Rand Co.*)

Design Alternatives Water can be brought into the pump intake upward in an axial manner or horizontally and from above, and distributed evenly around the area between the bell and the floor. The latter method requires less depth of excavation and is usually less expensive. Various design factors for this type are given below. The suction-bell diameter G (Fig. 7-11) is used for reference in design and is usually about twice the impeller-eye diameter. A different ratio will modify the design values listed below.

Bell clearance from the floor of the intake is perhaps the most important design variable. A clearance of $G/4$ provides the same flow area as the bell cross section and is the minimum depth to which the bell can be dropped and still retain accelerating flow. The best range for bell clearance, $G/2$ to $G/3$, should be used wherever possible. Pump efficiency is higher than with an infinite-pool inlet.

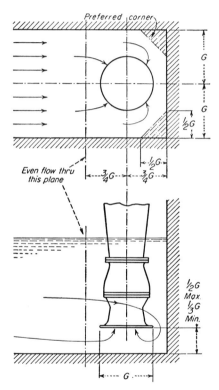

Fig. 7-11 Preferred dimensions for vertical pump intakes. (*Ingersoll-Rand Co.*)

When the flow approaches the intake from one direction, a wall behind the pump is beneficial. It helps prevent vortices from forming in the idle water behind the pump. Good practice locates this back wall $3G/4$ from the pump center.

For single-pump installations, the channel width is not too important if it equals or exceeds $2G$, and the pump is in the center of the channel. However, since space is usually at a premium and a large channel tends to induce secondary currents, it is seldom wise to exceed a width of $2G$. Whenever this is done, caution must be used.

The length of channel chosen should produce even water distribution just ahead of the pump. For reference, a plane $3G/4$ in front of the pump center line can be used. Flow through this plane should be normal to it and without irregularities. A channel length of 2.5 to 3 times the width upstream from the plane usually eliminates flow disturbances. This makes the over-all channel length about $7G$.

With several pumps on a site, installation of a number of units in one long channel should be avoided, if possible. This is because of the danger of mutual interference. But if a long channel must be used, keep the flow velocity at the same value as in a single-pump installation. Provision must be made to eliminate turbulence in the flow as the liquid passes from one pump to the next. The channel width at each pump should at least equal $W/G = 1 + (4G + 8GN)/3H$, where W = channel width, G = suction-bell diameter, H = water depth, and N = number of pumps downstream. Dimensions used in this equation should be in consistent units. Spacing between the pumps should be wide enough to stabilize and redistribute the flow as it passes any given pump. A distance of 2.5 to 3 times the channel width between reference planes is probably adequate.

On the basis of test work done by many pump manufacturers, the Hydraulic Institute has established acceptable size and arrangement parameters for a variety of sump operating conditions. Figure 7-12 shows recommended sump dimensions as a function of flow; Fig. 7-13 indicates proper arrangements as well as those specifically not recommended. Adherence to these general recommendations will assure satisfactory sump-pump operation and prevent the necessity of costly sump modification.

Pipe Size As a general rule, the suction pipe for any class of pump should never be of a smaller diameter than the pump inlet connection. If at all possible, the suction pipe should be two or more pipe sizes larger than the pump inlet connection. This ensures lower friction-head losses in the suction line.

Figure 7-14 shows a number of important factors to be kept in mind when planning suction lines for horizontal centrifugal pumps. Do not use arrangements like those shown in Fig. 7-14a or c because there is a likelihood of an air pocket forming. Use an eccentric reducer instead (Fig. 7-14b). Slope the suction pipe up toward the pump. Air pockets can form at high points, causing the pump to lose prime, even if fitted with a foot valve in the suction line. Avoid short-radius elbows (Fig. 7-14e) on double-suction pumps because more liquid will enter one side of the impeller than the other. This can reduce pump capacity and efficiency and may overload the thrust bearing, leading to rapid failure. Use a reducer as shown in Fig. 7-14f for double-suction pumps. Figure

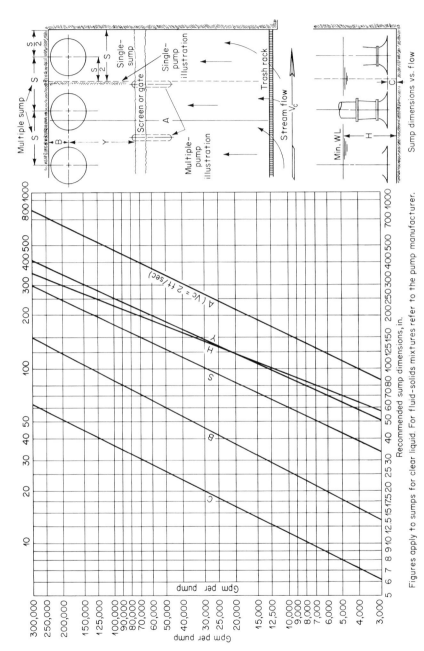

Fig. 7-12 Hydraulic Institute Standards set sump dimensions as a function of flow. (*Hydraulic Institute.*)

Figures apply to sumps for clear liquid. For fluid-solids mixtures refer to the pump manufacturer. Recommended sump dimensions, in.

MULTIPLE PUMP PITS

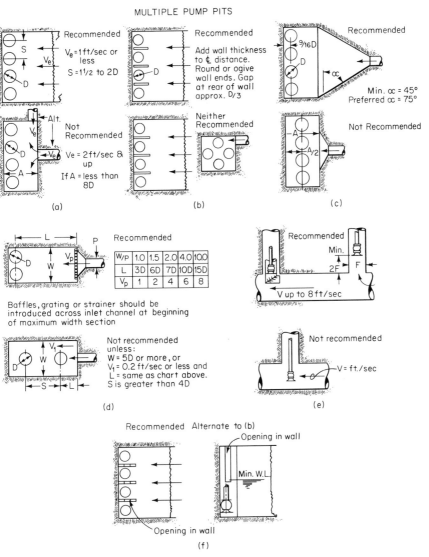

Baffles, grating or strainer should be introduced across inlet channel at beginning of maximum width section

NOTE: Figures apply to sumps for clear liquid. For fluid-solids mixtures refer to the pump manufacturer.

Fig. 7-13 Hydraulic Institute Standards show recommended sump arrangements as well as arrangements that are specifically not recommended. (*Hydraulic Institute.*)

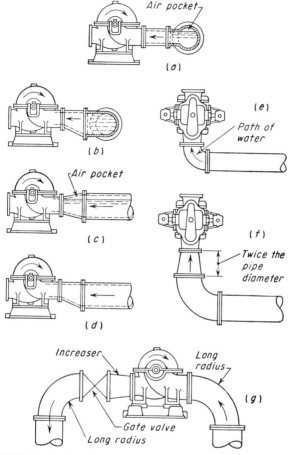

Fig. 7-14   Intake piping pointers for horizontal centrifugal pumps.

7-14*g* shows the perferred arrangement for piping horizontal centrifugal pumps. To prevent transmission of vibration to the pump casing,
both the suction and discharge piping should be independently supported at some point close to the inlet and outlet flanges.

For direct-acting steam pumps, several rules of thumb are often used
in sizing suction and discharge piping. These are

$$d_s = \sqrt{0.1g} \tag{7-1}$$

$$d_d = \sqrt{0.08g} \tag{7-2}$$

$$d_s \text{ or } d_d = 4.95\sqrt{g/v} \tag{7-3}$$

where d_s = internal diameter of suction pipe, in.; d_d = internal diameter of discharge pipe, in.; g = pump capacity, gal per min; and v = average velocity of liquid, ft per min. Equations (7-1) and (7-2) are based on a suction-line liquid velocity not exceeding 240 ft per min, a discharge-line velocity of 300 ft per min.

The importance of suction-pipe size is emphasized by Fig. 7-15a, showing a pump fed by an elevated roof tank and fitted with an upward loop in the suction. Even with the pressure tank, the pump was unable to deliver its rated flow because an air pocket was formed by the loop. Swinging the loop downward, as shown, corrected the trouble. Figure 7-15b shows the correct way to pitch a long suction line with respect to the center of the pump.

Liquid Velocity Figure 7-16 gives data on recommended liquid velocities in lines used in various types of industrial process plants. Based on extensive experience, it may be used for all classes of pumps in the usual plant. Note that, once the liquid is under pressure on the discharge side of the pump, higher velocities can be used without danger of trouble.

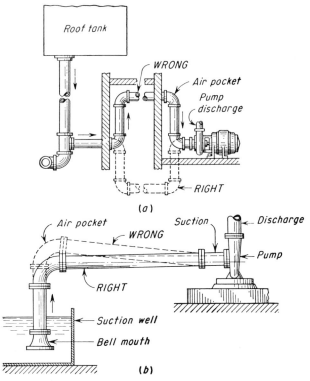

Fig. 7-15 (a) Upward loop in suction line caused an air pocket. (b) Slope a long suction line away from pump.

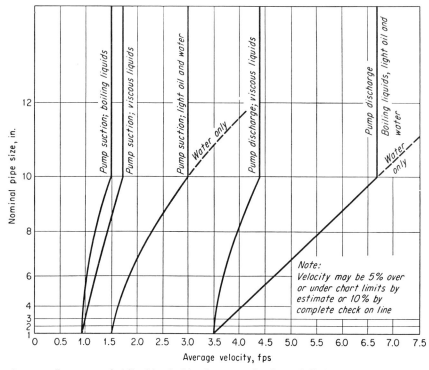

Fig. 7-16 Recommended liquid velocities for pump intake and discharge lines.

Figure 7-17*a* shows the recommended size of suction piping for rotary pumps handling viscous liquids. Sizes given are based on installations where no static suction lift exists and the line is of relatively simple configuration. Where a static lift exists, or the line has a number of bends and fittings, a larger-size suction pipe than shown must be used. Figure 7-17*b* shows the effect of altitude and temperature on the lift of a 6- by 4- by 6-in. valve-plate-design duplex direct-acting pump.

Vertical Pumps Figure 7-18 shows a number of applications of vertical centrifugal pumps for various industrial services. As can be seen, the pump body extends into the liquid being handled and there is no need for special suction arrangements. The vertical pump is finding many applications today for these and similar services because it has several outstanding advantages, including good performance and simple mechanical construction. The exact type used depends on the job requirements, but multistage-diffuser and mixed-flow types are the most common in use today. Much of the success being enjoyed today by these units is a result of their outstanding performance in deepwell service. Ground-water applications are discussed in a later chapter of this book.

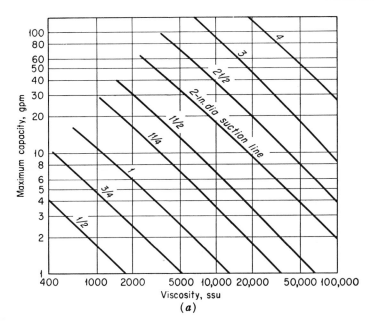

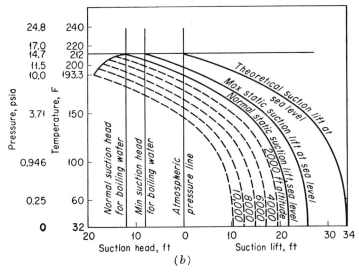

Fig. 7-17 (*a*) Recommended suction-line sizes for rotary pumps. (*b*) Effect of altitude and temperature on the lift of a direct-acting pump. (*Hydraulic Institute.*)

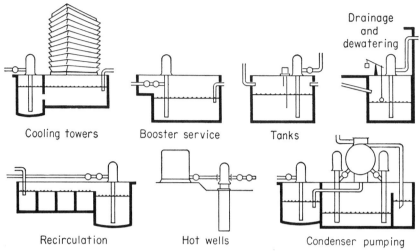

Fig. 7-18   Typical applications for vertical centrifugal pumps. (*Johnston Pump Co.*)

River Intakes   These (Figs. 7-19 and 7-20) may use either horizontal or vertical centrifugal pumps.   Reciprocating pumps are seldom used for this service because their flow characteristics are not the most desirable. Capacities of the usual rotary pumps are generally too low where large

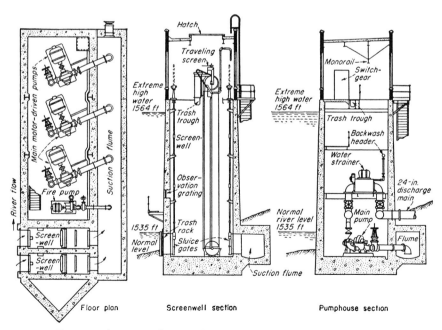

Fig. 7-19   Dry-type river pump house using horizontal centrifugal pumps.

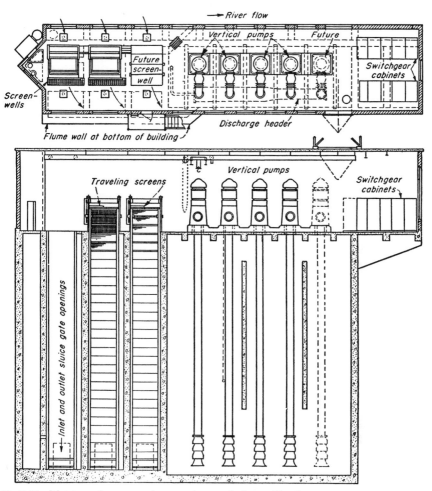

Fig. 7-20 Wet-type river pump house using vertical centrifugal pumps.

quantities of water are needed for industrial cooling or other similar services.

As Figs. 7-19 and 7-20 show, the typical river intake consists of some sort of housing for the pumps and their drives, an intake screen to protect the pumps from debris, an intake flume or channel, and a discharge header. Best practice locates the intake on the inside of river bends where less debris reaches it. Also, the intake should be as far as possible upstream from areas where plant wastes are discharged. Intake pipes should extend deep enough so they are always under the water. This requires determination of the water level during extreme droughts as well as during flood conditions.

The pump house should be as near the plant as practicable, accessible by truck during normal river stages, and easy to reach with a power line. Almost all pumps delivering water from rivers are motor-driven. For design calculations, river-bed contours are needed as well as careful subsurface study for the foundation.

All sites have their own problems, such as avoiding long lines for fire protection and taking advantage of different water pressures within the plant. For example, at a lakeside steel mill, water levels varied widely, adversely affecting pump suction lifts and therefore making mill-service pump-discharge pressures change unpredictably. The solution to this problem lay in the construction of a large, constant-level, in-plant reservoir. Figure 7-21a shows a plan view of the arrangement; Figs. 7-21b

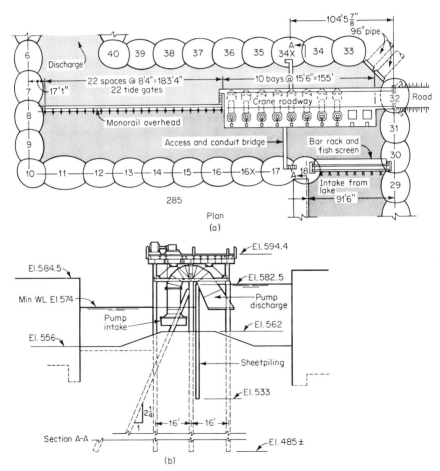

Fig. 7-21 (*a*) Plan view of constant-level reservoir. (*b*) Elevation of pumping plant. (*Hydrotechnic Corp.*)

Fig. 7-22 Exterior view of pumping station in Fig. 7-21. (*Hydrotechnic Corp.*)

and 7-22 show an elevation of the pump installation and the pump-drive arrangement. During periods of high lake level, tide gates hold the flow level to elevation 582.5 ft (Fig. 7-21*b*); as the lake level drops, pumps start as needed to hold this discharge level elevation constant. Thus in-plant pumps operate at all times with the same suction conditions, regardless of lake level.

The simplest type of pump house consists of a doghouselike covering for a single small unit. These, however, seldom find use except for emergency or temporary service, or in extremely small plants. Two general types of larger pump houses that have proved economical are of hollow-pier construction and are set in the stream. In a *wet* house, the pumps are submerged in the river water, whereas in the *dry* type they are housed in a deep dry room below water level. These two types are usually satisfactory, unless there are sound reasons for moving the house inland and feeding it through a conduit or open channel from the river.

Screening Most rivers and lakes require use of at least trash racks and traveling screens to protect the pumps, piping, valves, and fittings from debris in the water. Trash racks give a rough screening of logs, sticks, cornstalks, ice, etc., to protect the traveling screens (Fig. 7-23). The screens remove leaves, twigs, sticks, etc., and are installed vertically, preferably one per pump. In rivers of the middle Atlantic states, ³/₈-in. woven galvanized-wire mesh makes a common and satisfactory screen. Many other screen forms are available and find wide use in a number of different intake applications. Where required, strainers may be

Fig. 7-23   Fish collector above screen removes fish from intake water.

installed in addition to traveling screens.   The purpose of the strainers is to remove smaller particles which pass through the trash racks and traveling screens.   Usual traveling screens have a flow velocity through them of about 1 to 2.5 ft per sec with a head loss of about 2 ft of water.   The screen travels at velocities up to 10 ft per min.   Water jets at 70 psig are often used to knock debris off screens.

Fish Problems   During certain seasons of the year, fish may be drawn into river intakes, where they clog screens, pumps, and other equipment.   As Fig. 7-24 shows, fish swim upstream to avoid being drawn into a river intake.   Fish created such a problem at the intake in Fig. 7-25 that a collector (Figs. 7-23 and 7-26) was installed.   The fish take refuge behind the curtain wall (Fig. 7-26) from where they are removed by a bladeless-impeller-type pump (Fig. 7-27) and discharged to the stream.   Over 98 per cent of the fish returned to the river survive without ill effects.   The system handles fish up to 14 in. long.   It is estimated that the collectors in this system recover about 560,000 juvenile bass

Fig. 7-24 Fish in test flume swim upstream to avoid being pinned against screen, left.

annually, besides a number of their varieties, both juvenile and full-grown.

Figure 7-28 shows another method of keeping fish away from traveling screens. Installed in the intake canal, this electric fence shocks any fish swimming past. Though it is effective in stopping the flow of fish into the intake, the fence gives some trouble because debris catch on the suspended electrodes.

In many cases where large river dams prevent fish migration for spawning, so-called *fish ladders* permit the fish to bypass the dam. Here a structure resembling a large staircase is constructed alongside the dam

Fig. 7-25 Circulating-water intake using collector shown in Figs. 7-23 and 7-26.

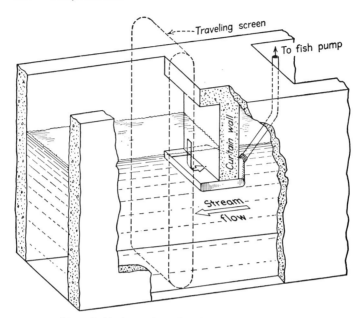

Fig. 7-26   Water behind curtain wall is slower-moving. Fish take refuge there, from where they are removed by fish pump. See Fig. 7-23, also.

spillway.   Water in sufficient quantity flows down the staircase, and each step is built to the height that a fish can jump.   The fish then jump up successive steps, against the flow as they would naturally do, until they reach the water level behind the dam and can continue upstream.   In order to create the necessary flow for a fish ladder to work, recirculating pumps are frequently used to conserve water.   These are usually medium-head pumps designed for high capacities.

Tidewater Areas   Pump houses may or may not be used in tidewater areas, depending on the existing hydraulic conditions.   For ocean-front

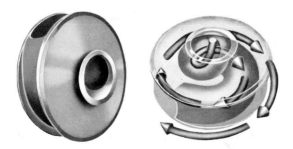

Fig. 7-27   Fish-pump impeller is a bladeless type.

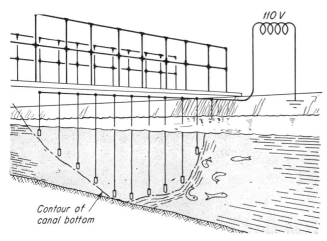

110 V

Contour of canal bottom

Fig. 7-28 Electric fence stops fish flow into intake.

intakes, channels and closed conduits are popular because they can be extended out far enough to eliminate the effects of the rise and fall of the tide. For bays, a siphon intake (Fig. 7-29) often proves economical. The siphon draws water into a pump-suction forebay. Vacuum pumps are used to evacuate the siphons for priming.

Vacuum Chambers Some direct-acting and power-type reciprocating pumps are fitted with a vacuum or surge chamber in the suction line (Fig. 7-30) to ensure that the pump cylinder will be completely full of liquid at each reversal of the piston. The chamber also provides an air cushion for the column of liquid in the suction pipe, when the movement of the liquid is suddenly arrested, because of the momentary stoppage of the piston at the end of each stroke. Suction surge chambers are more frequently required than discharge chambers.

Marine Pumps One of the peculiar problems met with the suction piping of marine pumps is related to the effect of rolling and listing of the vessel on the liquid level in a tank, bilge, or compartment. Where there is any possibility that the suction inlet will be deprived of liquid because of motion of the vessel, some means must be provided to ensure that liquid will reach the pump. This may be done by means of cross connections from one tank to another, use of structural members in such a way that liquid flows from one side or end of the ship to the other, etc. Marine pumps of various types, and their piping, are discussed in a later chapter.

Pump Priming Positive-displacement pumps—reciprocating and rotary—are self-priming for total suction lifts to about 28 ft when in good condition. But with long suction lines, high lifts, or other abnormal conditions, they must be primed.

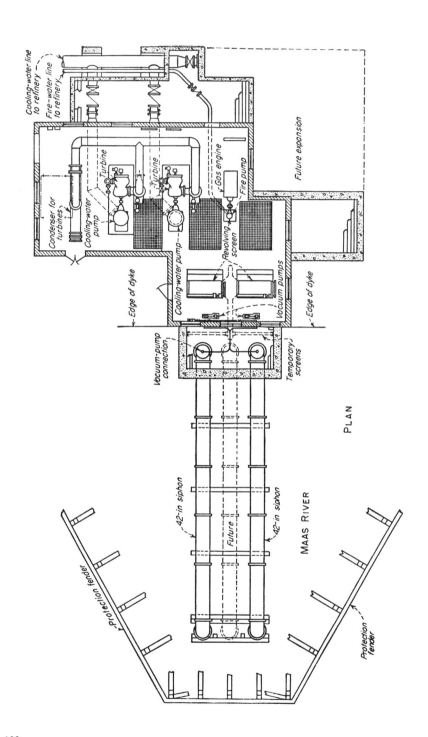

PLAN

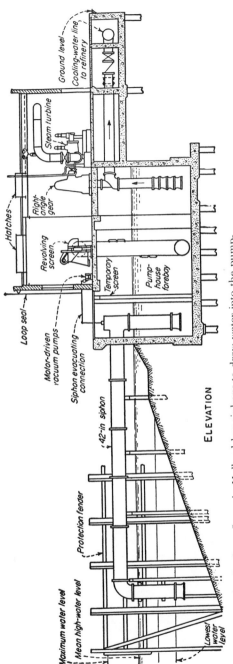

Fig. 7-29 Bay intake for refinery in Holland has siphons to draw water into the pump-house forebay.

Ground level

Cooling-water line to refinery

Steam turbine

Right angle gear

Hatches

Revolving screen

Loop seal

Motor-driven vacuum pumps

Siphon evacuating connection

Temporary screen

Pump-house forebay

42-in siphon

Protection fender

Maximum water level

Mean high-water level

Lowest water level

ELEVATION

187

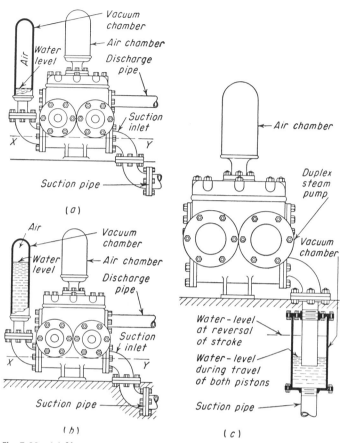

Fig. 7-30 (*a*) Vacuum or surge chamber at start of suction stroke. (*b*) Chamber at end of suction stroke. (*c*) Another form of vacuum chamber.

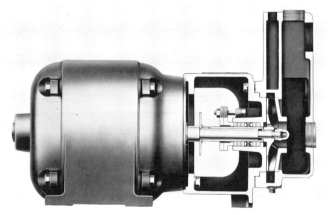

Fig. 7-31 Double volute in this pump recirculates liquid-vapor mixture during priming, separates vapor from liquid. (*Ingersoll-Rand Co.*)

Because centrifugal pumps are not self-priming, some priming means is required on a suction lift. Either special self-priming pumps may be used or auxiliary priming equipment may be installed. Figures 7-31 through 7-34 show a number of modern self-priming pumps. Though designs vary from one manufacturer to another, a liquid reservoir of some type on the pump discharge is common. The reservoir holds priming liquid and serves as an air separator. Other designs have a liquid reservoir on both the suction and discharge sides of the pump, liquid being circulated from the discharge to the suction during priming. Automatic valves or hydraulic action stop the circulation after the pump primes. Some pumps circulate the liquid continuously.

Fig. 7-32 Air separator and two discharge ports give priming action in this pump (*Lawrence Pumps, Inc.*)

Auxiliary equipment for pump priming includes ejectors, vacuum pumps, etc., used in hookups like those shown in Fig. 7-35. With a flooded suction *a*, opening the casing air-vent petcocks and the suction gate valve allows the incoming liquid to push the air out of the casing. A bypass around the discharge check valve *b* allows use of the liquid in the discharge line for priming of the pump. The food valve *c* holds liquid in the suction line and may be augmented by an auxiliary liquid supply. The separate pump *d* draws air from the casing of the main pump to give a priming action. Or an ejector *e* may be used to do the same job.

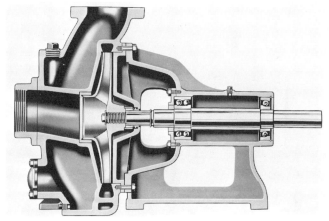

Fig. 7-33 Priming chamber here surrounds the straight-in suction; no valves are used. (*Gorman-Rupp Co.*)

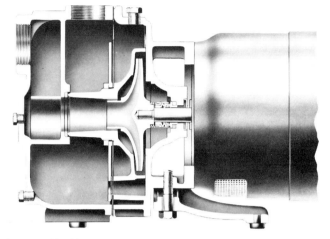

Fig. 7-34   Liquid flows into the lower volute and out the upper one during priming in this pump. (*Goulds Pumps, Inc.*)

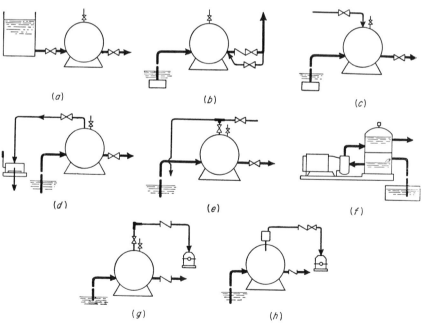

Fig. 7-35   Eight priming methods which may be used with many types of pumps.

A priming tank f holds a supply of liquid large enough to establish flow through the pump on starting. Vacuum pumps g and h are manually and automatically controlled to prime the main pump.

DISCHARGE PIPING

From the standpoint of the pump, the most important factors in dis-charge-piping design are pipe size, liquid velocity, length of run, number and type of fittings, and general nature of the piping layout—whether it is fairly straight or contains a number of bends, loops, or offsets. Also, over a period of time, it is important to know if the liquid handled is coating the inside of the pipe and reducing its internal diameter, because if this occurs, the frictional resistance of the pipe will increase.

Pipe Size With but few exceptions, the aim in designing a piping system is to secure the flow rate desired at the lowest over-all cost. The initial cost of the pipe and fittings is directly proportional to the pipe diameter. So, likewise, are depreciation and maintenance. The cost of pressure drop (i.e., the cost of pumping) is, however, inversely proportional to the diameter. So an economic balance can be struck, by proper analysis, at that diameter which will give the minimum sum for the initial, operating, and pumping costs.

Figure 7-36 can be used to determine the most economical diameter for a pipe in a pumping system when the liquid flow rate and density are known, and turbulent flow exists. When using this chart it is more economical to select the next standard pipe size above the actual diameter determined because standard sizes are cheaper and more readily obtained than special sizes. But it is worthwhile to note that in some instances pumping conditions, pipe size, and other economic factors combine to make it desirable to use a special diameter. This is most common in installations where the discharge piping is long—say, several miles—and the quantity of liquid handled is fairly large.

Allowable liquid velocity may be obtained from Fig. 7-16. Once the actual pipe size is known, the velocity in it can be found from Fig. 7-37. In general, the actual velocity of the liquid in the pipe should not exceed the values given in Fig. 7-16. See Chap. 10 for a comprehensive discussion of the economics of various types of pump installations.

Run Length To reduce the friction head on the pump discharge, the line should be run between the pump and equipment served in the most direct route possible. Every effort possible should be made to keep the number of valves, fittings, and bends to the minimum necessary for the installations. Loops or bends, if used, should be long-radius types, to keep friction losses to a minimum. Where pump capacity is regulated

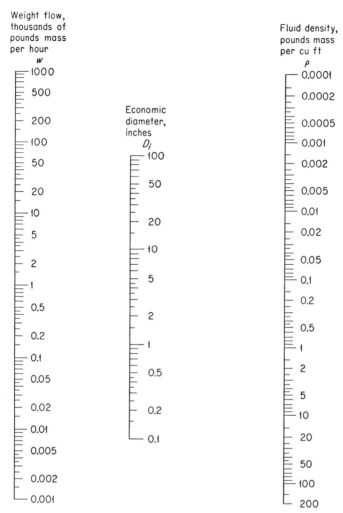

Weight flow,
thousands of
pounds mass
per hour
w

Economic
diameter,
inches
Dᵢ

Fluid density,
pounds mass
per cu ft
ρ

Fig. 7-36   Chart for determining economic pipe diameters.

by a throttling valve, the resistance at the position of maximum throttling must be used in head calculations.

There are some installations (Fig. 7-38) where the discharge piping can be arranged for a minimum run with easy bends and no fittings, but these are rare.   Other installations (Fig. 7-39) need only a minimum of valves in a short run.   To secure the minimum operating head, pumps arranged as in Fig. 7-38 should be piped so the outlet of the discharge pipe is submerged at all liquid levels.   However, the invert of the high point of the siphon must be above the high-water level on the discharge

VELOCITY IN FEET PER SECOND

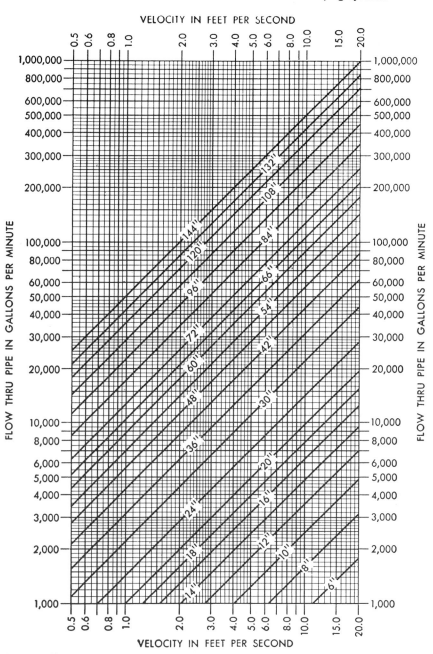

VELOCITY IN FEET PER SECOND

Fig. 7-37 Chart for determining liquid velocity in a pipe. (*Worthington Corp.*)

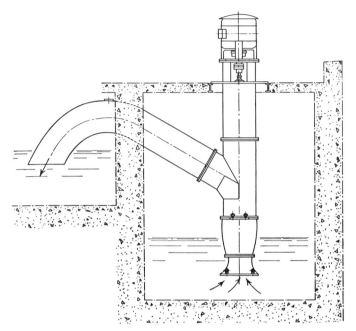

Fig. 7-38   Vertical pump with short siphon-type discharge. (*Worthington Corp.*)

side to break the siphon and prevent backflow of water when the pump is stopped.

Noise, Vibration   Pump and motor manufacturers today are tackling the problems of noise and vibration with greater intensity than ever before.   There are instances in recent practice where, of two equal boiler-feed pumps for a central-station job, the unit chosen was that having a lower noise level.   This may appear strange, in view of the high noise level in the usual central station, but there is a marked trend today toward reducing the noise in all types of plants.

A number of different methods are used to reduce noise and vibration in the piping and pump itself.   Flexible connectors (Fig. 7-40*a* and *b*) made of rubber or metal are useful in isolating noise and vibration originating in the pump or pipe.   Figure 7-41 shows other types of devices used to cushion the shock of the pump discharge stroke.   Discharge cushion chambers (Fig. 7-41*a*) are often built as part of the pump.   Pressure alleviators (Fig. 7-41*b*) find use with high-pressure power pumps to absorb shock from sudden stoppage of the liquid.   They usually consist of a spring-loaded plunger operating in a stuffing box.   Liquid does not escape from the piping system during pressure surges.   Some designs are also fitted with an air chamber.   In some large water-supply systems, surge tanks placed in either the suction or discharge piping, or both, are successful in eliminating the noise and damage caused by water hammer. Figure 7-42 shows a number of piping pointers helpful in reducing noise in rotary and other classes of pumps.

Reduced speed (Fig. 7-43) will lower the noise level of a rotary pump. The three lines X, Y, and Z represent the noise levels of three different pumps. Pump X when operating at rated speed is very noisy, point A. Slowing it down to 75 per cent of rated speed brings its noise level within the zone considered to be satisfactory for most installations, point B. At 50 per cent rated speed, point C, the pump approaches the quiet operating zone.

Pump Z, even at 100 per cent speed, operates in the usually satisfactory zone, point D, and at 75 per cent speed it is in the quiet zone, point

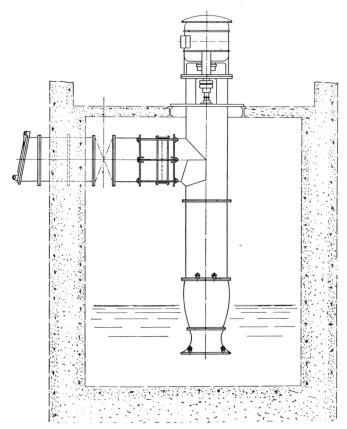

Fig. 7-39 Vertical pump with short discharge line and minimum number of valves. (*Worthington Corp.*)

(a)

(b)

Fig. 7-40 (a) Rubber sections in suction and discharge lines absorb noise and vibration. (b) Rubber expansion joint in pump discharge line.

E. Operating a rotary pump at reduced speed cuts its capacity by a like amount. For example, if a pump delivers 100 gpm at rated speed, its capacity will be only 75 gpm at 75 per cent speed and 50 gpm at 50 per cent speed. This should be kept in mind when choosing a pump—it should deliver the required capacity at a suitable operating speed. Figure 7-44 shows a rotary pump fitted with synthetic-rubber connectors

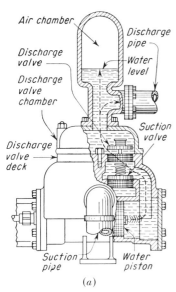

(a)   (b)

Fig. 7-41 (a) Discharge air or surge chamber. (b) Pressure alleviator. (*Aldrich Pump Division, Ingersoll-Rand Co.*)

A and *B* to prevent transmission of noise from the pump to its connecting piping.

Piping Arrangement The particular arrangement used for a given job depends on so many factors that it is difficult to give specific instructions applicable in every case. But many good practices can be found by study of typical installation photographs and drawings. A few are given in this chapter.

The four cooling-water pumps in Fig. 7-45 are mounted on a stepped base, permitting ready access for inspection and maintenance. Each pair of pumps takes suction from a common header and discharges to another header. The rising-stem gate valves used on both the suction and discharge lines are arranged for easy operation from the aisle in front of the pumps. Fittings used between the pump outlet and the discharge header include a union, an increaser, and a gate valve. This installation shows care in its planning; note the packing leak-off drain tubing connected to the bottom of each pump.

The three boiler-feed pumps in Fig. 7-46 serve multiple boilers. The piping is headered so any two of the pumps serve the normal load on the boilers. The third pump is for standby service. Motor valves and magnetic controllers give completely automatic operation of all three pumps.

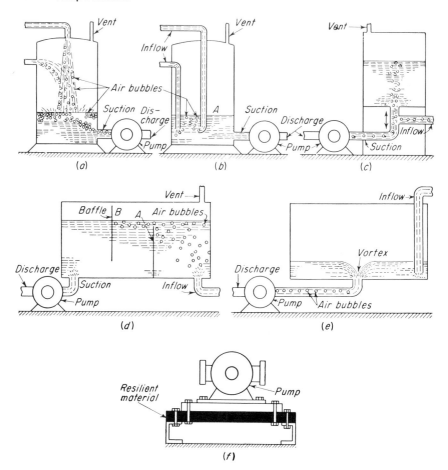

Fig. 7-42   (a) Incoming liquid entrains air. (b) Submerged supply reduces number of air bubbles. (c) Poor intake. (d) Baffles prevent air bubbles from reaching pump. (e) Vortex entrains air (f) Rubber mounting isolates noise.

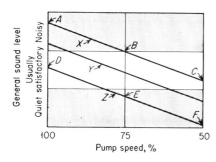

Fig. 7-43   Noise level in relation to pump speed.

Fig. 7-44 Synthetic-rubber connectors *A* and *B* isolate pump noise. (*Worthington Corp.*)

Fig. 7-45 Fig. 745 Four cooling-water pumps and their connecting piping. (*Fairbanks, Morse Pump Division, Colt Industries.*)

Fig. 7-46 Three multistage pumps and deaerating heater integrated for boiler-feed service. (*Schaub Engineering Co.*)

Fig. 7-47 Typical arrangement of an ash sump with the pump in a dry pit. (*The Allen-Sherman-Hoff Pump Co.*)

For an ash sump (Fig. 7-47) mounting the pump in a dry well permits use of a short, direct suction line containing only one bend. The discharge pipe rises vertically to a suitable level, from where a horizontal run can be made to the disposal point.

With some pumping systems little or no piping is required other than connections to the units served. Thus fuel-oil pumping and heating set can be completely piped, including strainers, pressure-regulating valves, bypasses, etc. To use a set of this type it need be connected only to the fuel tank, electrical supply, and the burners on the boiler served.

The two gear pumps in Fig. 7-48 are part of a group of eight used to maintain a constant flow of fuel oil to the burners on the boilers in a utility power plant. Note how each pump is fitted with a manually operated air-relief valve on its discharge side, arranged so the liquid bled off can be collected in a bucket or other receptacle. Suction legs are oversize, providing a reserve supply of liquid for the pumps. The discharge piping is fitted with a pressure-relief valve, pressure gage, sight-flow indicater, and strainer.

AUXILIARY PIPING

Many pumps require auxiliary piping of some kind—from simple drains for the base plate to oil, water, and smothering lines for lubrication,

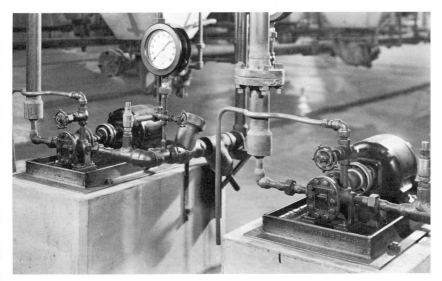

Fig. 7-48 Gear pumps for burner fuel-oil supply.

cooling, and reduction of packing-gland leaks. This piping may or may not be supplied by the pump manufacturer, depending on the contract under which the pump is purchased.

Oil Piping Figure 7-49 shows three arrangements for the oil piping for a horizontal barrel-type centrifugal pump. In Fig. 7-49a, only the pump bearings are supplied lube oil by the system. Both the pump and driver are supplied lube oil in Fig. 7-49b and c, with an auxiliary oil pump separately driven from the main pump being used in Fig. 7-49c.

Water Piping This is used to convey water to and from bearings,

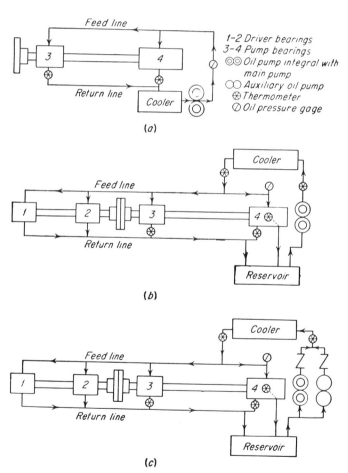

Fig. 7-49 (a) Pump bearings are supplied by lube-oil pump. (b) Both pump and driver bearings are supplied by lube-oil pump. (c) Hookup for auxiliary lube-oil pump.

stuffing boxes, oil coolers, shaft sleeves, lantern rings, and other parts of the pump requiring water for cooling or sealing purposes. Thrust bearings are more often fitted with water cooling than are line bearings because more heat is generated in thrust bearings. When clean low-temperature water is handled by a pump, the stuffing boxes are usually sealed by water supplied from the pump discharge through either external piping (Fig. 7-50) or internal passages (Fig. 7-51). In multistage pumps, the sealing water is generally taken from the waterway of the first stage.

Use of an independent seal-water supply is recommended where (1) the pump suction lift exceeds 15 ft; (2) the pump discharge pressure is less than about 10 psi (23-ft head); (3) the pump handles hot water (over 250 F) and does not have adequate cooling from other sources; (4) the pump handles muddy, sandy, or gritty water; (5) the pump serves a hotwell; and (6) acids, juices, molasses, or other similar liquids are handled with no special provision in the stuffing-box design for the liquid pumped.

Grease is used as a sealing medium for stuffing boxes when clear water

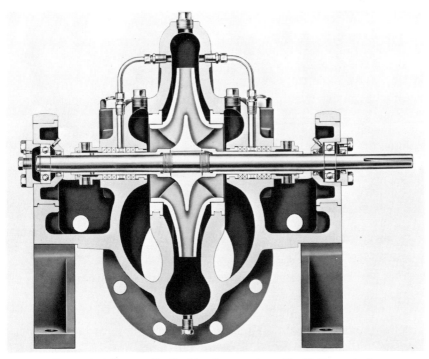

Fig. 7-50   External piping supplies liquid to lantern rings. (*Goulds Pumps, Inc.*)

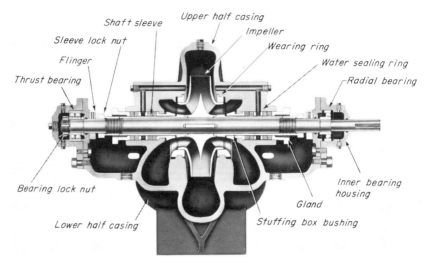

Shaft sleeve
Sleeve lock nut
Flinger
Thrust bearing
Upper half casing
Impeller
Wearing ring
Water sealing ring
Radial bearing
Bearing lock nut
Lower half casing
Inner bearing housing
Gland
Stuffing box bushing

Fig. 7-51 Internal drilled passages supply lantern rings. (*Weinman Pump Mfg. Co.*)

or a suitable clear liquid is not available or cannot be used. Grease sealing is common in sewage and low-head drainage pumps, and is also used in some designs of chemical pumps.

When stuffing-box leakage to the atmosphere is objectionable because it creates an explosion hazard or is dangerous to plant personnel, a smothering or quenching gland may be used. Water, or a hydrocarbon, supplied from external piping, is circulated through a channel between the packing gland and shaft sleeve.

Reciprocating Pumps Auxiliary piping used with reciprocating pumps includes steam-cylinder drains, lube-oil lines, steam supply for steam

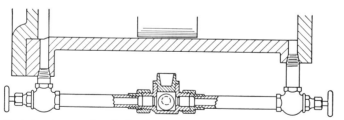

Fig. 7-52 Automatic cylinder drain for a direct-acting pump.

jacketing, and air bleeds. These may or may not be supplied by the pump manufacturer, depending on the purchase contract. Figure 7-52 shows an automatic steam-cylinder drain arrangement for a direct-acting steam pump. It contains a double-ended ball drain valve. The valve chamber is connected to a steam trap to permit condensate drainage without steam waste.

Other Auxiliary Piping   There are a number of special cases where additional auxiliary piping is required.   Figure 7-53 shows one example. The piping to and from the pump is fitted with couplings to which air or steam hoses can be connected for cleaning of the process lines and equipment.   The couplings permit quick attachment and removal of the air or steam hoses, reducing the time required for cleaning of the piping and equipment.

Water Hammer   This occurs in a closed pipe system when the liquid velocity is suddenly changed by (1) sudden starting, stoppage, or change in the speed of a pump or (2) sudden opening or closing of a valve or other flow-control device which alters the liquid velocity.   Water hammer is often accompanied by unpleasant noise, but the major result of a sudden velocity change may be a rapid pressure rise in the pipe.   If the pressure rise is excessive it can damage the pump and piping.   This may occur in either the suction or discharge piping of any class of pump.

The magnitude of the pressure rise can be computed from

$$h = vr/g \qquad (7\text{-}4)$$

where $h =$ pressure rise, ft of water; $r =$ reduction in liquid velocity, ft per sec; $g = 32.2$ ft per sec^2; $v = 4{,}660/(\sqrt{1 + KR}) =$ velocity of pressure

Fig. 7-53   Pipe hookup for cleaning lines with air or steam.

wave in pipe, ft per sec; K = (elastic modulus of water)/(elastic modulus of pipe material); R = (pipe diameter, in.)/(pipe wall thickness, in.). Values of K for common pipe materials are steel, 0.010; wrought iron, 0.0107; cast iron, 0.025; cement-asbestos, 0.088; wood, 0.20. Note that the pressure rise determined here is *above* the pressure existing in the pipe before hammer occurs.

The time required for the pressure wave to travel from one end of the pipe to the other is found from

$$t = 2l/v \qquad (7\text{-}5)$$

where t = time, sec, for the wave to travel the length of the pipe; l = length of pipe between pump and the device causing water hammer, ft.

example: What pressure rise will occur in a 10,000-ft-long cement-asbestos pipe if a valve in the middle is suddenly closed when the liquid velocity is 5.46 ft per sec and R = 14.7? How long will it take for the pressure wave to travel from the valve to the pump at the inlet end of the pipe?

solution: Velocity of the pressure wave is $v = 4,660/\sqrt{1 + (0.088)}$ (14.7) = 3,090 fps. Then, h = (3,090)(5.46)/32.2 = 521 ft of water = 226 psi above the pressure existing in the pipe before water hammer. The time for the pressure wave to travel from the valve to the pump is t = 2(5,000)/3,090 = 3.24 sec.

Preventing Water Hammer   The most common means used to prevent the destructive and annoying effects of water hammer act to reduce the pressure developed during the surge of the liquid stream. Pressure surges following sudden starting or stoppage of a pump can be held within allowable limits (1) by lengthening the flow-stopping time to several intervals of t [Eq. (7-5)], (2) by bleeding some water from the pipe, or (3) by a combination of these two methods. Designing the piping system for low flow velocity is also a help.

To lengthen the flow-stopping time, a flywheel may be used on the pump or an air chamber in the pipe near the pump or device causing water hammer. Or air may be injected by a compressor or drawn in through a special valve to form a cushion during a pressure surge. Air-relief valves can be installed to relieve air and water during a surge. To discharge water from the pipe, slow-closing check valves at the pump or one of several types of surge-suppressor valves may be used.

Air injection is used for medium-length pipes where the vacuum during a surge reversal does not exceed 15 in. Hg. The air pressure is about 10 psi less than the line pressure. Air aspiration is suitable for any high-velocity or low-velocity discharge pipe over 1,500 ft long. Air-relief valves keep a line relatively free of air and are used in conjunction with an air-check valve at the line high point near the pump.   Slow-closing check valves relieve water from the discharge pipe and are used when air aspiration or injection is unsuitable.   Suppressor valves are used where the pipe pressure does not fall below atmospheric during a surge.

CHAPTER EIGHT

Drives for Industrial Pumps

PROBABLY EVERY FORM OF PRIME MOVER AND POWER SOURCE, with some kind of power-transmission device, if needed, has been used for industrial pumps. Today, electric motors drive most pumps—be they centrifugal, rotary, or reciprocating. But steam, gas, and hydraulic turbines and gasoline, diesel, and gas engines are also used. Other power sources having limited popularity are air motors, air-expansion turbines, windmills, etc., but their use is usually confined to certain specialized applications. Power-transmission devices for pump drive include flexible couplings, gears, flat or V belts, chains, and hydraulic and magnetic couplings or clutches.

ELECTRIC MOTORS

For stationary applications, alternating-current (a-c) motors are the most common choice for pump drives. Some direct-current (d-c) motors are in service for driving pumps where for one or more reasons a-c motors are unsuitable. The d-c motor is extremely popular in marine service on all classes of vessels.

Load Characteristics Two characteristics of the driven unit are important in motor choice—the amount of starting torque required in normal operation and the speed requirements. Most centrifugal and rotary pumps are driven at constant speed, except some larger-size pumps where a variable-speed device may be used. Many reciprocating pumps run at constant speed, but in some applications the use of a variable-speed drive allows easy adjustment of pump capacity.

Electric motors for pump drive in the United States are generally full-voltage-starting squirrel-cage units, with some d-c motors also in use. Synchronous and wound-rotor motors are also used. Recent developments in canned and axial-air-gap sealed motors and electromagnetic pumps are already having an influence in the design of both large and small pumps. With some canned motors, straight-through flow is obtained, simplifying the pipe connections to the pump. Axial-air-gap motors shrink to almost pancake thickness, allowing easy installation and care. Both these types are leakproof, an important factor in process applications.

Alternating-current Motors While control and operating simplicity usually dictate a direct-connected constant-speed induction motor, wound-rotor induction motors offer four special advantages: (1) speed control with speed variation down to 50 per cent of full speed, 40 per cent of rated horsepower; (2) high starting torque on low kva for heavy loads; (3) high heat dissipation in the starting resistor, permitting large slip losses during starting without endangering the motor; and (4) cushioned peak loads provided by high-slip operation, giving a desirable flywheel effect on peak loads. Wound-rotor motors are often used when periodic operation at reduced speed is required.

Modern synchronous motors are double-duty units—they are an efficient means to drive pumps while at the same time they provide a practical way to improve plant power factor. A synchronous motor can be applied to any load that can be successfully driven by a NEMA design-B squirrel-cage motor. Other loads for which the synchronous motor is particularly well suited are those requiring low starting kva, controllable torque, or variable speed where a slip coupling is permissible. At 3,600 rpm, synchronous motors may be used for loads from 2,000 to 5,000 hp. Above this horsepower range they should be the first choice. At 1,800 rpm it is questionable whether synchronous motors show any advantages above 1,000 hp. In lower ratings they may be used to improve the plant power factor.

In the range from 500 to 1,200 rpm, any motor slated for continuous duty, and rated 700 hp and above, can well be synchronous. From 200 to 700 hp in this speed range, the choice depends on the value of power-factor improvement, energy cost, and the number of operating hours. Because of their low efficiency and power factor, induction motors should rarely be used for operation below 500 rpm. Synchronous motors are built to operate at unity and leading power factor and with good efficiencies down to 72 rpm. For direct connection in ratings above 200 hp and speeds below 500 rpm, the synchronous motor should be first choice for large reciprocating pumps. Figure 8-1 shows typical speed and power ranges for synchronous and induction motors.

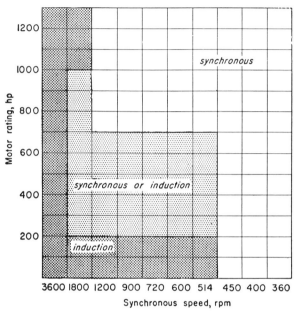

Fig. 8-1   Application ranges for various a-c motors

Vertical hollow-shaft synchronous and induction motors are used for some deepwell pumps. The electrical characteristics of these units are the same as for a horizontal drive but the mechanical features are somewhat different. Reverse-overspeed-protection requirements for the motor may be higher than standard. For example, when the motor stops, the liquid in the discharge pipe may drain through the pump, producing a backspin of the motor. To prevent this, a ratchet coupling or other device is generally used. It prevents reversal of the motor or limits its reverse speed to a safe value. It is also used with other types of motors.

For low-horsepower applications with small pumps, the capacitor-type single-phase induction motor fits many needs. All single-phase units, other than the universal type, must have an auxiliary means for developing starting torque. The basic types of single-phase motors include shaded-pole, universal, split-phase, capacitor, repulsion-start, and synchronous. With the exception of the universal, synchronous, and some forms of the repulsion design, they all run as induction motors.

Direct-current Motors The d-c motor is used where an a-c motor is not satisfactory. D-c motors are premium-priced products, especially in the larger sizes. But they offer easily adjustable speed and effective and simple control of torque, acceleration, and deceleration. Even with the limitations of a commutator, they can and do handle difficult duty cycles. A rule of thumb states that the maximum economical size of a d-c motor is not exceeded so long as the product of horsepower and speed is less than 1.5 million. From a voltage standpoint, 250 volts is used for units up to 500 hp, 600 volts for 600 to 1,000 hp, and 700 or 900 volts above 1,000 hp.

Figure 8-2 shows the speed, torque, and horsepower characteristics of the three types of d-c motors—series, shunt, and compound. Many centrifugal pumps are driven at either 1,800 or 3,600 rpm; both these values are synchronous speeds of a-c equipment. With d-c motors, any intermediate or higher or lower speed can be obtained, if suitable controls are used. However, it is better to operate the motor at its standard

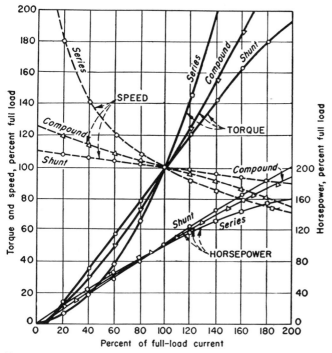

Fig. 8-2 Speed, torque, and power characteristics of d-c motors.

speed. This ranges from 50 rpm for 1,000- to 8,000-hp motors to 3,500 rpm for 1.5- to 40-hp motors.

Enclosures   The type of enclosure specified for a motor driving a pump is particularly important because pumps are located in a variety of areas—outdoors, indoors, in mines, ship pump rooms, etc. Typical enclosures used with pumps include dripproof, splashproof, pipe-ventilated, weather-protected, etc. For a complete listing of available enclosures for motors up to 200 hp, see appropriate NEMA publications. Pipe-ventilated motors having an outside blower are shown in Fig. 8-3, while self-ventilated units in a pipeline pumping station are shown in Fig. 8-4. In all ventilated motors it is important that the supply air be free of hazardous gas and dust.

Integral Auxiliaries   Motor manufacturers now incorporate power-transmission units in the frames of many motors. Typical are the gear-motor (Fig. 8-5) and the fluid-drive motor. Gearmotor popularity hinges on the fact that the cost and size of electric motors increase considerably as the speed for a given horsepower decreases. It is generally more economical to reach low driver speeds through use of a standard 1,750-rpm motor, with its output speed reduced by some means. Covering a horsepower range between $\frac{1}{6}$ and 200, many gearmotors are designed for output speeds from 5.7 to 780 rpm. Where pump speed is 900, 1,200, or 1,800 rpm synchronous, a motor with 8, 6, or 4 poles may be used without a speed reducer. In the 700- to 900-rpm range,

Fig. 8-3   Pipe-ventilated motor driving a boiler-feed pump.   (*Worthington Corp.*)

Fig. 8-4 Self-ventilated motors in pipeline pumping station. [*Standard Oil Co. (N.J.).*]

however, some engineers prefer an 1,800-rpm motor in combination with flat or V belts. At about 700 rpm and below, the practice is to use a chain drive, a belt, a separate gear reducer, or a gearmotor. Geared or belted variable-speed drives for many different output speeds are also popular.

Torque Requirements For the starting and running torque requirements of a particular pump, consult its manufacturer. Starting torque usually varies considerably from one type of pump to another. With typical centrifugal pumps the starting torque is 15 to 20 per cent of the normal operating torque. Rotary and reciprocating pumps have higher starting torques—up to 150 per cent when not fitted with a bypass.

Drive Arrangement The motor, turbine, or engine can be arranged to drive the pump directly through a flexible coupling (Fig. 1-15), spacer-

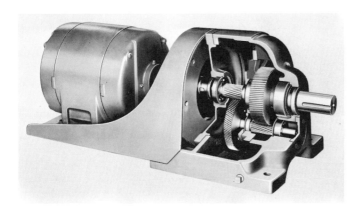

Fig. 8-5 Double-reduction gearmotor. (*Link-Belt Division, FMC Corp.*)

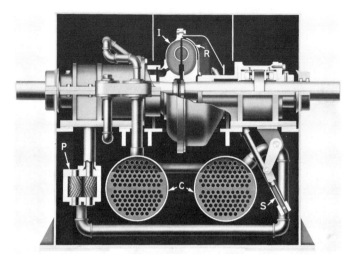

Fig. 8-6 Fluid drive for pump application. (*American Standard Industrial Division.*)

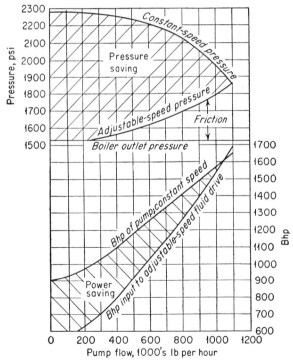

Fig. 8-7 Pressure and power savings with fluid drive for a boiler-feed pump. (*American Standard Industrial Division.*)

type coupling, gears (Fig. 3-4), belts, fluid drive (Fig. 8-6), or a combination of these. The type chosen depends on a number of factors, discussed below.

Fluid Drives These (Fig. 8-6) are popular for boiler-feed pump drive because they permit a substantial reduction in the power input to the pump and the discharge pressure. The unit shown consists of an impeller I connected to the motor or input shaft, a runner R connected to the output or pump shaft, a gear-type oil pump P, oil coolers C, and a speed controller S. As Fig. 8-7 shows, the pressure developed by a typical pump driven by a fluid drive is less than a constant-speed pump at low flows. This differential decreases as the flow increases and will be different for each installation, but the pressure saving exists throughout the entire flow range, except at the point of maximum flow. The power saving which results from the lower pressure and improved pump operation is a major item. Fluid drives are also used to drive many other types of pumps, including refinery, deepwell, process, etc.

STEAM TURBINES

There are a number of installations where steam turbines offer simple speed control and the possibility of improving the plant heat balance by use of exhaust steam, or by exhausting to heaters of some type. Power plants, refineries, chemical process, and pipeline pumping stations are examples of plants using steam turbines to drive pumps. The most common class of pump driven by steam turbines is the centrifugal, but rotary and reciprocating pumps are also occasionally found. Figure 8-8 shows a typical turbine designed for centrifugal-pump drive. Units of this type are generally known as mechanical-drive turbines and may drive the pump directly or be connected to it through a set of step-down or step-up gears, depending on the desired speed of the pump.

Governors Several different types of governors can be used for steam turbines driving pumps—constant-speed, constant-pressure, and differential-pressure governors. Figure 8-9 shows pump characteristic curves obtained with various types of drives and turbine governors. Note how the shutoff head of the centrifugal pump driven by these units varies from one type of drive and governor to the next. To take full advantage of a plant heat balance, some pumps may be driven by either a motor or a turbine. When one drive is operating, the other spins idly with the pump.

Staging There are no fixed rules with regard to the number of pressure stages to be used in a turbine driving a pump. However, below 100 hp, single-stage turbines are generally used, with multistage units for

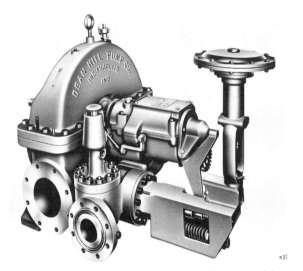

Fig. 8-8 Mechanical-drive steam turbine for pump drive.

larger outputs. But where space is limited or simplicity of construction is required, single-stage units are sometimes used at ratings above 100 hp. Some marine installations use single-stage turbines for large horse-powers because excessive space and complexity are often a disadvantage.

Turbine Steam Rates Figure 8-10 gives data for determining the steam

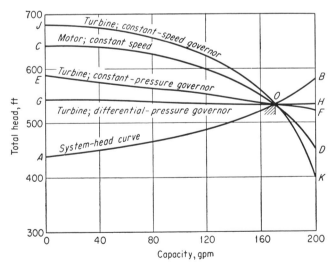

Fig. 8-9 Characteristic curves obtained from same pump with different drivers.

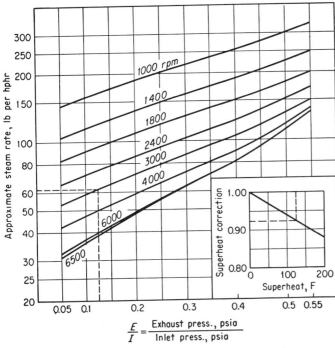

Fig. 8-10   Approximate steam rates for 25-in.-diameter single-stage tur-bines. (*General Electric Co.*)

rate for one size of turbine wheel for single-stage units rated at 10 to 1,200 hp, 1,000 to 6,500 rpm.   Initial steam pressure ranges up to 850 psig, temperature to 750 F, exhaust pressure to 75 psig.

example: Find the steam rate for a 400-hp 3,000-rpm steam turbine driving a pump if the initial steam pressure is 200 psia, initial temperature is 125-F superheat, and the exhaust pressure is 25 psia, when a 25-in.-diameter wheel is used.

solution: $E/I = 25/200 = 0.125$.   Enter the bottom of Fig. 8-10 and project vertically upward to the 3,000-rpm speed curve.   At the left read 43 lb per hp-hr. Correct for superheat by entering the small chart (Fig. 8-10) and projecting to the correction curve.   At the left read the correction factor as 0.895.   Steam rate for the 25-in. wheel is $43 \times 0.895 = 38.5$ lb per hp-hr.   Values found by this method are approximate but may be used in all estimates of steam needs when the manufacturer's guarantees are not available.

Turbine Application   Steam turbines should be considered for pump drive when one or more of the following requirements must be met: (1) quick starting of the pump, as in emergency or standby installations; (2) where low-pressure exhaust steam is needed for plant process or other equipment; (3) in explosive atmospheres where motors or internal-combustion engines are unsuitable (the turbine must be fitted with non-sparking accessories and control apparatus); (4) in hot damp areas; (5) where variable-speed drive of the pump is necessary; and (6) for high-speed pumps requiring shaft speeds to 12,000 rpm.

STEAM ENGINES

With the exception of direct-acting steam pumps, reciprocating steam engines are rarely used today for pump drive in stationary applications. They are, however, still rather widely used in some marine applications —tugboats and dredges. Simple or compound engines are used in tugs and similar vessels for pump drive. In suction dredges, the uniflow engine is often used to drive the main pump. Steam rates of usual uniflow engines range from a low of about 9 lb per ihp-hr to a high of about 22 lb per ihp-hr. Direct-acting steam pumps of various types are discussed and illustrated in Chap. 3. The variable length of the discharge pipe and the variety of materials handled has made the uniflow engine popular for suction dredges because it combines flexible control of power and speed with low steam rates.

INTERNAL-COMBUSTION ENGINES

Recent studies of United States industrial and service establishments show that about 86 per cent of the centrifugal pumps in use are motor-driven, 44 per cent of the reciprocating pumps, 96 per cent of the rotary pumps, and 95 per cent of the deepwell pumps. Steam engines, turbines, and internal-combustion engines drive about 13 per cent of the centrifugals. Steam is outstanding for larger reciprocating units, driving almost half of those in use. Though internal-combustion engines can seldom best electric motors on a purely economic basis, they are extremely important for a number of different types of installations. Probably the most common use is in isolated areas where electricity is not available. But internal-combustion engines are also extremely important for portable pumping units, emergency sets, certain types of pumping stations, irrigation service, and oil wells.

Types of Engines Used Diesel, gasoline, dual-fuel, gas, and low-compression oil engines are all used for pump drive. The type of engine chosen depends on the amount of power required, type of fuel most readily available at low cost, type and number of operators that can be employed, and the class of installation the engines will serve—permanent or temporary. Many permanent installations are in the petroleum industry for crude- or refined-oil transportation through pipe-lines. The construction and mining industries use portable pumps. Sewage plants have engines utilizing sewage gas for their fuel. Some vessels—for example, shrimp boats, tugs, dredges, and other harbor craft—use internal combustion engines for pump drive. Here small size and weight are important.

Petroleum Pipelines In many pumping stations the engines driving the pumps use the oil being handled for their fuel, reducing transpor-tation costs for the fuel and eliminating much of the fuel-storage capac-ity that would otherwise be required. Saving in tankage reduces the first cost of the station.

Figure 8-11 shows a typical station of this type. This station is some-what unusual in that the diesel engine driving the 6-stage flammable-liquid centrifugal pump is located in the same room with the pump. In most stations it is common practice to separate the pump from the engine by means of a gastight wall and shaft seal (Fig. 8-12). Here, however, the pump is fitted with double mechanical seals, and air from the vicinity of the seals is constantly drawn through copper tubing to a combustible-gas analyzer and alarm on the instrument panel (arrow, Fig. 8-11). As many as 23 different combustible liquids have been piped through this pump at one time. The diesel drives the pump through a variable-speed drive. Engine speed is primarily controlled by the pump suction and discharge pressures.

Fig. 8-11   Typical diesel-powered petroleum pumping station showing pump, engine, and combustible-gas analyzer (arrow). (*Mine Safety Appliances Co.*)

Figure 8-12 shows the interior of another station using diesel engines for pump drive. Each unit shown is a 4-cycle 2,000-hp supercharged dual-fuel engine driving a centrifugal pump through a speed-increasing gear. As is customary in many stations of this type, a vaportight wall

(a)

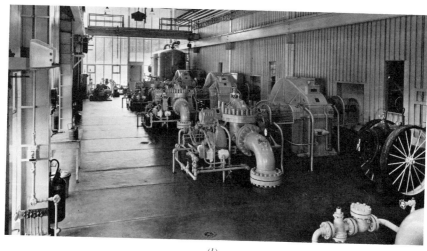

(b)

Fig. 8-12 Two views of a petroleum pipeline pumping station show the diesel-engine drives (a) separated by a fire wall from the speed increasers and pumps (b). (*Nordberg Mfg. Co.*)

Fig. 8-13   Nine radial engines rated at 2,125-hp each directly drive large sewage-lift pumps beneath the engine room. (*Nordberg Mfg. Co.*)

separates the gear and pumps from the engines. Figure 8-13 shows several large radial diesel engines direct-connected to vertical sewage-lift pumps on the floor below.

Flood Control, Irrigation, and Sewage   Figure 8-14 shows the interior of an internal-combustion-engine-powered flood-control station.   Each of

Fig. 8-14   Interior of flood-control station showing diesels and gears for pump drive. (*Fairbanks, Morse Pump Division, Colt Industries.*)

the six 116-in.-diameter 360,000-gpm 11.1-ft static head horizontal axial-flow propeller pumps in this station (Fig. 5-3) is driven by a 1,600-hp opposed-piston diesel engine. The combined capacity of the six pumps is 2,160,000 gpm. Only a few higher lift stations such as at Grand Coulee Dam and Mill Creek, Ohio, can pump more water.

Diesel engines were chosen to drive these pumps primarily to ensure continuity of operation during storm periods, when the plant would be most urgently needed, and when electric power-transmission lines would be most vulnerable to failure. Also, the power and efficiency characteristics of diesel engines suited them to driving axial-flow pumps at variable speeds under various load conditions. Another factor was the operating economy of the diesels in a plant where the annual service equivalent of 52 days of full-capacity pumping is anticipated.

Sewage-disposal plants often have large quantities of sewage gas available. The gas usually has a heating value of 600 to 700 Btu per cu ft, with 3 to 6 cu ft of gas being obtained per day per capita from digestion tanks. Engines used for sewage gas are usually high-compression units. Both the spark-ignite and diesel types are used in modern plants.

Fire Protection   The pumps and drives installed in the pumping station of the John F. Kennedy International Airport in New York provide an interesting study of good application practices. To give as flexible and as dependable a system as possible, two banks of pumps are used. Both banks are identical, each containing five centrifugal pumps. Two pumps in each bank are rated at 2,500 gpm, while the other three are rated at 4,000 gpm. Five diesel engines drive the pumps in one bank, and five electric motors the pumps in the second bank (Fig. 8-15). In addition, a 100-gpm make-up pump maintains a pressure of 165 psi in the entire water-supply system, which contains 10 miles of piping.

Marine Installations   Internal-combustion-engine drives for pumps in marine service find most use in motorships, though some steam vessels use them for emergency service. Sets combining three auxiliary ser-

Fig. 8-15   Interior of fire-fighting station, John F. Kennedy International Airport, New York.

Fig. 8-16   Diesel-driven set for marine use has pump, compressor, and generator.

vices—pumping, power generation, and air compression—are common for smaller motor vessels, like tugs and fishing boats. One typical set is shown in Fig. 8-16. The pump (lower center, Fig. 8-16) is belt-driven from the engine shaft through a clutch. The generator is direct-connected to the diesel engine, while the air compressor is driven in the same way as the pump. Smaller sets have the same components but may not be fitted with clutches.

In some large vessels, especially those driven by steam, an emergency pumping unit may be used, driven by a gasoline or diesel engine. Newer dredges are often fitted with large diesels to drive the main pump. Reduction gears and a hydraulic coupling are commonly used between the engine and the pump. For offshore drilling of oil wells, diesel-electric generating sets mounted on skids are very popular. One or more sets supply power to motor-driven mud pumps, drawworks, and control equipment. The mud pumps and drawworks can be operated singly or in combination.

Rating Standards   The capacity of a diesel engine is affected by the temperature and barometric pressure of the intake air and hence by the altitude at which the engine operates. To make corrections unnecessary for the average installation, the standard sea-level rating specified by the

Diesel Engine Manufacturers Association (DEMA) holds up to 1,500 ft altitude at atmospheric temperatures to 90 F and barometric pressures not less than 28.25 in. Hg. Engines to operate at altitudes above 1,500 ft must be derated. The DEMA publishes a derating curve for altitudes up to 15,000 ft. At 5,000 ft an engine will turn out about 85 per cent of its sea-level rating; at 10,000 ft about 67 per cent. Where the engine must operate at intake temperatures in excess of 90 F the manufacturer should be asked to make his guarantee on the bas s of the actual temperature.

To estimate the fuel consumption for a projected pumping plant, the best procedure is to calculate the number of horsepower-hours' operation expected at various loads and multiply by either the actual guarantee given by the engine builder or by typical fuel-consumption data. Small diesels use about 6 gal of fuel per 100 hp-hr; large engines use about 5 gal per 100 hp-hr.

Gasoline Engines   Industrial-type gasoline engines for pump drive are generally rated for continuous duty at maximum output. Since the price of fuel for a gasoline engine is generally much higher than for a diesel, the gasoline engine does not find too much use in ratings above about 150 hp for pump drive. Today, gasoline engines are popular for portable applications where fuel cost is secondary to light weight, ease of operation, and freedom from fuel-supply problems in isolated areas where gasoline is usually more readily available than diesel fuel. Other uses of low-compression spark-ignited engines, using gasoline, kerosene, or natural gas for fuel, include oil-well pumping (Fig. 8-17), mud-pump drive, etc. Where there is danger of ignition associated with fuel storage, diesel or kerosene engines are preferred over gasoline and natural-gas engines.

The fuel consumption of typical 4-cycle gasoline engines ranges between 0.45 and 0.70 lb per hp-hr, depending on engine size, method of cooling, mechanical condition, etc. Engine weight varies from 10 to 50 lb per hp, compared with 10 to 150 lb per hp for diesels. Spark-ignited natural-gas engines use 8 to 14 cu ft of gas per hp-hr, depending on the heating values of the gas, amount of moisture in it, etc. Values given here are suitable for estimating when manufacturer's data are not available.

Comparative Data   Gasoline engines serving continuous loads have a fuel cost three to four times that of a diesel driving the same pump. Where the fuel cost, cents per gallon, for a diesel equals or is less than the electric-power cost, mills per kwhr, the diesel is usually a worthwhile investment. For example, in an area where fuel costs 4 cents per gallon, the cost of driving a pump with a diesel engine using this fuel is equiva-

Fig. 8-17 Engine-driven pumping rig in Kansas lifts oil from a
3,700-ft depth, runs unattended 5 days a week. (*Fairbanks, Morse
Pump Division, Colt Industries.*)

lent to buying electric power at 4 mills per kwhr. Spark-ignition
natural-gas engines often prove profitable when the cost of gas ranges
between 20 and 30 cents per 100 cu ft. While these rules of thumb are
not exact for every installation, they are helpful in preliminary estimates.
The capacity of gasoline-engine-driven pumps decreases about 1.5 per
cent per 1,000 ft above sea level; head falls off about 2.5 per cent per
1,000 ft above sea level.

GAS TURBINES

In recent years, gas turbines (Figs. 8-18 and 8-19) have become popular
for a wide variety of pump-drive services. Packaged with an integral
reduction gear, any output speed between 30,000 and 900 rpm (at 100
per cent power-turbine speed) can be provided at powers of well over
100 hp. Usable fuels for gas-turbine drives cover a wide range: natural
gas; LPG, including liquid propane and liquid butane; distillates, including light diesel fuel; kerosene; and JP-series fuels.

Exhaust products from gas turbines are usually clean, dry, and leave
the engine at about 850 F. Where *both* shaft horsepower and exhaust
heat are put to work, thermal efficiencies of over 70 per cent are often
attained. Gas-turbine heat rates vary from a low of about 9,600 Btu per
hp-hr for regenerative, intercooled two-shaft open units to about 22,900
Btu per hp-hr for simple, open single-shaft units.

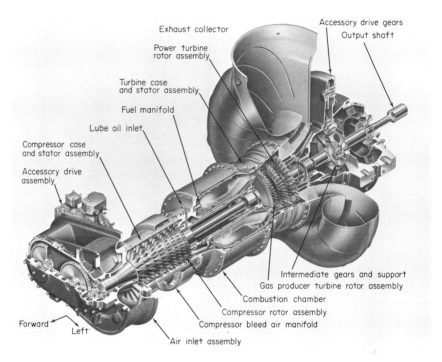

Exhaust collector

Accessory drive gears
Output shaft

Power turbine
rotor assembly

Turbine case
and stator assembly

Fuel manifold

Lube oil inlet

Compressor case
and stator assembly

Accessory drive
assembly

Intermediate gears and support

Gas producer turbine rotor assembly

Combustion chamber

Compressor rotor assembly

Forward

Left

Compressor bleed air manifold

Air inlet assembly

Fig. 8-18 Gas-turbine engine for direct or integral-gear industrial drive. (*Solar Division, International Harvester Co.*)

Fig. 8-19 Gas-turbine-driven multistage centrifugal pump on petroleum pipeline service. (*Solar Division, International Harvester Co.*)

MISCELLANEOUS POWER SOURCES

The variety of miscellaneous power sources used to drive all types of pumping equipment is limited only by the engineer's imagination.   In many cases, these power sources reflect the availability of an unusual power source—compressed air or abundant, high-pressure water, for example.   Compressed air is probably the most widely used power source.   Typical pumps using compressed air from the plant air system develop high pressures while handling small volumes of liquid.   They are used to test tubing, valves, and pressure vessels and to operate small molding presses and similar equipment.   Other types of pumps operated by air include sewage, food-handling, hydraulic-press, and metering pumps.

Many ingenious hydraulic power sources have been developed to take advantage of the availability of high-pressure water to boost low-pressure water.   In Central Park, New York City, two single-stage centrifugal pumps are coupled together to do just this.   One, taking available

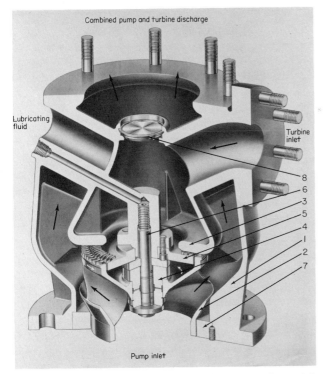

Fig. 8-20   Liquid turbine-driven centrifugal pump can be used for booster service and can act as a blender. (*Worthington Corp.*)

high-pressure water, acts as a water turbine to drive the other. Since the high-pressure water would have to be throttled before entering the city water system, the power source is virtually free. A more modern adaptation of this installation is shown in Fig. 8-20. Here an integral water turbine takes high-pressure water and drives a hydraulically conventional mixed-flow pump. The exhaust water from the turbine combines with the pumped flow. The unit is entirely enclosed in an in-line spool piece, requiring no external stuffing box or other sealing means.

While manual operation of pumps is often thought of as ancient, Fig. 20-1 shows a typical priming application of a manually operated pump. The vertical reciprocating hand pump is used to prime the horizontal close-coupled centrifugal pump rated at 275 gpm. Priming, lube-oil drum emptying, and hydrostatic testing represent the major service of hand pumps today.

CHAPTER NINE

Pump Selection

PROBABLY THE BIGGEST PROBLEM facing an engineer designing a pumping system is choice of the class, type, capacity, head, and details of the pump or pumps to be used in the system. There are such a variety of pumps available, and so many applications are possible for each, that it is often difficult to narrow the choice to one specific unit. This chapter is aimed at reducing many of the difficulties met in pump selection. Using the methods presented here, the engineer can start with the hydraulic conditions to be met, and proceed, by means of a few simple steps, to the pump best suited for the liquid conditions. Then, using an economic analysis, as discussed in Chap. 10, he can arrive at the most economical unit for his plant.

Selection Methods Pumps are usually chosen by one of three methods: (1) the prospective purchaser supplies one or more manufacturers with complete details of the pumping conditions and requests a recommendation and bid on the units which appear best suited for the conditions; (2) the purchaser makes a complete calculation of the pumping system and then chooses a suitable unit from current catalogs and rating charts; or (3) a combination of these two methods is used to arrive at the final selection. Regardless of which method is used to select pumping equipment, responsibility for accurate determination of operating conditions rests with the purchaser. Even though he supplies a manufacturer with all pertinent data to calculate required flow, total head, and suction conditions, he must satisfy himself that the resultant operating condition determinations are correct. The manufacturer's sole responsibility is to provide equipment that meets the stated operating conditions in all respects.

Manufacturer's Choice This method is used for large pumps, for jobs having unusual conditions, and in instances where the engineer does not have the time or desire to choose the pump himself. Though it appears to relieve the engineer of much of the responsibility in the choice of the pump, it actually does not. Recommendations and bids must be evaluated, and to do this requires a complete knowledge of the pumping problem, the relative merits of various designs, and the economics of the installation.

Data for Manufacturer Table 9-1 summarizes the essential data required by any pump manufacturer before a recommendation and bid can be prepared. Many manufacturers have forms which the engineer can fill out when requesting a recommendation. These can be extremely useful because they help prevent omission of important data.

When supplying data to a manufacturer, extreme care must be used to see that all the facts concerning the installation are given. Incomplete data can lead to a poor or wrong recommendation because the engineer choosing the unit may make the wrong assumptions. Thus, the plant engineer requesting a pump recommendation or bid has a definite responsibility toward the manufacturer, and if this is neglected, the engineer can hardly expect to secure the right pump.

Proposal Most manufacturers combine their recommendation and bid into a single document called a proposal. The usual proposal contains the following information: pump model number, class, type, construction details and materials, type of drive for which the pump is designed, performance curves or tabulation, unit weight, price, availability of pump for shipment after receipt of the order, and legal agreements concerning drawings, warranties, unit installation, shipping date, terms of payment, taxes, insurance, transportation, etc. Enclosed with the usual proposal is a drawing of the pump and a catalog. If the pump is to be specially built for the purchaser, the catalog may not be included because the manufacturer may not have bulletins on it available.

To evaluate a proposal it is necessary to review all the steps made in choosing a pump from a given set of hydraulic conditions. These steps are given in detail below.

Calculations in Pump Choice Basically, there are five steps in choosing any pump—be it large or small, centrifugal, reciprocating, or rotary. These steps are: (1) sketch the pump and piping layout, (2) determine capacity, (3) figure total head, (4) study liquid conditions, and (5) choose

TABLE 9-1   Summary of Essential Data Required in Selection of Centrifugal Pumps*

1. Number of Units Required

2. Nature of the Liquid to Be Pumped
   Is the liquid:
   a. Fresh or salt water, acid or alkali, oil, gasoline, slurry, or paper stock?
   b. Cold or hot and if hot, at what temperature? What is the vapor pressure of the liquid at the pumping temperature?
   c. What is its specific gravity?
   d. Is it viscous or nonviscous?
   e. Clear and free from suspended foreign matter or dirty and gritty? If the latter, what are the size and nature of the solids, and are they abrasive? If the liquid is of a pulpy nature, what is the consistency expressed either in percentage or in lb per cu ft of liquid? What is the suspended material?
   f. What are the chemical analysis, pH value, etc.? What are the expected variations of this analysis? If corrosive, what has been the past experience, both with successful materials and with unsatisfactory materials?

3. Capacity
   What is the required capacity as well as the minimum and maximum amount of liquid the pump will ever be called upon to deliver?

4. Suction Conditions
   Is there:
   a. A suction lift?
   b. Or a suction head?
   c. What are the length and diameter of the suction pipe?

5. Discharge Conditions
   a. What is the static head? Is it constant or variable?
   b. What is the friction head?
   c. What is the maximum discharge pressure against which the pump must deliver the liquid?

6. Total Head
   Variations in items 4 and 5 will cause variations in the total head.

7. Is the service continuous or intermittent?

8. Is the pump to be installed in a horizontal or vertical position? If the latter,
   a. In a wet pit?
   b. In a dry pit?

9. What type of power is available to drive the pump and what are the characteristics of this power?

10. What space, weight, or transportation limitations are involved?

11. Location of installation
    a. Geographical location
    b. Elevation above sea level
    c. Indoor or outdoor installation
    d. Range of ambient temperatures

12. Are there any special requirements or marked perferences with respect to the design, construction, or performance of the pump?

* Worthington Corporation.

class and type. For convenience in quick estimates these five steps can be remembered as relating to *size, class,* and *best buy.*

Sketch Layout: Base sketch on the actual job. Single-line diagrams (Figs. 4-3 and 4-4) are usually satisfactory. Show all piping, fittings, valves, equipment, and other units in the system. Mark the length of pipe runs on the sketch. Be sure to include all vertical lifts. Where the piping is complex, an isometric sketch is often helpful.

Determine Capacity: Job conditions fix the capacity required. For example, the maximum steam flow from the exhaust of a turbine, along with steam conditions, determines the minimum amount of cooling water needed 'at a given temperature. Seasonal changes, safety factor desired, etc., influence the actual capacity chosen. Data in Chap. 5 and in later chapters covering specific applications are helpful in determining the pumping capacity required for a given set of conditions.

Figure Total Head: Use the data in Chap. 4 to compute the head on the pump. As a double check, it is worthwhile to submit a complete sketch of the system layout to the manufacturer when requesting a proposal. Then his engineers can also compute the total head on the pump, verifying the calculations made here. This is one additional means of assuring a more accurate selection of the pump.

Study Liquid Conditions: Liquid specific gravity, temperature, vapor pressure, viscosity, chemical characteristics, etc., must be carefully considered. See Chap. 6 for a discussion of these and other factors, and their effect on pump performance.

Choose Class and Type: Studying the layout tells what *size* (capacity and head) pump is needed. This furnishes the first clue as to what class of pump is suitable. For example, where high-head small-capacity service is required, Table 1-1 shows that a reciprocating pump would probably be suitable. Reviewing the liquid characteristics furnishes another clue to class because exceptionally severe conditions may rule out one or another class right at the start. Sound economics (Chap. 10) dictate choosing the pump that provides the lowest cost per gallon pumped over the useful life of the unit.

Operating factors deserving recognition when deciding on the class of pump include type of service (continuous or intermittent), running-speed preferences (high-speed pumps may cost less), future load expected and its effect on pump head, possibility of parallel or series hookup, and many other conditions peculiar to a given job. These factors deserve as much study as head and capacity because they are just as important.

Select Number of Pumps: Depending upon the need for part-capacity operation and for a spare pump to insure operating continuity, the total capacity may be advantageously split into two half-capacity pumps with, perhaps, a third duplicate pump as an installed spare. This would be particularly good if periods of operation at less than half the total capacity were planned; the need to operate only one pump would require far less power than a full-capacity pump operating at less than half capacity. Generally, several single- and multiple-pump schemes should be compared against planned operating conditions to arrive at the optimum number of pumps to use.

Once class and type are known, a rating table (Table 1-2) or rating chart (Fig. 1-11) can be checked to see if a suitable pump is available from the particular manufacturer whose unit is to be purchased. This assumes, of course, that a complete set of his bulletins and other data are on hand. Where the required hydraulic conditions fall between two standard models, it is usual practice to choose the next larger size of pump, unless there is some reason why an exact capacity and head are required of the unit. Where one manufacturer does not have the particular class and type of pump available, or a unit which meets the desired hydraulic conditions, refer to the data of one or more other manufacturers. One important fact to keep in mind is that some pumps are custom-built for a given job or plant. Under these conditions the pump manufacturer performs most of the steps listed above, basing his design on data supplied by the plant engineer.

Power Input   The power required to drive any class or type of pump can be computed from

$$P = fhs/3{,}960e \qquad (9\text{-}1)$$

where P = power input, hp; f = liquid flow rate, gpm; h = total head on the pump, ft of liquid handled; s = liquid specific gravity; e = pump efficiency, expressed as a decimal. This equation is suitable for all liquids having a viscosity equal to that of water; for other viscosities use Fig. 9-1 or the correction factors given in Chap. 6.

example: What power input is required to deliver 1,000 gpm of water at 68 F (having a viscosity of 32 SSU and a specific gravity of 1.0) against a total head of 120 psi [total head is $(120)(2.31)/1.0 = 277.2$ ft] with an efficiency of 70 per cent?

solution: $P = (1{,}000)(277.2)/(3{,}960)(0.70) = 100$ hp.   To use Fig. 9-1, connect 1,000 gpm and 120 psi by line A–B. Then connect the fluid horsepower intersection C with 70 per cent efficiency and extend brake horsepower D. Connect point D with viscosity index (32 SSU for water at 68 F) and extend to brake horsepower, point E.

Change of Performance   Altering the speed or impeller diameter of a centrifugal pump changes the performance of the unit. There are three rules relating performance with change in speed and three for change in diameter. With a constant-diameter impeller, (1) pump capacity varies directly as speed, (2) head varies as the square of the speed, and (3) horsepower input varies as the cube of the speed. At a constant speed, (1) capacity varies directly with the impeller diameter, (2) head varies as the square of the impeller diameter, and (3) horsepower varies as the cube of the impeller diameter. These rules hold, approximately, for all types of centrifugal pumps. An example will show their use.

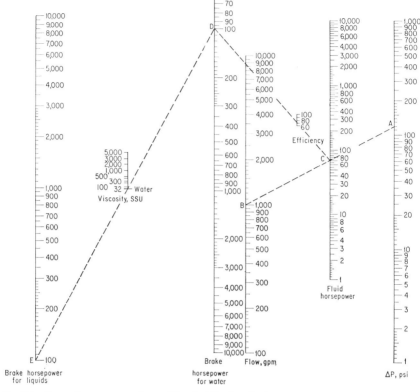

Fig. 9-1 Pump horsepower chart. (*Chemical Engineering.*)

example: A pump delivering 500 gpm at 1,150 rpm and 50-ft head requires 10 hp. What will be the capacity, head, and power input of this unit if its speed is increased to 1,750 rpm?

solution: New capacity is in the same ratio as the speeds, or $(1,150/1,750) = (500/x \; gpm)$, and $x = 760$ gpm. New head is in the ratio of the speeds squared, or $(1,150)^2/(1,750)^2 = (50)/(y \; ft)$, and $y = 116$ ft. New horsepower is in the ratio of the speeds cubed, or $(1,150)^3/(1,750)^3 = (10)/(z \; hp)$, or $z = 35.2$ hp.

Rules for impeller diameter are used in a similar way. By computing the performance of the pump at a number of points along its characteristic curve, a new set of curves can be plotted. These will usually agree closely with actual test curves.

Flexible Couplings To choose a flexible coupling for a pump the horsepower to be transmitted must be known, as well as the speed of rotation and the applicable *service factor*. This last item compensates for shock loads and other variations in the power input. Couplings are usually rated in horsepower per 100 rpm, determined from

$$C = 100(PF)/S \qquad (9\text{-}2)$$

where C = coupling rating, hp per 100 rpm; P = horsepower input to pump; F = coupling service factor; S = coupling speed, rpm. Service factors vary from one coupling manufacturer to another and with the type of drive used for the pump. For example, one manufacturer uses the following service factors: turbine-driven centrifugal pump, 1.25; motor-driven centrifugal pump, 1.5; motor-driven duplex and triplex pumps, 3.5; engine-driven centrifugal pumps, 3.0; engine-driven duplex and triplex pumps, 5.5.

example: What rating coupling should be chosen for an engine-driven triplex pump rated at 600 hp at 1,100 rpm?
solution: $C = 100(600)(5.5)/1,100 = 300$ hp per 100 rpm. When choosing an actual coupling be sure to use the correct service factor because the values vary from one manufacturer to another.

Specific Speed It is a wise practice to check the specific speed of a proposed pump to see that it comes within the usual limits for the type of pump chosen. See Chap. 4.

example: What is the specific speed of a 3-stage horizontal 1,750-rpm centrifugal pump handling 900 gpm if the total head developed is 300 ft?
solution: Specific speed, N_S = (gpm)(rpm)/$h^{0.75}$, where h = head per stage, ft of liquid. $N_S = (900)(1,750)/(100)^{0.75} = 1,640$. From Fig. 1-7, it can be seen that a typical centrifugal pump has an efficiency of about 82 per cent at this specific speed and capacity.

Net Positive Suction Head The Hydraulic Institute condensate-pump npsh curve can usually be used beyond its plotted ranges if, for a definite npsh, the product (rpm)($\sqrt{\text{gpm}}$) is constant, but the manufacturer should be advised.

example: A 1,150-rpm condensate pump has a capacity of 360 gpm at 2 ft npsh. What will its capacity be at 3,450 rpm?
solution: Using the above relation, $(1,150)(\sqrt{360}) = (3,450)(\sqrt{x \text{ gpm}})$, or $x = 40$ gpm. This relation also holds for the Hydraulic Institute curves for hot-water pumps when the speed is within ±25 per cent of the charted values.

Friction Loss To interpolate in pipe-friction tables requires use of a simple square relation. Results obtained are approximate.

example: The friction loss per 100 ft of 6-in. pipe is 1.5 psi when the flow is 1,200 gpm. What is the loss for 1,600 gpm?
solution: Loss at higher flow rate is $(1,600/1,200)^2(1.5) = 2.67$ psi.

Horizontal vs. Vertical Pumps This consideration is becoming of greater importance today because vertical pumps of a number of different designs are extremely popular. From the standpoint of floor space occupied, required npsh, priming, and flexibility in changing the pump duty, vertical pumps may be preferable to horizontal designs. But where headroom, corrosion, abrasion, and ease of maintenance are factors, horizontal pumps may be preferred. Recent vertical-pump design advances eliminate many operating problems related to abrasive liquids. Study of a number of typical close-coupled and vertical turbine pumps shows the ratio of the floor area of the first to the second is about 1.5 in capacities through 500 gpm. With a horizontal double-suction split-case pump the ratio is about 3.1. Height ratios for these pumps are the reverse because the vertical turbine pump is higher, and the values are 3.4 and 1.9, respectively. There is, of course, some variation in these from one design to another.

Single- vs. Double-suction Designs Figure 9-2 shows the approximate ranges in which single- and double-suction pumps are used. It also

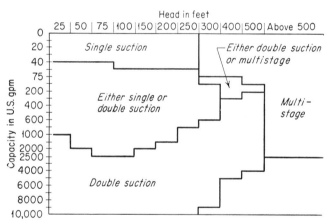

Fig. 9-2 Approximate head and capacity ranges for centrifugal (single- and double-suction) pumps.

shows the general ranges for applying single- and multistage pumps. Within these ranges, however, the choice between double- and single-suction pumps may be limited by suction conditions. For a given pump capacity and speed, a double-suction pump will normally be capable of a higher suction lift (lower npsh) than a single-suction pump because of the greatly increased impeller-eye area involved.

Packing vs. Mechanical Seals Rapid advances in mechanical-seal design and manufacture make these units available for many routine services. Seals (Fig. 9-3) have almost zero leakage and may be used for

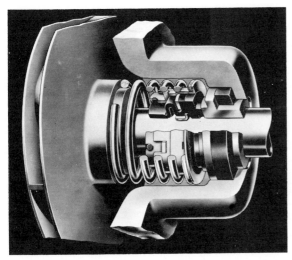

Fig. 9-3   Mechanical seal for a centrifugal pump. (*Weinman Pump Mfg. Co.*)

centrifugal and rotary pumps.   The labor saving of seals over packing often tips the scale in their favor.   Mechanical seals, however, do require expert installation and maintenance.   If treated by operators like a packed stuffing box, they can be troublesome and may require early, expensive replacement.

Packing, however, is not outmoded and will continue to be used in almost every class and type of pump.   At temperatures above 300 F, the packing boxes of boiler-feed and other hot-water pumps should have cooling jackets.   Smothering-type glands are also often recommended. Chemical-type pumps may use extra-deep stuffing boxes.   An external source of lubrication or cooling water may also be used.   Condensate and heater-drain pumps operating under a vacuum should have a seal cage (Fig. 1-29) and a connection to an external supply of clear water. Hot-oil pumps having sealed stuffing boxes often use oil as the seal liquid.   Seal-oil pressure should be at least 25 psi above the stuffing-box pressure.   Smothering-type glands should be used.   Figure 9-4 shows a typical packed pipeline pump with water-cooled stuffing boxes and smothering glands.

Volatile-liquid pumps often use mechanical seals, as do some hot-oil pumps.   Packed volatile-liquid pumps may have a smothering-type gland, water jacketing, oil or grease seal, or a bleed-off connection.   A large number of packing materials are available today for all types of pumps.   These include cotton, flax, viscose rayon, leather, asbestos, Teflon, phenolic resins, silicone, rubber, and metals.

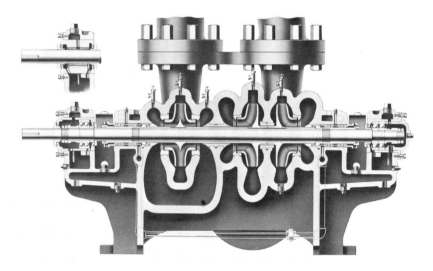

Fig. 9-4   Typical 3-stage pipeline pump. (*Goulds Pumps, Inc.*)

Cooling Mechanical Seals   There are a number of reasons for cooling certain mechanical seals: (1) prevent high temperatures from destroying liquid film between sealing faces; (2) stop vaporization of liquid at the seal faces; (3) protect the seal; (4) reduce the fire hazard when handling flammable liquids; and (5) prevent corrosion by high-temperature liquid.

The coolant used may be the liquid handled—bled from an intermediate stage in a multistage pump, or from the discharge in a single- or multistage pump.   If the pressure differential is high, an orifice or bushing should be fitted in the cooling-liquid supply pipe to reduce the pressure and restrict flow.   Almost any clear cool nonflammable liquid handled by the pump is suitable for cooling the seal.

Double mechanical seals handling acid-treated oils, phenol solutions, acids, and other chemicals often have an auxiliary cooling system.   Consisting of a small pump, a heat exchanger, connecting piping, pressure regulator, and gage, the system circulates a light lube oil, distillate, or water to cool the seal as well as lubricate it.   Consult the seal manufacturer for the correct hookup.

Another common cooling system uses cored glands through which the liquid is circulated.   Cored sealings rings, either carbon or metal, are also used.

Shaft Deflection   Before choosing centrifugal pumps it is often wise to check the shaft deflection to be expected and the approximate first criti-

cal speed. Figure 9-5 shows a typical chart supplied by the pump manufacturer to permit calculation of deflection speed.

example: What are the shaft deflection and approximate first critical speed of a
pump if the combined weight of its impellers is 23 lb and the data in Fig. 9-5 are
applicable to it?

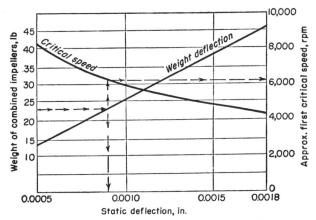

Fig. 9-5 Static shaft deflection and critical speed. (*Goulds Pumps, Inc.*)

solution: Enter Fig. 9-5 at the left at a combined weight of 23 lb and project to
the right to intersect the weight-deflect curve. At the bottom of the chart read
the deflection as 0.009 in. at full speed. Project vertically upward from the
weight-deflection curve to intersect the critical-speed curve. At the right read
the approximate first critical speed as 6,200 rpm. Note that this chart is suitable
for only one particular pump. The chart for a pump being considered for a
given job must be obtained from the manufacturer.

In using any chart or calculation to determine shaft deflection, it must
be remembered that several variables can radically affect actual values.
Wearing rings, especially if running clearances are low, can act as water-
lubricated bearings and have a dampening effect. Also, packed stuffing
boxes act to dampen shaft deflection, and this effect can vary widely
depending upon the stuffing-box adjustment.

Typical Capacities Figure 9-6 shows the capacity and pressure ranges
available with two groups of horizontal centrifugal pumps of one manufacturer. For volume pumping, 19 sizes of the single-stage double-suction unit *B* provide capacities to 6,400 gpm, heads to 280 ft. For higher
pressures, 5 sizes of the two-stage pump *A* provide heads to 1,000 ft,
capacities to 1,200 gpm.

Figure 9-7 shows the capacity ranges of three models of one make of

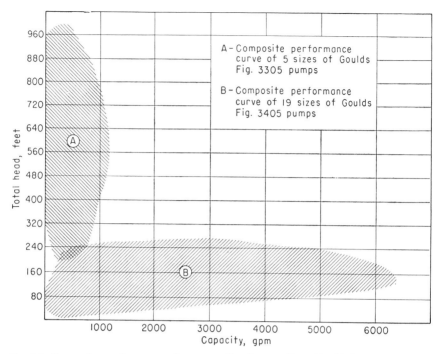

Fig. 9-6 Composite performance of two centrifugal pump lines. (*Goulds Pumps, Inc.*)

rotary pump. Typical head and capacity ranges of 1,750-rpm regenerative-turbine pumps are shown in Fig. 9-8. Table 9-2 shows the recommendations of one manufacturer for vane selection in rotary pumps. Typical capacities of horizontal duplex direct-acting plunger pumps are given in Table 9-3. While these data on pump head and capacity are specific, they give some idea of a few of the ranges available. For

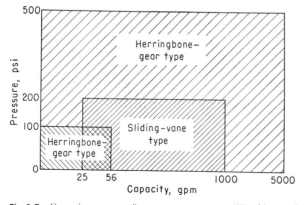

Fig. 9-7 Capacity ranges of some rotary pumps. (*Worthington Corp.*)

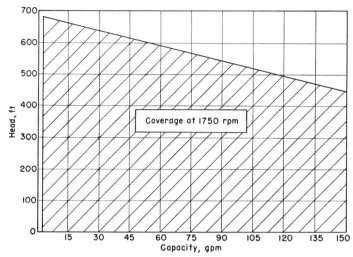

Fig. 9-8 Head and capacity range of 1,750-rpm regenerative turbine pumps. (*Aurora Pump, a Unit of General Signal Corp.*)

further information on the many other pumps available and their head and capacity ranges, consult one or more manufacturers. See the Standards of the Hydraulic Institute for further recommendations on pump standardization.

TABLE 9-2 Typical Vane Selection for Swinging-vane Rotary Pumps

Service No.	Service	Type vanes	Vane material	Reference notes
1	Volatile liquids	Sliding	Metal	*a* and *d*
2	Nonvolatile and non-viscous liquids	Swinging	Metal	*a* and *d*
3	Viscous liquids	Sliding with push rods	Metal	*a* and *b*
4	Liquids containing solids	Swinging	Metal	*a*
5	Tank truck	Sliding	Composition No. 1	*c*
6	Dry cleaning	Sliding	Composition No. 2	*c*

a. The metal used would be iron or bronze according to past practice for type of liquids pumped.

b. This group of liquids would be of 15,000 SSU or above when pump runs at full or rated speed; or 7,000 SSU and up when pump is running at reduced speed.

c. Composition vanes are satisfactory for these services. With wide change in characteristics of liquids pumped, it might require a change in composition materials.

d. Occasionally it is desired to use a pump alternately for services No. 1 and No. 2. In this case sliding vanes are recommended up to 5,000 SSU. Above 5,000 SSU sliding vanes require push rods; otherwise swinging vanes would be necessary.

TABLE 9-3 Capacities of Typical Horizontal Duplex Plunger Pumps

Size, in.	Boiler-feed capacity		Water and liquids, under 250 SSU visc		Liquids, 250 to 500 SSU visc		Liquids, 500 to 1,000 SSU visc		Liquids, 1,000 to 2,500 SSU visc		Liquids, 2,500 to 5,000 SSU visc		Liquids, 5,000 to 10,000 SSU visc	
	Gpm	Piston speed, ft per min*	Gpm	Piston speed, ft per min	Gpm	Piston speed, ft per min	Gpm	Piston speed, ft per min	Gpm	Piston speed, ft per min	Gpm	Piston speed, ft per min	Gpm	Piston speed, ft per min
3 × 2 × 3	7	22	12	37	11	35	10	33	9	29	8	24	4	12
4½ × 2¾ × 4	17	28	29	47	28	45	26	42	22	36	19	31	9	15
5¼ × 3½ × 5	32	32	53	53	50	51	47	47	41	41	34	35	17	17
6 × 4 × 6	47	36	78	60	75	57	69	53	60	46	51	39	25	19
7½ × 5 × 6	73	36	123	60	117	57	110	53	95	46	80	39	40	20
9 × 5¼ × 10	101	45	169	75	162	72	151	67	130	58	110	49	55	25
10 × 6 × 10	132	45	220	75	211	72	196	67	170	58	144	49	72	25
10 × 7 × 10	180	45	300	75	288	72	268	67	232	58	196	49	98	25

* Piston speed for boiler feed is based on water at 212 F.

Design Pointers   Here are a number of miscellaneous pointers helpful in specifying and purchasing pumps of all types:

Volatile liquids can be prevented from flashing in the pump suction line by introduction of cold liquid from an auxiliary source. Control of cold-liquid flow can be either manual or automatic. Vent vapor from the eye of the first-stage impeller to the vapor space of the suction vessel, except where the pump suction nozzle is in the vertical position. Recirculate a portion of the liquid handled from the discharge to the suction to prevent overheating the pump during periods of low-capacity operation. When the packing gland is allowed to leak to provide shaft lubrication, check to see that the amount of liquid lost is not excessive. Where liquid is injected to the packing, an outside-packed pump is often used, especially for oil. Be sure this liquid does not contaminate the material being handled by the pump. Fit a pressure regulator to the packing-box sealing system set to relieve a few psi above the casing pressure. Double mechanical seals, used for problem liquids, also require a circulating-type sealing system fitted with a small pump. Consult seal manufacturer for hookup, cooling arrangement, and capacity needed. Wherever possible, use a single seal instead of the double type.

Standby units must be carefully chosen if they are to handle more than one liquid. Choose head and capacity for the most severe requirements. Check the construction materials to see that they are suitable for all liquids to be handled.

Available npsh should be as much as can be readily provided without adding too greatly to system cost.

Existing pumps can sometimes be used for new jobs in a plant. To compute the effect of new conditions on the pump, use the rules given earlier in this chapter.

Viscous liquids are handled more easily in heated pipes, pumps, strainers, and other equipment. Steam, hot-water, or electric tracing can be used for this purpose.

Friction losses recommended for discharge and suction lines vary. Here are some typical values: discharge lines, 1.5 to 6 psi per 100 ft for flows of 0 to 150 gpm; 1 to 4 psi per 100 ft for 151 to 500 gpm, 0.5 to 2 psi per 100 ft for flows over 500 gpm. Size suction lines for 0.05 to 1 psi per 100 ft, depending on the available npsh.

Pump center line is usually 1.5 to 3.0 ft above the floor level, depending on foundation height and pump size.

Pump bids should contain six copies of the pump-performance curves, six copies of a cross-section drawing of the pump, six copies of the pump-outline drawing, mechanical-seal data (if seals are used), and complete flange ratings.

Critical speed of pump shaft should be at least 20 per cent above, or 30 per cent below, the maximum operating speed of the pump.

Ball bearings used for pump shafts should have a minimum service life of at least 10,000 hr.

Auxiliary connections for centrifugal pumps should be $\frac{3}{8}$ in., or larger, except drains and vents, which can be $\frac{1}{2}$ in.

Base plate should have a drain rim, drain holes, anchor-bolt holes, and grout holes. Either fabricated steel or cast iron is acceptable.

Glands for pumps handling flammable liquids should be made of spark-proof material or lined on the interior with it.

Flushing-oil systems for mechanical seals should have a strainer, pressure gage, and thermometer.

Nameplate and rotation arrow should be furnished with pump. The metal nameplate should contain pump manufacturer's name, serial number of pump, size and type, design head, capacity, speed and temperature, and hydrostatic test pressure.

Pump piping is safest when designed to conform with the ASA Code for Pressure Piping, ASA B31.1.

CHAPTER TEN

Pumping-system Economics

EVALUATION OF THE HYDRAULIC CONSIDERATIONS IN a pumping system seems to receive more attention than the many economic factors which are an inseparable part of every pumping problem. From the standpoint of the user of a pump both factors are of prime importance—he seeks the desired pressure and capacity at the lowest cost per gallon of liquid pumped. Once a suitable class and type of pump have been chosen for a given application, the engineer is faced with a number of decisions. These relate to the first cost of the pump and its driver, the installation and operating costs, the estimated life of the unit, its probable maintenance cost, the "cost" of money invested in the pump, the return on the investment, and the possible salvage value, if any, of the pump and its driver.

Many of these factors are covered, in a general way, in texts on engineering economy. However, their application to pumping problems cannot be covered in too great detail in most engineering-economy texts because so many basic concepts must be discussed. Here, in this book, basic concepts will be touched on but lightly, allowing major emphasis to be given to the successful use of economic analyses in practical pumping problems.

COMPARING DIFFERENT SCHEMES

While not every pumping installation is susceptible to use of more than one scheme for handling the liquid, many are. As a result, it is often possible, and necessary, to consider more than one class or type of pump, piping arrangement, and driver. The aim of any comparison of this type is to secure the lowest annual cost per gallon of liquid pumped, all other factors like dependability, ease of maintenance and repair, and flexibility also being taken into consideration. At times these so-called intangible factors may swing the balance away from the unit or units giving the lowest annual cost to ones having higher annual cost. Evaluation of intangibles can be discussed only in general terms—each installation has its peculiar characteristics which yield only to experienced judgment. Many of these will be discussed in the following examples.

example: Two schemes (Fig. 10-1) have been proposed for supplying water to an industrial plant. Scheme A uses two vertical turbine pumps run in parallel and discharging into existing 6-in. cement-asbestos pipe. Either motor- or diesel-engine drive will be used for both pumps. Scheme B (Fig. 10-1) has four pumps, in two stages. In each scheme, all pumps run continuously during the day, when the plant requires 360 gpm of water. At night, only half the number of pumps installed will run, supplying about 180 gpm to the plant. Which scheme is best? What are the pipe losses during maximum flow and when only one pump operates? What auxiliary devices, if any, should be installed?

solution: Since the supply pipe between the pumps and plant is already in existence, it is unnecessary to make a comparison of the losses in lines of various sizes. However, if the pipe size were not already fixed, several sizes would have to be chosen and the loss in each analyzed to determine the best size for the

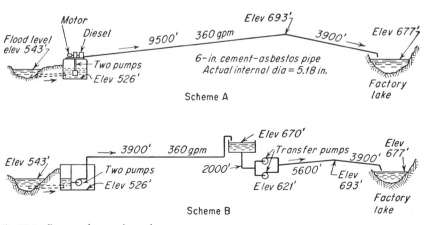

Fig. 10-1 Proposed pumping schemes.

installation. The second example in this chapter contains a comprehensive solution for various sizes of pipe, and the methods used there could be applied here if the pipe size was not fixed.

1. With such long lines (Fig. 10-1) water hammer should be studied, to see that the pipe is suitable for the shock pressures that might be met. See Chap. 7 for a discussion of water hammer. In some plants the class of pipe needed to withstand the shock pressures from water hammer is much heavier than for the normal operating pressure. Surge suppressors or other equipment might be necessary.

2. Even with the class of pipe fixed, it is worthwhile to note that the 3,900-ft-long line from elevation 693 ft to the factory lake (Fig. 10-1) could be made of lighter material than the 9,500-ft run between the pumps and the high point. This is because the water runs freely from the high point to the lake and its pressure is lower than at any point in the 9,500-ft run. Since there is a 16-ft drop (= 693 − 677) between the high point and the lake, it would be desirable to have the friction of the pipe sufficient to cause the line to be full of water at all foreseeable flow rates. Also, a slight positive pressure should exist at elevation 693 ft, the high point. This pressure is needed to rid the pipe of air at the high point. An economical velocity from a friction standpoint would not be great enough to wash air out of the pipe. Automatic evacuating equipment would not be advisable at this point in the line because it would complicate the installation. Instead, automatic relief or vent valves would be cheaper and more economical.

3. With the line full of water and no flow, the maximum vacuum that could exist at elevation 693 ft is 16 ft of water, the distance between the high point and the lake. Cement-asbestos pipe would not be affected when subjected to this vacuum. Some types and classes of pipe may collapse under vacuum; so all conditions must be studied before installing the line. There is a possibility in this installation of separation of the water column at the high point when a pump is stopped or the power supply fails. So vacuum breakers (automatic type), to admit air when the line pressure falls below atmospheric, will be needed at the high point, in addition to the vent valves.

4. Compute the friction loss for various flow rates between 0 and 500 gpm, using data from Chaps. 4 and 6. Plot this loss (Fig. 10-2).

5. Two 180-gpm pumps are proposed, to operate against the head resulting from a 360-gpm flow. As the capacity delivered through any pipe is that at which the system head equals the head developed by the pump or pumps, running one pump alone gives only 180 gpm if the head is all static. But since the head in this system is partly static and partly friction, single-pump operation gives increased capacity and reduced head (point A, Fig. 10-3). The capacity at point A depends on the shape of the pump HQ curve and the system-head curve. If system head is mostly static with little friction, the head at A is a little less than that for two-pump operation, while the capacity is a little more than the rated value.

With a large friction component, as in Fig. 10-3 and as the first scheme involves, the head at A is considerably less than rated and the capacity is considerably more. If most pumping is done with a single pump, a pump having best efficiency at A is desirable. This could not be obtained with a centrifugal pump

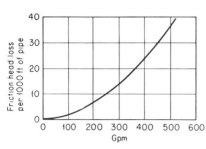

Fig. 10-2 Pipe-friction loss. (*Worthington Corp.*)

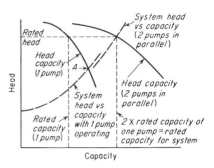

Fig. 10-3 Characteristic curves. (*Worthington Corp.*)

unless a design giving a capacity near that obtainable with two-pump operation is acceptable.

If most running is with two-pump operation, a unit with best efficiency at the rated head is desirable. Single operation of such a design at reduced heads may cause complications because of suction conditions. Two-pump operation is normal in this installation during the day. Single-pump operation is necessary most of the night. So a high efficiency is desirable both at *A* and at the rated head.

6. Assume elevation 543 ft is the normal water level and there is no loss in the conduit feeding the suction pit. Find the hydraulic gradient at the high point, elevation 693 ft, to see if a siphon will exist with the minimum flow of 180 gpm. Loss in 3,900 ft of 5.18-in. ID pipe at 180 gpm is 3,900(5.55)/1,000 = 21.6 ft, using data from Fig. 10-2. Adding an exit loss of 0.1 ft at the factory lake gives a total loss of 21.7 from the high point to the lake. As the drop in elevation is 16 ft there would have to be 21.7 − 16 = 5.7 ft positive head (5.6-ft pressure, 0.1-ft velocity head) at the high point with 180-gpm flow. So unless there is a loop in the profile of this 3,900-ft line there is no danger of siphoning or of the pipe's being only partly full.

Next, investigate the suction conditions of the transfer pumps in scheme B. This is done to ensure that the scheme does not involve impossible suction conditions because of friction loss in the 2,000-ft line from the reservoir to the transfer pumps. Loss with 360-gpm flow is 2,000(20)/1,000 = 40 ft, using data from Fig. 10-2. Static suction head is 49 ft; so there should be 9-ft positive suction head at the station entrance. So even with some friction in the suction piping to individual pumps there would be a positive suction head. No adverse suction conditions are involved.

7. The static lift in scheme A is 134 ft, pipe length 13,400 ft. Scheme B has the same static lift in two steps and 15,400 ft of pipe. Immediately, it is seen that scheme B uses 2,000 ft more pipe. Offhand it appears that scheme A is better because it saves 2,000 ft of pipe, a second pumping station with its transmission line, and the cost of the reservoir. But scheme B has these advantages: (1) If the reservoir has enough capacity, pumps for the supply station can be picked for daily or weekly average consumption and run almost 24 hr per day, 5 or 7 days

per week. (2) Diesel-engine standby drive is not needed if the reservoir is large enough because water can be drawn from it during power failures. Equipment cost is lower. (3) Pressure in the pipe near the supply station is much lower in scheme B than in A, allowing use of lighter, less costly pipe. (4) As heads are lower, the pumps chosen may be a more efficient type.

In view of these facts it is advisable to investigate the relative merits and net operating costs of both schemes before deciding which to use. Vertical turbine pumps are being considered for scheme A and horizontal centrifugal pumps for scheme B. Proper analysis includes a study of both types for both schemes. Also, other possible pumping equipment should be studied.

8. Based on friction values in Fig. 10-2 and 134-ft static head, the system head, less station-piping loss for scheme A, is as shown in Fig. 10-4. System head for the first or supply station in scheme B is also shown in Fig. 10-4, as is the head for the transfer station.

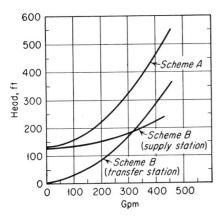

Fig. 10-4 Head loss for two schemes. (*Worthington Corp.*)

9. Lacking information on what station piping and valves are needed, assume a 9-ft piping loss for the vertical-type pumps and 11-ft for the centrifugal pumps. These losses may be higher than in the actual plant but they will serve to show how the problem is solved. Since the load factor affects pumping-equipment choice it is necessary to assume certain flow rates. With a 5-day 10-hr demand at 360 gpm, flow is roughly 1 million gal. With 5-day 14-hr demand during the night at 180 gpm, plus a 24-hr demand at this rate during Saturday and Sunday, flow is roughly 1¼ million gal. Total weekly flow is 2¼ million gal, an average of 223 gpm. Average working-day consumption is 255 gpm.

Pumping equipment suitable for scheme A with 5.18 ID pipe includes: (1) Two 180-gpm 410-ft total-head vertical turbine pumps to handle 1¼ million gal weekly with one pump running. (2) Two 180-gpm 412-ft total-head centrifugal pumps selected the same as in (1). As before, single-pump operation gives over 180-gpm capacity. (3) One 360-gpm 410-ft total-head vertical turbine pump for days, and one 180-gpm 216-ft total-head vertical turbine pump for nights and week ends. (4) One 360-gpm 412-ft total-head and one 180-gpm 218-ft total-head centrifugal pump. (5) Two 180-gpm 216-ft total-head vertical turbine

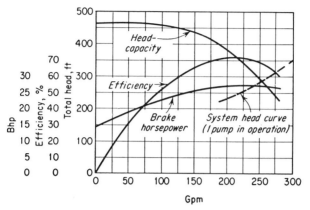

Fig. 10-5   Pump and system characteristics. (*Worthington Corp.*)

pumps with one 360-gpm 200-ft total-head centrifugal booster pump piped so either turbine pump can run alone giving 180 gpm, or so both can run in parallel and discharge to the booster, giving 360 gpm.   Head loss of 6 ft is allowed for booster piping and valves.   (6) Two 180-gpm 218-ft total-head centrifugal pumps and one 360-gpm booster.

If recommendations from various pump manufacturers are requested, they will differ considerably. The pumps listed in Table 10-1 are units built by one firm.

10. Figure 10-5 shows the characteristics of a vertical turbine pump good for recommendation (1) above.   Flow is 252 gpm with one-pump operation.   When demand is only 180 gpm, the pump is started and stopped to keep factory-lake level within desired limits. The pump runs about 70 per cent of the time. Figure 10-6 shows the characteristics of a centrifugal pump that gives 278 gpm on one-pump operation.   It runs about 65 per cent of the time when 180 gpm is needed.

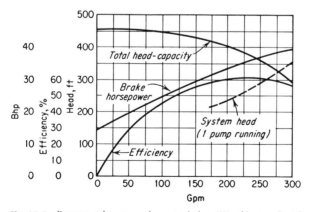

Fig. 10-6   Pump and system characteristics. (*Worthington Corp.*)

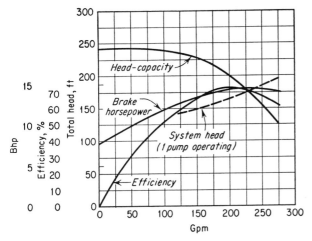

Fig. 10-7 Pump and system characteristics. (*Worthington Corp.*)

11. Results obtainable with the various plans for scheme A are summarized in Table 10-1. Motors are 220/440-volt 3-ph 60-cps units. Power costs are based on an assumed rate of 15 cents per kilowatthour. Plan A3 is lowest and is taken as the base from which differences in column (14) are figured. Assuming capitalization as 15 per cent, extra power cost is figured and tabulated in column (15). This amount is added to the first cost of all plans other than A3. While the importance of standby pumping equipment is not stated in the problem, its costs are computed in column (18), assuming 360-gpm flow with one pump in the station out of service. Summing items in columns (15) and (18) shows plan

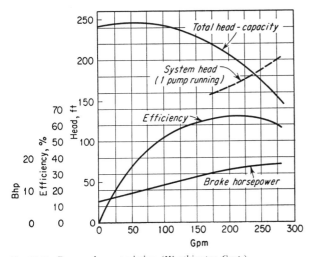

Fig. 10-8 Pump characteristics. (*Worthington Corp.*)

TABLE 10-1 Scheme A: Summary of Results Obtainable with Various Plans

Plan	(1) Type of pump	(2) Min No. required	(3) Unit type (see Table 10-3)	(4) No. run	(5) Gpm per pump	(6) Total head, ft	(7) Motor hp	(8) Over-all efficiency, %	(9) Kwhr per 1,000 gal	(10) Weekly pumpage 10^6 gal	(11) Kwhr per week	(12) Total kwhr per week	(13) Power cost 52 weeks at 1½¢ a kwhr	(14) Extra power cost over $2,580 per year base	(15) Extra annual power cost at 15%	(16) Cost of unit	(17) No. units with spares	(18) Total cost of units (pumps and motors)	(19) Adjusted cost of plan
A1	Vtp	2	a	2	180	410	30	63.0	2.04	1.00	2,040	3,865	$3,020	$ 440	$2,930	$1,940	3	$5,820	$ 8,750
				1	252	188		69.0	1.46	1.25	1,825								
A2	Cent.	2	b	2	180	412	40	52.9	2.45	1.00	2,450	4,850	3,790	1,210	8,060	875	3	2,625	10,685
				1	278	326		53.2	1.92	1.25	2,400								
A3	Vtp	1	c	1	360	410	50	67.8	1.90	1.00	1,900	3,310	2,580	0	0	1,770	2	4,355	4,355
			d	1	180	216	15	59.8	1.13	1.25	1,410					815	1		
A4	Cent.	1	e	1	360	412	60	63.5	2.04	1.00	2,040	3,560	2,700	120	800	1,030	2	2,615	3,415
			f	1	180	218	20	56.3	1.215	1.25	1,520					555	1		
A5	Vtp	2	d	1	180	216	15	59.8	1.13	1.25	1,410	3,560	2,700	120	800	815	3	3,935	4,735
				2	180					1.00	1,130								
			g	1	360	200	30	61.7	1.02	1.00	1,020					745	2		
A6	Cent.	2	f	1	180	218	20	56.3	1.215	1.25	1,520	3,755	2,930	350	2,430	555	3	3,155	5,585
				2	180					1.00	1,215								
			g	1	360	200	30	61.7	1.02	1.00	1,020					745	2		

A4 is best, if the above assumptions are correct. This neglects the differences in station and installation costs. These must be considered before a final decision is made.

Plan A3 does not have any suction piping because vertical turbine pumps are used. Saving in piping and station costs might make A3 more desirable. More expensive power favors plan A3, less expensive A4. Plan A1, which was originally proposed, is fifth highest in first cost, based on the above assumptions.

12. A similar study is made for scheme B, considering pumps for both the supply and booster stations. Since the booster station has a positive suction head, only centrifugal pumps are studied for it. Head is mostly friction; so, with one-pump booster operation, head will be much lower than rated and the logical choice is one 360-gpm pump for days, one 180-gpm pump for nights. Parallel operation of 180-gpm pumps to give 360-gpm delivery gives results B1 and B2 in Table 10-2. Plans B3 and B4 assume the reservoir has sufficient capacity to permit constant pumping at a 260-gpm rate.

Using the power consumption of plan A3 as base, it is found that the cost of the four plans for scheme B are considerably more than A4. Since scheme B uses more piping, a reservoir, and a second station, it is obvious that the cost is higher than plans A3 and A4. Scheme B can be dropped from further consideration.

This analysis assumes the factory lake is relatively small and has little storage capacity. If storage capacity or level variation sufficient for 75,000 gal or more could be obtained, the best plan would be scheme A with 260-gpm pumps delivering at constant rate for practically 24 hr. The centrifugal pump for this modified scheme A would be a single-stage 3,500-rpm unit with a rated head of 292 ft. Yearly power cost is $2,700. Cost of the two units with motors is about $1,490. The additional cost of standby diesel drive will vary with the capacity needed. Three 30-hp engines would be used with plan A1, if 360 gpm must be delivered. In plan A3, with only one 360-gpm pump on standby, one 50-hp diesel is needed. Engines in both plans would use drive gears.

PRELIMINARY COST ESTIMATES

The above problem was based upon 1956–1957 cost data supplied by one or more pump manufacturers. At the end of this chapter are a table (Table 10-10) and discussion that will permit updating these and other cost data presented in this chapter to the 1967–1968 period. By using the method given above, updated appropriately, a firm decision based upon economic grounds can be made between various pumping schemes.

In many instances, however, exact cost data may not be available. To provide basic information for these situations, two methods are presented here for determining the economic pipe size for a plant using different types of pumps. Again, the cost-updating factors tabulated at

TABLE 10-2 Scheme B: Summary of Results Obtainable with Various Plans

Plan	Station	Type of pump (1)	Min No. required (2)	Unit type (see Table 10-3) (3)	No. run (4)	Gpm per pump (5)	Total head, ft (6)	Motor hp (7)	Over-all efficiency, % (8)	Kwhr per 1,000 gal (9)	Weekly pumpage 10⁶ gal (10)	Kwhr per week (11)	Total kwhr per week (12)	Power cost 52 weeks at 1½¢ a kwhr (13)	Extra power cost over $2,580 per year base (14)	Extra annual power cost at 15% (15)	Cost of unit (16)	No. units with spares (17)	Total cost of units (pumps and motors) (18)	Adjusted cost of plan (19)
B1	Supply	Vtp	2	h	2	180	214	15	61.4	1.095	1.00	1,095	3,925	$3,060	$480	$3,200	$1,265	3	$5,820	$9,020
	Supply					228	175		61.9	0.89	1.25	1,110								
	Booster	Cent.	1	i	1	360	242	40	62.6	1.212	1.00	1,212					785	2		
	Booster		1	j	1	180	73	5	56.3	0.406	1.25	508					455	1		
B2	Supply	Cent.	2	k	2	180	216	20	56.3	1.205	1.00	1,205	4,195	3,270	690	4,600	555	3	3,690	8,290
	Supply					239	183		56.5	1.015	1.25	1,270								
	Booster	Cent.	1	i	1	360	242	40	62.6	1.212	1.00	1,212					785	2		
	Booster		1	j	1	180	73	5	56.3	0.406	1.25	508					455	1		
B3	Supply	Vtp	1	l	1	260	178	20	64.5	0.865	2.25	1,945	3,665	2,860	280	1,865	1,415	2	4,855	6,720
	Booster	Cent.	1	i	1	360	242	40	62.6	1.212	1.00	1,212					785	2		
	Booster		1	j	1	180	73	5	56.3	0.406	1.25	508					455	1		
B4	Supply	Cent.	1	m	1	260	180	20	65.0	0.87	2.25	1,950	3,670	2,865	285	1,900	555	2	3,135	5,035
	Booster	Cent.	1	i	1	360	242	40	62.6	1.212	1.00	1,212					785	2		
	Booster		1	j	1	180	73	5	56.3	1.015	1.25	1,270					455	1		

TABLE 10-3 Identification of Pumps in Tables 10-1 and 10-2

Code	Description	Curve No.
a	1,770-rpm 15-stage vertical turbine	Fig. 10-5
b	3,500-rpm 2-stage horizontal split-case centrifugal	Fig. 10-6
c	3,500-rpm 9-stage vertical turbine	
d	3,500-rpm 4-stage vertical turbine	
e	3,500-rpm single-stage horizontal split-case centrifugal	
f	3,500-rpm single-stage horizontal split-case centrifugal	
g	3,500-rpm single-stage horizontal split-case centrifugal	
h	1,770-rpm 8-stage vertical turbine pumps	Fig. 10-7
i	3,500-rpm single-stage horizontal split-case centrifugal	Fig. 10-8
j	1,770-rpm single-stage horizontal split-case centrifugal	
k	3,500-rpm single-stage horizontal split-case centrifugal	
l	1,770-rpm 7-stage vertical turbine	
m	3,500-rpm single-stage horizontal split-case centrifugal	

Vertical-turbine-pump prices, Tables 10-1 and 10-2, based on 20-ft approximate length.

the end of the chapter will prove valuable. These methods can be applied to any plant, particularly those in the process industries.

In the first method, average cost and utility data, presented in charts and tables in this chapter, are used to prepare an individual economic analysis of each pump type and pipe size chosen. Tabulations of economic liquid velocities, also presented in this chapter, are used in the second method to determine the most desirable pipe size for a given set

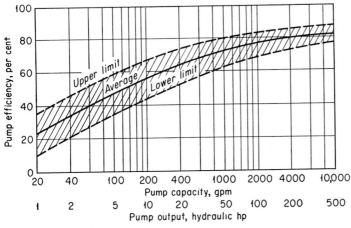

Fig. 10-9 Efficiencies of centrifugal pumps.

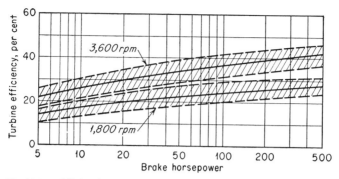

Fig. 10-10   Efficiencies of single-stage steam turbines.

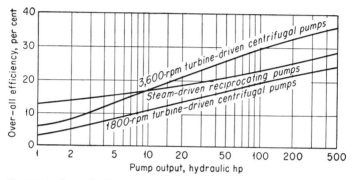

Fig. 10-11   Over-all efficiencies of pumps and drivers.

of conditions. Correction factors permit adjustments in the cost of steam and electric power, steam rate of turbine drives, payout time, and the differential pressure head on the pump. While the methods are basically for the discharge piping of a pump, they may also be used for the suction piping, except where the flow is by gravity or differential head. Then the line should be sized on the basis of npsh. The effects of erosion and pipe vibration are neglected. If the reader intends to solve many problems using the methods given here, he should refer to the original article for a comprehensive discussion of the theory and derivations of the equations used.

example: A refinery charge pump and its spare handle 400,000 lb per hr of 27.5 ° API gas oil. The length of pipe run between the pump and the cracking unit it serves is 400 ft; there are thirteen 300-lb ASA valves in this run. Heat exchangers and a control valve cause a 105-psi pressure loss. Differential head on the pump, exclusive of pipe friction, is 270 psi. Standard-weight pipe and fittings will be satisfactory, with the first 125 ft uninsulated, the next 75 ft having

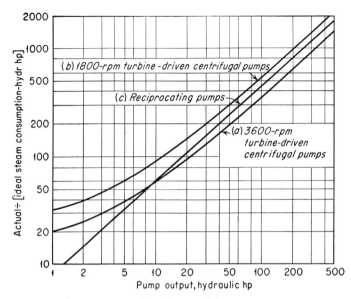

Fig. 10-12 Steam consumption of steam-driven pumps.

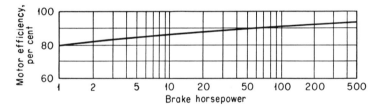

Fig. 10-13 Efficiencies of 3-phase induction motors.

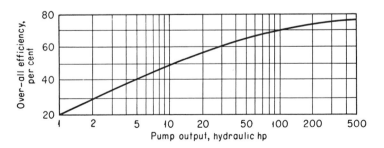

Fig. 10-14 Over-all efficiencies of motor-driven centrifugal pumps.

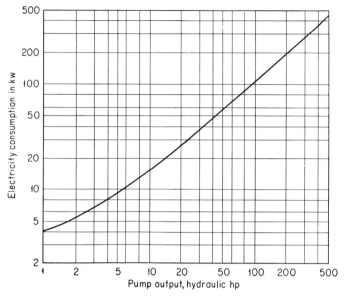

Fig. 10-15   Electrical consumption of motor-driven pumps.

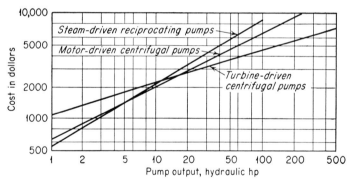

Fig. 10-16   Cost of pumping equipment.

270 to 350 F insulation, and the last 200 ft 400 to 500 F insulation. The average temperature of the gas oil is 300 F, specific gravity 0.8, and viscosity 2.0 centipoises at this temperature. At 300 F, 1,010 gpm of gas oil is flowing, while at 100 F the flow is 920 gpm. Compute the economic pipe size for a 2-year payout time for: (Case A) 3,600-rpm steam-turbine-driven centrifugal pumps when steam costs $0.75 per 1,000 lb and exhaust steam is worth $0.10 per 1,000 lb. Steam conditions are 150 psig and 450 F at the inlet, 20 psig at the exhaust, for which the ideal steam rate is 19.9 lb per hp-hr. (Case B) Motor-driven centrifugal pumps when electricity costs $0.008 per kilowatthour.

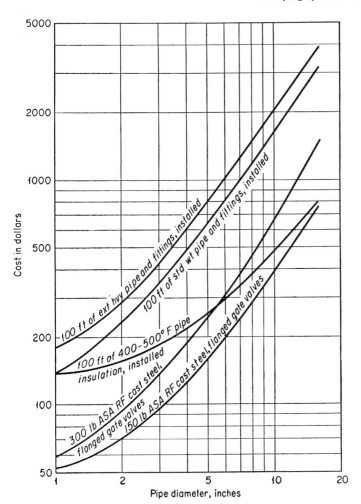

Fig. 10-17 Cost of piping and accessories.

solution: With a 1,010-gpm flow rate, a pipe size somewhere between 4 and 10 in. will probably be satisfactory. The calculations for solving this problem by the first method are detailed in Table 10-9 and Fig. 10-24.* They are easily followed if the steps are performed in the same order as listed in Table 10-9. The total investment required shows that an 8-in. line would be used with a steam-turbine-driven centrifugal pump and a 6-in. line size would be used with a

---

* In Fig. 10-24, curve A1 refers to case A investment, A2 to yearly cost of steam, A3 to total yearly fixed and operating costs. Curves B1, B2, and B3 are the corresponding values for case B.

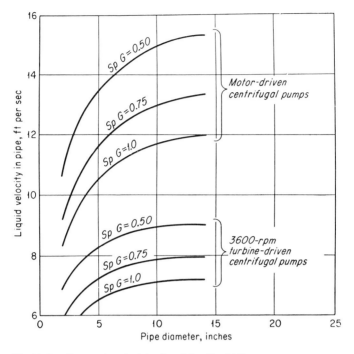

Fig. 10-18 Economic velocities for sizing liquid lines.

TABLE 10-4   Cost of Pipe and Fittings
(Basis: Fabricated flanged and welded, erected in place, expressed as dollars per 100 ft of pipe)

Nominal diam, in.	Standard weight	Extra heavy
2	$ 223	$ 273
3	362	448
4	508	622
6	781	987
8	1,155	1,449
10	1,648	1,994
12	2,125	2,553

TABLE 10-5   Cost of Gate Valves
(Basis: RF cast carbon steel, flanged, OS and Y)

Nominal diam, in.	150-lb	300-lb
2	$ 74	$102
3	92	135
4	128	175
6	201	311
8	278	458
10	377	639
12	505	861

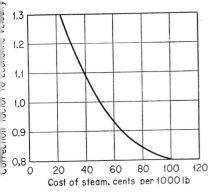

ig. 10-19 Correction factors for steam-cost variations.

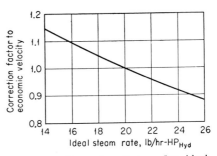

Fig. 10-20 Correction factors for ideal steam-rate variations.

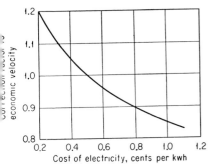

ig. 10-21 Correction factors for electricity-cost variations.

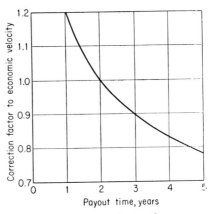

Fig. 10-22 Correction factors for payout-time variations.

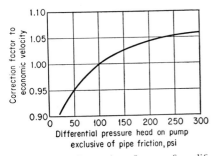

Fig. 10-23 Correction factors for differential-pressure variations.

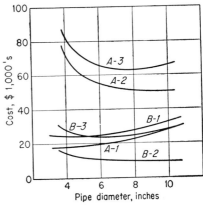

Fig. 10-24 Results of sample problem.

TABLE 10-6   Cost of Pipe Insulation

(*Basis: Installed cost per 100 ft of straight pipe. For approximate cost of insulating valves and fittings as well as straight pipe, multiply values in table by 1.35. M denotes 85% magnesia, S denotes Superex or high-temperature block insulation; number immediately following denotes thickness in inches*)

Nominal pipe diam, in.	Temp range, F						
	200–270	270–350	350–400	400–500	500–600	600–700	700–800
2	M1, $113	M1, $113	M1½, $137	M2, $156	M2, $156	S1½ + M1½, $289	S1½ + M2, $356
3	M1, $124	M1, $124	M1½, $152	M2, $181	M2, $181	S1½ + M1½, $327	S1½ + M2, $388
4	M1⅛, $135	M1⅛, $135	M1½, $164	M2, $216	M2, $216	S1½ + M1½, $387	S1½ + M2, $458
6	M1⅛, $158	M1½, $206	M2, $267	M2, $267	M2½, $353	S1½ + M2, $476	S1½ + M2½, $567
8	M1¼, $191	M1½, $233	M2, $319	M2½, $404	M2½, $404	S1½ + M2, $556	S1½ + M2½, $660
10	M1¼, $227	M1½, $274	M2. $379	M2½, $479	M2½, $479	S1½ + M2, $626	S1½ + M2½, $857
12	M1½, $302	M2, $430	M2½, $535	M3, $606	M3, $606	S1½ + M2, $706	S1½ + M2½, $993

motor-driven centrifugal pump. In either case a 6-in. line involves a $3,500 lower initial investment, but with a turbine-driven pump the yearly cost of steam would be $3,500 higher for the 6-in. than for the 8-in. line. With the motor-driven centrifugal, the 6-in. line increases the yearly cost of electricity by only $400 over the cost for an 8-in. line. It is important to note that for the 6-in. size the yearly cost of steam for the turbine-driven pump is $54,000, against only $10,180 for the cost of electricity with the motor-driven pump. Investment cost for the motor-driven pump is $24,493, compared with $19,493 for the turbine-driven unit.

To use the second method, economical velocities are chosen from Tables 10-7 and 10-8. These are then corrected as detailed below for each case.

Case A: The economic velocity from Table 10-7 for a specific gravity of 0.8 is: 6-in. pipe, 7.3 ft per sec; 8-in. pipe, 7.6 ft per sec. The correction factor for the cost of steam at $0.65 per 1,000 lb is 0.9 from Fig. 10-19; for the ideal steam rate, 1.0 from Fig. 10-20; for a 2-year payout time, 1.0 from Fig. 10-22; for the differential head on the pump, 1.06 from Fig. 10-23. To correct for completely insulated extra-heavy pipe with 300-lb ASA valves and 400 to 500 F insulation, instead of uninsulated standard-weight pipe, a factor of 1.17 is used for a turbine-driven pump, 1.19 for a motor-driven unit. But since the piping here is only partially insulated and has 300-lb ASA valves, a factor of 1.10 will be used. Corrected economical velocity is found by applying all the above correction factors to the velocities obtained from Tables 10-7 and 10-8. For the 6-in. pipe, corrected economic velocity = $(7.3)(0.90)(1.06)(1.10) = 7.7$ ft per sec; 8-in.

TABLE 10-7 Economic Pipe Velocities, Motor Drive

(Basis: Liquid flow in ft per sec, using motor-driven centrifugal pumps)

Nominal pipe diam, in.	Specific gravity of liquid		
	0.50	0.75	1.0
2	10.7	9.2	8.4
3	12.0	10.3	9.4
4	12.9	11.1	10.1
6	13.9	12.0	10.9
8	14.6	12.6	11.4
10	15.0	13.0	11.7
12	15.2	13.2	11.9

TABLE 10-8 Economic Pipe Velocities, Turbine Drive

(Basis: Liquid flow in ft per sec, using 3,600-rpm turbine-driven pumps)

Nominal pipe diam, in.	Specific gravity of liquid		
	0.50	0.75	1.0
2	7.0	6.0	5.5
3	7.6	6.6	5.9
4	8.0	7.0	6.3
6	8.5	7.4	6.7
8	8.8	7.7	7.0
10	8.9	7.8	7.1
12	9.0	7.9	7.1

pipe, $(7.6)(0.90)(1.06) = 8.0$ ft per sec. Table 10-9 shows that a 6-in. line with an economic velocity of 7.7 ft per sec will not handle the required 1,010 gpm, whereas an 8-in. line is more than adequate.

Case B: Using the same steps as above, the 6-in. line is found to be suitable. In both cases, the pipe-size selections made by this method agree with those made by the previous method. The most accurate procedure here would be to plot on a single graph a curve showing the economic velocity vs. pipe diameter, similar to those of Fig. 10-18, but corrected for the particular conditions which apply, and another curve of required velocity vs. pipe diameter. Where the two curves intersect is the economic diameter.

Note several items about these methods: In the average process plant the cost of insulating bent pipe is 1.35 times that of straight pipe. Apply this factor to the data in Table 10-6 and Fig. 10-17. Standard-weight carbon-steel pipe and fittings with 150-lb ASA valves are assumed to be used. The installation is assumed to operate 8,150 hr per year, for an operating factor of 83 per cent. For further data, see the original source of this material.

MODERNIZING EXISTING INSTALLATIONS

More thought is being given today to replacing old pumps with newer, more efficient units. Recent advances in pump design often make replacement an extremely attractive scheme.

example: How large an investment can be justified or written off by a new pump which operates 4,500 hr per year if it uses 8 hp less than an older unit and power costs $0.01 per kilowatthour?

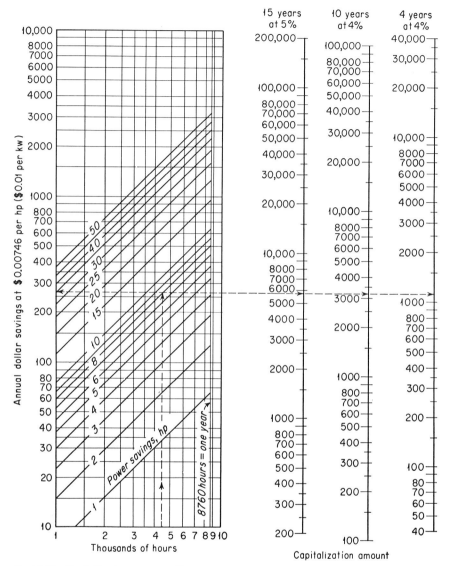

Fig. 10-25   Capitalization amounts for pumps.

TABLE 10-9   Summary of Calculations for Sample Problem

Nominal pipe diam, in.	4	6	8	10
Case A				
1. Velocity (1,010 gpm), ft per sec.......	25.5	11.2	6.5	4.1
2. Pipe-friction drop, psi per 100 eq ft....	19.0	2.5	0.70	0.23
3. Total pipe-friction pressure drop, psi = (400/100) (1 + 0.186D) (item 2)......	132.5	19.2	7.0	2.5
4. Differential head on pump, psi........	402.5	289.2	277.0	272.5
5. Pump output, hydraulic horsepower = (item 4) (920)/(1,714)...............	216.0	155.2	148.7	146.3
6. Cost of standard weight pipe and fittings = (Table 10-4) × 400/100.......	$2,032	$3,124	$4,620	$6,592
7. Cost of 300-lb valves = (Table 10-5) × 13.............................	$2,280	$4,040	$5,960	$8,310
8. Cost of insulation = (Table 10-6) (L/100) (1.35) 270–350 F insulation = (Table 10-6) (75/100) (1.35).....................	$136	$209	$236	$277
400–500 F insulation = (Table 10-6) (200/100) (1.35).....................	$583	$720	$1,090	$1,293
9. Total cost of pipe and piping accessories	$5,031	$8,093	$11,906	$16,472
10. Yearly cost of pipe and piping accessories = (item 9) (0.06 + 1/2.0).......	$2,820	$4,530	$6,670	$9,230
11. Cost of two steam turbine-driven centrifugal pumps = (Fig. 10-16) × 2.....	$12,800	$11,400	$11,200	$11,200
12. Yearly cost of pumps and drivers = (item 11) (0.06 + 1/2.0).............	$7,170	$6,390	$6,270	$6,270
13. Yearly fixed costs = (item 10) + (item 12)................................	$9,990	$10,920	$12,940	$15,500
14. Steam consumption, lb per hr = (Fig. 10-12) (19.9).....................	13,350	10,300	9,650	9,550
15. Yearly cost of steam = (item 14) (8,150) (0.65/1,000)...............	$70,800	$54,600	$51,100	$50,600
16. Yearly fixed and operating costs = (item 13) + (item 15)...............	$81,790	$65,520	$64,040	$66,100
17. Total investment = (item 9) + (item 11).............................	$17,831	$19,493	$23,106	$27,672
Case B (Steps 1 through 10 are identical to case A)				
11. Cost of two motor-driven centrifugal pumps...........................	$19,300	$16,400	$16,000	$16,000
12. Yearly cost of pumps and drivers......	$10,800	$9,190	$8,960	$8,960
13. Yearly fixed cost = (item 10) + (item 12)................................	$13,620	$13,720	$15,630	$18,180
14. Electric power consumption, kw = (Fig. 10-15)............................	212	156	150	148
15. Yearly cost of electricity = (item 14) (8,150) (0.008).....................	$13,800	$10,180	$9,780	$9,650
16. Yearly fixed and operating costs = (item 13) + (item 15)...............	$27,420	$23,900	$25,410	$27,830
17. Total investment = (item 9) + (item 11).............................	$24,331	$24,493	$27,906	$32,472

solution: Enter the bottom of Fig. 10-25 at 4,500 hr per year and project vertically upward to intersect the 8-hp power-savings curve. At the left read the annual saving as $270. Project horizontally to the right and read the justifiable investment for different periods at various interest rates. Results are accurate to within about 2 per cent. Where the power rate is higher or lower than $0.01 per kilowatthour, multiply the chart result by the ratio of the actual cost to $0.01.

Write-offs used vary from one industry to another. Municipal installations often use 15 years at 5 per cent. Central-station plants may use 8 to 10 years, while chemical and refining plants often use 2 to 4 years. Note that these periods may or may not be the same as those approved by Federal income-tax authorities for the useful life of plant equipment.

Other Equipment   It is possible to make similar economic analyses for other equipment used with a pump installation. Figure 10-26 shows an

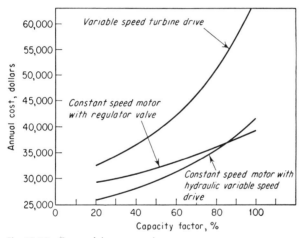

Fig. 10-26   Pump-drive comparisons.

annual cost comparison of several drives for a boiler-fed pump. As can be seen, the capacity factor at which the plant normally operates will strongly influence pump operating costs.

USING COST INDEXES

For general cost-estimating work, the reader should not overlook the many helpful cost indexes published by various magazines. These include the *Engineering News-Record* construction cost indexes and others. In addition, the Bureau of Labor Statistics of the United States Department of Labor makes available a wide variety of comparative cost

TABLE 10-10   Price Adjustment Factors for Wholesale Price Index (1957–1968)*

Equipment or service	1957–1958	1967–1968
Pumping equipment	100	128.2
Small, integral-horsepower electric motors (10 hp)	100	43.1
Medium, integral-horsepower electric motors (50 hp)	100	90.6
Large, integral-horsepower electric motors (250 hp)	100	111.4
Steam turbines	100	102.0
Internal-combustion engines	100	109.8
Cast-iron pipe	100	100.8
Steel pipe	100	107.1
Valves	100	118.8
Insulation materials	100	92.2
Electric power	100	101.0

* Condensed from data from the Bureau of Labor Statistics, U.S. Dept. of Labor.

indexes on the basis of wholesale commodity prices. For the specific purpose of permitting the reader to update problems in this chapter, Table 10-10 provides a general indication of applicable cost changes from 1957–1958 to 1967–1968.

For similar problems to be worked out in the years ahead, the reader can obtain consistent cost indexes directly from the Bureau of Labor Statistics, United States Department of Labor, or actual up-to-date cost figures from manufacturers or other sources.

PART THREE

Pump Application

CHAPTER ELEVEN

Power-plant Services

FOSSIL AND NUCLEAR STEAM, internal-combustion engine, and hydro power plants use one or more pumps in their cycles. Of the three, steam plants use many more pumps and present more diverse application problems. Steam-power-plant discussions in this chapter pertain exclusively to fossil-fuel-fired (coal, oil, and gas) steam power plants. Nuclear steam-supply-system pumping equipment is covered in Chapter 12.

FOSSIL-FUELED STEAM POWER PLANTS

Pumps commonly used in fossil-fueled steam power plants include boiler-feed, boiler-circulation, condensate, condensate-booster, vacuum, hotwell, fuel-oil, ash-handling, gland-seal, lube-oil, seal-oil, water-supply, sump, circulating, and chemical feed. Careful selection is important because plant efficiency is directly affected by pump efficiency.

Boiler-feed Pumps Figure 11-1 shows the types of feed pumps popular in low- and medium-pressure plants. The ranges shown are general and should not be considered fixed. Reciprocating feed pumps for low-pressure steam plants (up to about 400 psi) may be horizontal (Fig. 3-1) or direct-acting vertical units. They are available in a number of

271

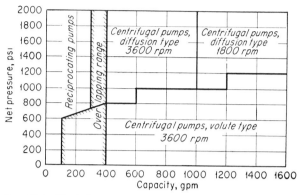

Fig. 11-1 Typical economic operating ranges for boiler-feed pumps. The *diffusion* and *volute* designations may vary between manufacturers. (*Worthington Corp.*)

different designs, some being fitted with condensate receivers having float controls to start and stop the pump. Regenerative turbine pumps (Figs. 11-2 and 1-26) find many applications for feed service in low-pressure plants. They may or may not have a receiver, depending on the requirements. Horizontal single- or multistage volute-type centrifu-

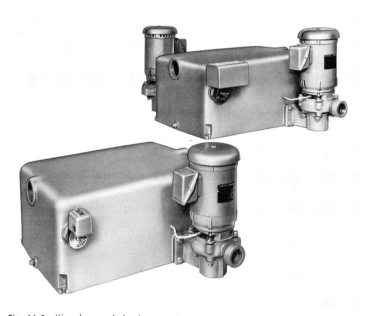

Fig. 11-2 Simplex and duplex condensate return units for plants where lines are close to the floor. (*Aurora Pump, A Unit of General Signal Corp.*)

gal pumps (Figs. 1-14 and 1-15) find some use in low-pressure installations where the flow exceeds about 100 gpm.

Medium-pressure plants, 400 to 1,500 psi, use horizontal multistage split-case volute-type centrifugal pumps with two to six stages (Fig. 11-3*a* and *b*).

Power-type reciprocating pumps (Fig. 11-4) are also used in feed service for medium-pressure steam plants. They overlap into the high-pressure field and are often used for desuperheater feed. Inverted designs in Figs. 11-4 and 3-4 can be built with two, three, five, or more cylinders. Power pumps are easily controlled automatically in stepless straight-line fashion from 0 to 100 per cent of rated capacity.

High-pressure plants (1,200 psi and up) use split-case multistage twin-volute and diffuser pumps up to about 1,500 psi, with some designs of a few manufacturers being suitable for somewhat higher pressures. Power-type reciprocating pumps are also popular in this range. Beyond about 2,000 psi, the double-case barrel-type multistage centrifugal pump (Fig. 11-5) is the most common choice. Above 1,600 psi, keeping a tight seal in horizontally split pumps is often a problem. So diffuser- and twin-volute-type barrel pumps with up to 12 stages are applied because sealing their end or ends is somewhat easier. Pressures developed by units of this design range up to about 6,000 psi; temperature of the water handled in feed service may be 700 F, or even slightly higher.

Power-plant sizes in the 1950s and 1960s more than doubled to well over 1,200,000 kw. This rapid growth rate revolutionized boiler-feed-pump technology and produced several successive drive and speed trends. Early in the period, direct-motor drive (or motor drive through a variable-speed coupling) gave way to turbine-generator drive (either from the generator or governor end). This eliminated a costly, high-horsepower motor that was sometimes difficult to start at normal power-station auxiliary voltage. At almost the same time, true high-speed pumps (7,000 to over 10,000 rpm) became commercially available. This facilitated the wide use of turbine drives. In very large power plants today (500,000 to 1,300,000 kw) turbine-driven boiler-feed pumps predominate. Horsepowers can range to the total power-plant capacity of but a few years ago. Here, all the advantages of variable-speed operation (power saving, minimum pump-discharge pressure at light loads, and reduced pump wear) are retained with the added heat-balance saving available through use of cold-reheat steam to drive the turbines.

High-speed boiler-feed pumps physically resemble their 3,600-rpm counterparts. Fewer stages (four to six) supplant the up to twelve stages needed at 3,600 rpm. Nevertheless higher heads are developed per stage, and impeller diameters are smaller. High-speed diffuser and twin-volute boiler-feed pumps develop 600 to 1,000 psi per stage vs. about 300 psi per stage at 3,600 rpm.

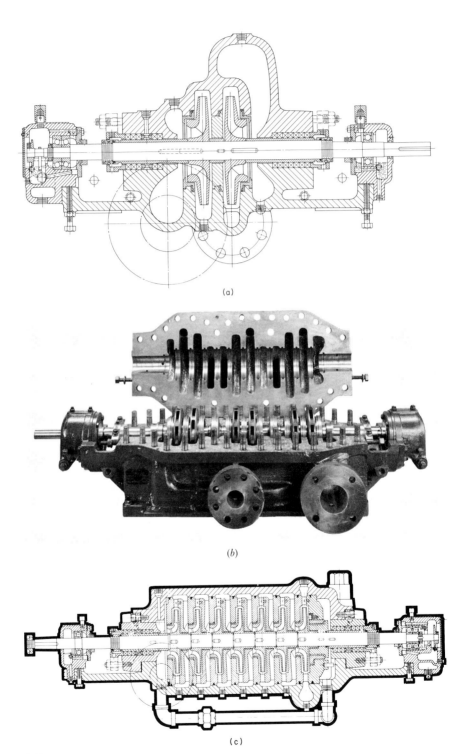

(a)

(b)

(c)

Increased boiler-feed-pump reliability and power-plant capital cost economics led to single-pump installations and the elimination of full-capacity spares. In large power plants, a one-third capacity start-up pump (motor-driven) serves well as the spare during an emergency plant shutdown caused by a highly unlikely boiler-feed-pump failure. Experience has shown that the modern boiler-feed pump can operate economically well over 100,000 hr before requiring major overhaul.

Feed-pump Selection   *Capacity:* This should be based on the maximum allowable overload rating of the boiler. This may range from 110 to 250 per cent or higher and can be obtained from the boiler manufacturer. With two half-size pumps, the sum of their capacities should at least equal the maximum steam capacity of the boiler, at the maximum overload rating. Generally, it is advisable to use a safety factor of 20 per cent, or more, for pump capacity. *Head* to be developed must equal the boiler pressure *plus* friction losses in the superheater, boiler, economizer, heaters, valves, and piping, *plus* a 6 per cent excess pressure so the pump can supply the boiler when the safety valve is open. The final head chosen is usually 15 to 25 per cent higher than the rated boiler pressure. *Materials:* See Table 6-4. Selection numbers 4, 5, 8, and 14 in this table are common for medium- and high-pressure pumps. *Drives:* See Chaps. 8 and 10. Both motors and turbines drive low-, medium-, and high-pressure pumps, with splashproof motors presently having the edge.

NPSH: See Standards of Hydraulic Institute and Chap. 4. *Minimum Flow:* See manufacturer's recommendation on bypass arrangement and flow required. *Control:* This varies with pump class and type. But the economy of variable-speed operation is receiving greater recognition today and is finding more applications in medium- and high-pressure plants. Low-pressure plants (and some medium- and high-pressure plants) commonly use throttling-type feed-water regulators for units other than reciprocating pumps. By locating the regulating valves at the pump discharge, the heaters and piping are subjected to much lower pressures than if the valves are at the steam-drum inlet. *Packing:* See Chap. 9. Mechanical seals, recently introduced for medium- and high-pressure feed pumps, will probably find greater use in coming years; they provide better sealing with less upkeep.

11-3 3,600-rpm centrifugal boiler-feed pumps for low, me-
m, and high pressures. (*a*) A 2-stage pump for low-pressure
vice. (*b*) Horizontally split case with opposed impellers for axial
ance. (*c*) Pump with radially split inner element and horizon-
y split outer casing. Impellers are arranged sequentially, and
ernal line balances internal pressure differential.

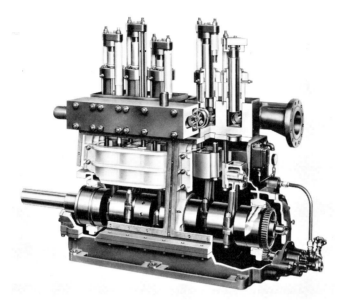

Fig. 11-4   Quintuplex power-type feed pump. (*Aldrich Division, Ingersoll-Rand Co.*)

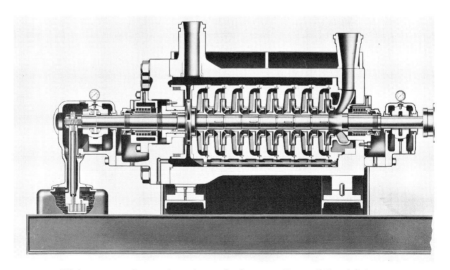

Fig. 11-5   High-pressure 8-stage barrel-type feed pump. (*Ingersoll-Rand Co.*)

Arrangement: Vertically mounted split-case volute pumps and double-case diffuser pumps find wider application today for medium and high pressures where floor space and foundation construction are factors.

Number of Pumps: One operating pump per boiler is now almost standard for large, base-loaded plants (a small start-up, standby, motor-driven pump is usually included, especially if the main boiler-feed pump is turbine-driven). For peaking plants or others where load swings are anticipated during most of the plant's life, two or more pumps operated in parallel are usually more economical. When the total capacity is too small to be economically served by two pumps in parallel, a single unit should be used no matter what the load is.

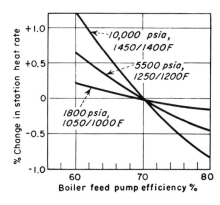

Fig. 11-6 Effect of feed-pump efficiency on station heat rate.

Efficiency: Figure 11-6 shows the effect of feed-pump efficiency on a typical heat balance. The importance of correct selection is apparent.

Boiler-circulation Pumps Forced-circulation boilers, at operating pressures over 2,000 psig, require a pump to remove water from the boiler drum and to force it into the evaporating section. Single-stage and two-stage volute-type pumps (Fig. 11-7) are used, as are canned-motor pumps and large vertical, controlled-leakage designs (Chap. 12). *Capacity* is usually two to three times the boiler steam capacity. Pressure developed ranges to 350 psi, suction pressure being equal to the rated drum pressure of the boiler. Temperatures range to 700 F. *Materials:* See Table 6-6. Special high-temperature construction is required. *Drives:* Usually motor. *Number of pumps:* Two or more per boiler, with one spare.

Condensate Pumps *Low-pressure Plants:* In heating and other plants where separate condensate pumps are not needed, centrifugal and regenerative turbine condensation sets fitted with a receiver (Chap. 19) serve as both the condensate and boiler-feed pump. Reciprocating condensate pumps are rarely used today. Where a separate condensate

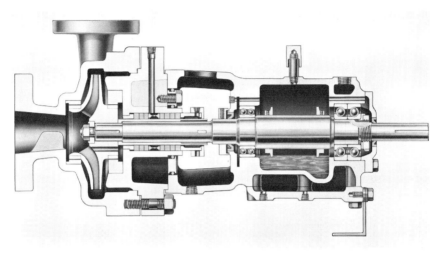

Fig. 11-7 Boiler circulating pump. (*Worthington Corp.*)

pump must be used, and the head is low or moderate, single- or two-stage double-suction general-purpose volute-type centrifugal pumps are satisfactory. Suction passages should be tapped on both sides of the impeller to vent any vapor present to a lower-pressure area. Stuffing boxes fitted with cages for sealing by an external water source are generally recommended. Other designs of centrifugal pumps used in this service include close-coupled, vertically-split-casing, and regenerative-type single- or two-stage pumps. These units are usually marketed as low-pressure condensate pumps.

Medium-pressure Plants: Two-stage opposed-impeller volute-type horizontal centrifugal pumps and multistage regenerative pumps are used in medium-pressure plants. It is common practice to vent the suction chamber of volute-type pumps. Stuffing boxes are normally under discharge pressure, as in the unit in Fig. 11-8. Seal cages are used in the stuffing boxes (Fig. 1-32) because there is chance of air in-leakage when the pump is idle.

High-pressure Plants: Multistage horizontal volute-type pumps (Fig. 11-8) are a common choice for higher heads and larger capacities. Though the pump in Fig. 11-8 has four impellers, it is actually a three-stage unit because the two center impellers form the first stage, providing a larger inlet area and permitting the pump to operate at lower npsh. The suction passage is vented at the top to prevent vapor binding. External sealing of the stuffing boxes is again used. In some designs, three stages are provided by using two pumps on one shaft. Providing the required npsh for horizontal condensate pumps often necessitates locating the unit in a pit below floor level. This makes maintenance more difficult and introduces the possibility of damage to the driving motor should the pit be flooded. To overcome these disadvantages, vertical multistage turbine pumps, called canned pumps (Fig. 11-9), have been developed for both medium and high pressures. The pumping element is enclosed in a shell which forms the suction well. The shell is installed below the floor level, supporting the pump and its drive. The shell and discharge column can be furnished in a length suitable for the npsh required. Single-stage units in this design are also available. Self-venting, these pumps are furnished with open or closed impellers. A hollow-shaft motor is commonly used as the drive. In these, the condensate passes through one or more stages, leaves the pump, and goes

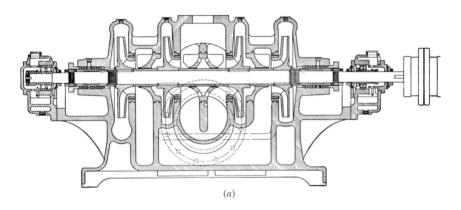

(a)

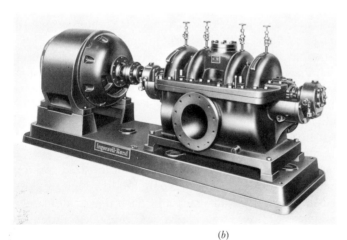

(b)

Fig. 11-8 Typical 3-stage condensate pump with two first-stage impellers operating in parallel. (a) Cross section. (b) Exterior. (*Ingersoll-Rand Co.*)

Fig. 11-9 Vertical multi-stage canned pump for heater-drain or condensate service.

through one or more heaters. It then reenters the pump, passes through one or more stages, and is discharged to the next heaters in the cycle.

Condensate-pump Selection: Capacity: Base capacity on maximum rating of plant or unit the pump serves. Horizontal pumps are often oversized 25 per cent to allow for reduced submergence. *Head:* Npsh is of extreme importance because the pump usually takes its suction directly from the condenser hotwell. See Chap. 5. *Caution:* Vacuum on the pump suction (when the unit takes its suction from a condenser hotwell) *reduces* the apparent static suction head. *Materials:* Condensate at low temperature is usually less corrosive than high-temperature feed water. Bronze-fitted pumps are therefore often used. *Drives:* Generally, electric-motor drive is chosen, but steam turbines are also popular. *Control:* Submergence, condensate recirculation, and discharge throttling are common control methods. Submergence control is simple and gives slight power saving at the exact head and capacity required by system. However, discharge throttling, with some type of recirculation, is generally thought to be better than submergence and condensate-recirculation control. *Packing:* Use seal cages with an external water supply to prevent air in-leakage. *Number of Pumps:* Either one full-capacity or two half-capacity pumps may be used, depending on whether the station is base-loaded or not. Provide standby capacity by installing an additional pump or by interconnecting with other pumps in the plant. Condensate-booster and hotwell pumps perform the same duties as condensate pumps and should be chosen in the same manner. Booster pumps may be located between heaters, with the hotwell pump taking its suction from the condenser and discharging to the boosters.

Heater-drain Pumps While many small and medium-sized power plants cascade regenerative heater drains (let them flow through traps to the next lower-pressure-heater steam space and flash there, ultimately flowing to the condenser), most large plants find added economy in pumping these drains ahead in the feed-water cycle.

Heater-drain service is perhaps one of the most severe to be found in a modern power plant—it combines most, if not all, of the problems of condensate and boiler-feed services. Usually, the available npsh is very low and variable; so vertical canned pumps (Fig. 11-9) are popular. Where oil-lubricated, horizontal or vertical pumps (Fig. 11-10) are hydraulically possible, however, they tend to be less troublesome.

While heater-drain-pump available npsh tends to be low (like condensate service), temperatures and pressures can be quite high (as in boiler-feed service). Thus, all stainless-steel construction is often required, especially in high-pressure heater-drain service. Also, pressure, temperature, flow, and available npsh conditions vary widely and inconsistently with power-plant load changes—all adding to operating difficulties. Like condensate pumps, heater-drain pumps must be effectively vented to the suction source to continuously bleed off vapor that can form in front of the first-stage impeller.

Circulating-water Pumps Also called *circulators*, these supply cooling water to the steam condenser. Many larger condensers have two half-capacity circulators and a means for reversing flow through the condenser. Each pump serves one water box. Single-stage horizontal double-suction volute-type pumps (Fig. 1-14) are popular for circulator duty, but vertical propeller, turbine, and mixed-flow pumps (Fig. 20-2C) have become popular in recent years for the large flows and moderate heads characterizing condenser cooling. Bronze-fitted pumps are usually suitable for this service. *Pull-out-type* circulators are vertical units arranged for easy removal for maintenance.

Fuel-oil Pumps These, and ignition-oil pumps, are usually some form of rotary type. Figure 11-11 shows a typical two-screw pump handling 80 gpm at continuous pressures of 275 psig. Direct-acting steam pumps are used for fuel-oil service in smaller low- and medium-pressure plants but their use appears to be declining. Some high-pressure centrifugal fuel-oil pumps are used in large stations. See Chap. 13.

Other Steam-plant Pumps A number of other pumps are used in steam power plants. These are briefly discussed below.

Ash-handling Pumps: The single-stage, centrifugal ash-handling pump (Fig. 11-12) is made of alloy iron and is fitted with water seals. Other designs are packingless and have flat-blade impellers and special arrangements for the suction inlet. While the pump is important, so are other details of the system.

Chemical-feed Pumps: Reciprocating metering and controlled-volume pumps of various types (Chap. 3) are standard for this service. They are used for pH control, coagulation, activated-silica and raw-water preparation, demineralization and lime-soda system, CO_2 and O_2 control, etc.

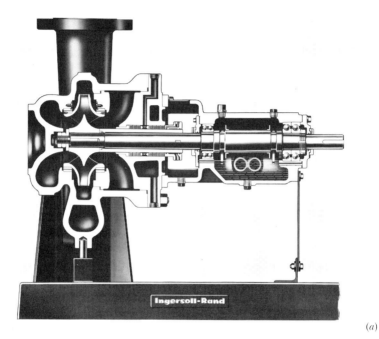

(a)

(b)

Fig. 11-10 (a) Horizontal single-stage heater-drain pump has double-suction impeller for low required npsh. (b) Vertical close-coupled design is pipe-supported; low impeller reduces required npsh. (*Ingersoll-Rand Co.*)

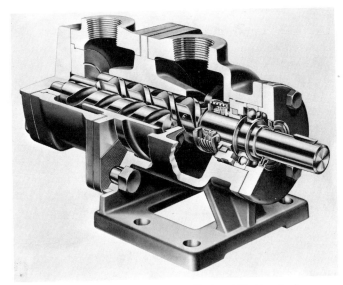

Fig. 11-11 Screw-type fuel-oil pump. (*De Laval Turbines Inc.*)

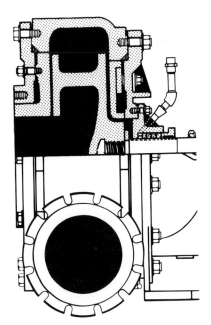

Fig. 11-12 Ash-handling pump. (*The Allen-Sherman-Hoff Pump Co.*)

Sump Pumps: See Chap. 18. *Water-supply Pumps:* See Chap. 17. Ash-service water, hydrogen circulating water, fresh-water, salt-water booster, gland-seal, river-water, fire, condensate-storage, city-water, and demineralizing make-up pumps are usually single-stage horizontal single- or double-suction bronze-fitted, all-bronze or all-iron volute pumps. However, vertical turbine and mixed-flow pumps have become popular for many of these services in recent years. They may be either motor- or turbine-driven. Acid and caustic pumps are often similar to the above, except that special materials are used in their construction. Some are packingless types. *Boiler-drain Sump Pumps:* See Chap. 18. *Hydrostatic Test Pumps:* See Chap. 3. *Steam-turbine lube-oil supply* and *turning-gear oil-supply* pumps are usually rotary units, driven by either motors or turbines, or both. *Seal-oil* pumps may be centrifugal or rotary units.

Cooling-tower Circulation: Both horizontal and vertical pumps are used for this service, with vertical single- or multistage turbine pumps being extremely popular and showing much promise of greater use in the future. Either semiopen or closed impellers are used in vertical turbine pumps. Since most pumps for cooling-tower circulation are outdoors, they are fitted with weatherproof drive motors. Horizontal pumps for cooling towers are either single- or double-suction single-stage motor-driven volute units.

Rotary vacuum pumps are often the cam-and-piston type (Fig. 2-1). Advantages include less maintenance, elimination of hogging jets for startup, and reduction of ammonia contamination.

Pump Materials Recent studies of central-station feed pumps show that cast iron and steel are unsuitable for high-pressure installations. In general, 5 per cent chromium steels provide suitable corrosion-erosion resistance. Higher alloy concentrations do not provide a marked increase in resistance. With a relative service life of 1.0 for iron and steel, 5 per cent chrome steel has a life of 100, 18-8 a life of 100 to 105. Stainless or K monel shafts are recommended for these pumps.

Where dissolved gases are present, either all-bronze or 13 per cent chrome steel should be used for heater-drain pumps. Circulating pumps handling polluted harbor water may require bronze casings and stainless impellers. Shafts should be stainless or monel. When pumps are interconnected to provide standby capacity, as is often done today, the effect of the liquid must be considered.

Process-condensate Pumps A number of different units are available today to handle condensate returning from process equipment. The general objectives for which they are designed are: (1) prevent system waste, (2) keep condensate in liquid form, (3) rid the return system of air, and (4) return condensate to the boiler at the temperature and pressure at which it was formed. Units for this service generally consist of a

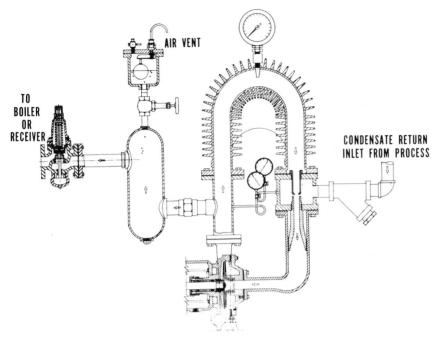

AIR VENT

TO
BOILER
OR
RECEIVER

CONDENSATE RETURN
INLET FROM PROCESS

Fig. 11-13 Condensate-drain pump. (*Cochrane Corp.*)

receiver and one or more pumps connected to it (Fig. 11-2). Condensate returning to the receiver is held there until the level reaches a predetermined height, when the pump is automatically started. Many of these units use a regenerative-type turbine pump because it has highly desirable characteristics for low- and medium-pressure feed.

A tankless system (Fig. 11-13) uses a combined jet and centrifugal pump. The centrifugal pump draws condensate from the finned priming loop and discharges to the loop and an outlet point. The amount of condensate discharged to the outlet equals the quantity of liquid and steam entering the venturi-shaped jet tube on the suction side of the finned loop. From the loop outlet the condensate flows to an intermediate point in the boiler make-up system or to the feed-water header.

INTERNAL-COMBUSTION ENGINES

Diesel-, gasoline-, and gas-engine plants use pumps for several services —cooling-water circulation, lube-oil supply and circulation, and fuel-oil supply. Horizontal single-stage single-suction volute-type close-coupled centrifugal pumps are popular for cooling-water circulation. Since these pumps are often mounted in the power-plant basement where there is usually a large floor area available, vertical pumps are not too popular.

Rotary-gear pumps are the usual choice for lube- and fuel-oil supply and circulation. The capacity and head required vary with engine size, length of piping, suction lift, etc. As a typical example, a 960-bhp diesel engine in a municipal generating station is fitted with a 50-gpm prelubrication pump to supply oil to the engine before it is started, and a 10-gpm lube-oil transfer pump for engine operation. In many stationary installations the pumps serving the diesel engine are not supplied with it but must be purchased separately. Portable engines, however, come equipped with the necessary pumps as part of the original equipment. With a 15-F rise in jacket-water temperature, 0.4 gpm per rated engine horsepower must be circulated.

HYDRO PLANTS

While a true hydroelectric power plant involves merely a water turbine directly connected to an electric generator, several modified plants have recently been built for *pumped storage.* Here, artificially constructed upper and lower ponds provide for cyclic drawdown and fill to meet peak electrical load conditions.

When electrical loads are high, pumped-storage plants operate as true hydroelectric plants. Water from the upper pond drains to the lower pond through the water turbine to generate power. But when electrical loads are light (at night, for example), the process is reversed. Now the generator operates as a motor, driving the water turbine as a pump. This fills the upper pond, using relatively inexpensive power, and prepares the system for later generation of needed power during high-load, or "peaking," conditions.

These two modes of operation (pump and turbine) in a single piece of equipment complicate control and design. The pump-turbine must be designed to be highly efficient in either mode. In some cases, this is accomplished by using variable-position impeller blades.

Peaking unit sizes are usually very large (up to several thousand kilowatts), as opposed to true hydroelectric generating units which may vary anywhere from a few kilowatts in a small stream to enormous, multiunit power plants such as at Niagara Falls. Thus design and application of pump-turbines are highly specialized tasks and are best done by the manufacturer. The combined motor-generator requires careful design because both are comb ned into a single frame.

Other pumps used in typical hydro plants include sump, dewatering, lube-oil, servomotor-oil, etc. They are discussed later in this book.

CHAPTER TWELVE

Nuclear-energy
Applications

EVER SINCE THE END OF WORLD WAR II, peaceful uses of nuclear energy have been pursued on a wide front. Perhaps the most immediately successful area thus far has been in the generation of electric power. In a nuclear-power plant, the nuclear reactor essentially replaces the conventional fossil-fueled boiler (Chap. 11); the remaining portions of the plant—turbine-generator, feed-water system, and condensing-cooling system—are the same.

During the 1950s and early 1960s several prototype, light-water-cooled nuclear-power plants came on the line. Initial troubles were overcome, and commercial success was within reach. Other coolants, too, were tried. Sodium-cooled plants anticipated breeder-reactor development, and gas-cooled plants demonstrated increased thermal efficiencies.

In the mid-1960s, utilities became convinced that nuclear power offered many operating advantages, especially in geographic areas of high fuel cost, and, beginning in 1966, nuclear-power plants purchased represented nearly 50 per cent of the total kilowatts ordered. Plant sizes grew startlingly—up to well over 1,000 Mw.

AEC Program Early in the design stages of nuclear-power plants, the Atomic Energy Commission requested development of two hermetically sealed pumps. The pumps requested were: (1) a 4,000-gpm unit to develop 100 psi when handling high-temperature water at high pressure (2,000 psi), and (2) a 150-gpm unit to develop 120 psi when handling high-temperature high-pressure water. The first unit was the main coolant pump and the second a hydraulic service pump for a nuclear-power plant.

From these, and other studies and tests, have come a number of successful pumps for various types of reactors and nuclear units. Pressures up to 5,000 psi at temperatures to 1200 F have been developed by some of the pumps. Liquids handled include heavy water, radioactive water, and liquid sodium, among others.

Service Conditions   The head and capacity requirements of nuclear applications are not severe—heads are moderate (in the 100- to 200-psi range), while the capacity needs are moderate to very large (up to about 75,000 gpm, at present). What introduces major design and manufacturing problems is the zero-leakage or controlled-leakage requirement. To meet this requirement, many firms have a standard line of pumps available for nuclear-energy and other related, or unrelated, applications.

To meet nuclear-energy service conditions, at least six design variations have been studied and developed into commercially available units: (1) canned-motor pump, (2) submersed-motor pump, (3) gas-filled-motor pump, (4) oil-filled-motor pump, (5) controlled-leakage pump, and (6) electromagnetic pump.

The first four units are zero-leakage pumps fitted with special close-coupled drive motors, with the entire assembly hermetically sealed in a suitable housing. Initially, these zero-leakage pumps found favor in most nuclear applications. However, as unit sizes grew, their inherent inefficiencies and high costs forced them out of use.

The fifth is a lower-cost, higher-efficiency, controlled-leakage pump having a mechanical seal to limit the leakage along the drive shaft of the unit. In large nuclear-power plants today, this design, or a variation of it, predominates.

The last pump, also a zero-leakage pump, is limited to applications where the liquid pumped is electromagnetic.

Canned-motor Pump   Figure 12-1 shows a typical design of this type. Pumped liquid is allowed to fill motor cavity, but it is excluded from the rotor and stator windings by sealing jackets or cans in the magnetic gap. No external shaft seals are used, and the suction and discharge nozzles are designed to be welded into the pipeline. The liquid pumped completely fills all the space inside the sealing tube and provides lubrication for the bearings. This tube, within the magnetic gap, also seals the motor stator from the liquid pumped. The only auxiliary requirement is an external supply of ordinary clean cooling water.

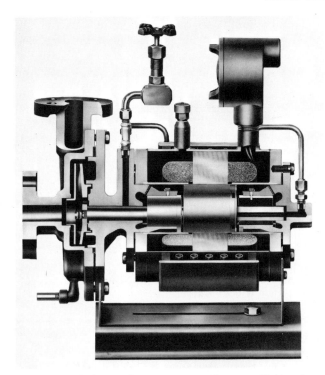

Fig. 12-1   Typical canned-motor pump. (*Chempump Division, Crane Co.*)

These pumps are designed to handle radioactive water with no leakage. Though originally designed to circulate high-pressure high-temperature water between a reactor and steam boiler, these pumps are suitable for many other clear nonabrasive liquids. Special designs are available for handling liquid metal at temperatures to 1000 F. These pumps can also be designed for liquids at system pressures to 10,000 psi. Single-stage, they are volute-type pumps, with all parts contacting the pumped liquid made of corrosion-resistant alloys (steel or Inconel).

Variations in the design of canned-motor pumps include a high pressure type having the stator liner supported by laminations and taking the full discharge pressure of the pump and a high-pressure type fitted with a balancing system to equalize the liquid pressure across the stator liner, eliminating the liner stress regardless of the pump suction pressure. An external heat exchanger to cool the motor can be used with some designs, if desired.

Submersed-motor Pump   In this design (Fig. 12-2) the pumped liquid surrounds and contacts the stator laminations and winding, the shaft bearings, and the motor rotor. Waterproof insulation is required for

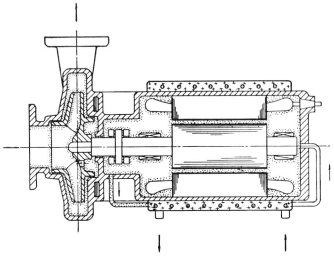

Fig. 12-2   Submersed-motor pump rated 250 hp operates with suction pressures to 2,000 psi and temperatures to 650 F. (*Byron Jackson Pump Division, Borg-Warner Corp.*)

the stator winding.   Many units of this type are used as boiler circulating pumps with suction pressures up to 2,000 psi and temperatures to 650 F. Because radioactive water might attack the motor insulation, this pump has not yet been used for reactor circulation service.

Gas-filled-motor Pump   In this design, the liquid pumped is not allowed to enter the motor (Fig. 12-3).   An inert gas, nitrogen or helium, sur-

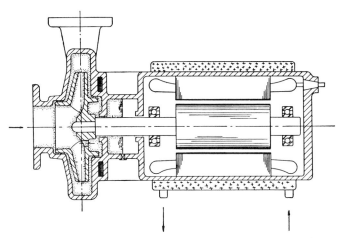

Fig. 12-3   Gas-filled-motor pump rated 10 hp handles uranium slurry at temperatures to 1000 F. (*Byron Jackson Pump Division, Borg-Warner Corp.*)

rounds the rotor and stator. Oil- or grease-lubricated ball bearings are used in the motor. The gas-filled design successfully handles abrasive uranium slurry.

Oil-filled-motor Pump   These have a number of advantages, including a high-efficiency motor, effective motor cooling, and oil-lubricated bearings. A shaft seal of either the mercury or mechanical type separates the oil in the motor from the liquid pumped. The mercury seal is a zero-leakage type, while the mechanical seal has an average leakage of about 4 gal per year. When a mercury seal is used (Fig. 12-4) contami-

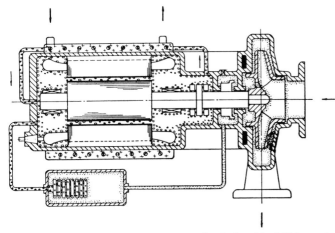

Fig. 12-4   Oil-filled-motor pump has mechanical seal in 200-hp unit, mercury seal in 400-hp unit. (*Byron Jackson Pump Division, Borg-Warner Corp.*)

nation of the mercury by oil or radioactivity presents problems. The power rating of these units varies from 10 to 400 hp, with the submergence ranging up to 1,200 ft. A low differential pressure across the seal and a low shaft peripheral velocity are required for successful use of the mercury seal.

Table 12-1 compares the various pump designs discussed above. While canned-motor pumps are at present the most common type for reactor service, their motor efficiency is low. To keep rotor friction and liner losses low in liquid-filled motors, a long slender motor is often used.

Controlled-leakage Pump   As previously mentioned, controlled-leakage pumps now predominate in light-water-reactor circulating service. Size-for-size, their cost is only 60 per cent of a canned-motor pump, and their efficiency is 10 to 15 per cent higher. Figure 12-5 shows the

TABLE 12-1 Comparison of Efficiencies of Various Motors*

(300-hp 1,800-rpm vertical units for 2,000 psi)

	Motor type			
	Canned	Submersed	Oil-filled	Gas-filled
Electrical losses, kw	14	18	12	14
Rotor-friction loss, kw	10	10	16	3
Liner loss, kw	36			
Motor efficiency, %	79	89	89	93
Zero-leakage feature	Good	Good	Fair	Good
Bearing lubrication	Fair	Fair	Good	Good
Motor-winding cooling	Fair	Good	Good	Fair
Motor-terminal design	Good	Fair	Fair	Fair
Design weakness	Liner	Contamination	Oil leaks	Liquid-level control

* Byron Jackson Pump Division, Borg-Warner Corp.

Fig. 12-5 A 9,000-hp primary reactor circulating pump; 6-ft figure shows immense size. (*Bingham Pump Co.*)

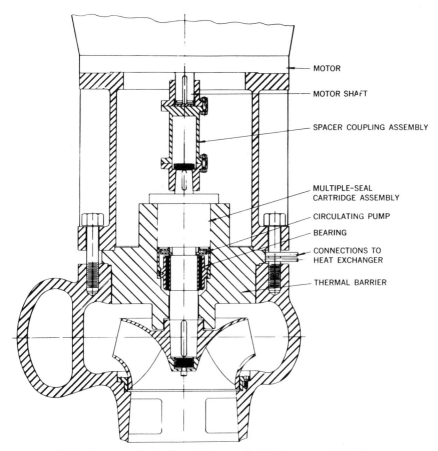

Fig. 12-6 Typical large mechanically sealed controlled-leakage pump for light-water-reactor service. This pump is rated at 1,000 hp, 36,000 gpm. (*Byron-Jackson Pump Division, Borg-Warner Corp.*)

immense size of modern nuclear-steam system circulators. Flows here can range up to about 75,000 gpm, but total heads are relatively low—100 to 200 psi. Suction pressures, however, are characteristically high—1,000 to 2,500 psig. Hence shaft-sealing arrangements are complex.

Figure 12-6 shows a typical large controlled-leakage pump for light-water-reactor service. It incorporates a cartridge-type mechanical-seal assembly readily accessible for maintenance. A single water-lubricated, self-aligning, lower radial bearing in the pump has a large area for very light loading. Axial thrust is carried by the motor by a pivot-shoe thrust bearing. In this vertical unit, major emphasis is placed on precise alignment of the motor shaft, pump shaft, seal assembly, and seal bearing. The complete motor-pump is allowed to "float" on the piping.

Electromagnetic Pump   These units have no moving parts, which means that seals of all types are eliminated. As a result, there is no danger of shaft leakage. But only liquid metals having a high enough conductivity to be satisfactorily pumped can be handled. Typical metals having suitable conductivity include magnesium, aluminum, sodium, potassium, and alloys of these metals.

With an a-c source, the principle of the electromagnetic pump is the same as an induction motor. The alternating current builds up a moving magnetic field in a suitable winding. This induces an electric current in the liquid metal flowing through the winding. The induced current has its own magnetic field, which reacts with the winding field and forces the individual liquid-metal particles to move with the moving magnetic field. Figure 12-7 shows a flat-bed induction-type electro-

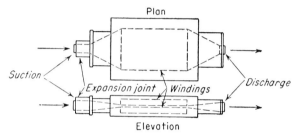

Fig. 12-7   Flat-bed induction-type electromagnetic pump develops 37 psi when handling 1,200 gpm of 500-F liquid sodium.

magnetic pump which develops 37 psi when handling 1,200 gpm of liquid sodium at 500 F. The unit is fitted with an expansion joint at one end.

A d-c Faraday-type pump is shown in Fig. 12-8, while a reverse-flow induction-type pump is shown in Fig. 12-9. This is also known as the Einstein-Szilard pump. The over-all efficiency of typical electromagne-

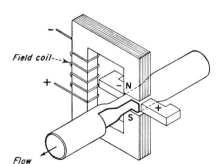

Fig. 12-8   D-c Faraday-type pump.

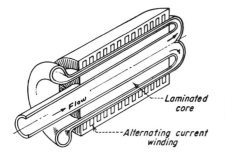

Laminated core

Flow

Alternating current winding

Fig. 12-9 Reverse-flow induction-type pump.

tic pumps is about 50 per cent; cost is about 33 per cent less than a canned-motor pump of the same capacity.   Head developed by these pumps ranges up to about 100 psi, with most designs being in the 50-psi range.   Some foundries are using electromagnetic pumps to move liquid metals in the molten state from the furnace to the molding area.   The greatest use, however, is in nuclear-energy plants.

Sodium Pumps   One of the coolants being actively considered for the breeder-reactors of the future—reactors that breed, or make, more fuel than they consume—is liquid sodium.   Sodium's unique characteristics require very special pumps.

Figure 12-10 shows the largest sodium pump built to date.   It is rated at 12,000-gpm sodium at 330-ft total head and 1000 F.   Its height from the face of the bottom discharge nozzle to the top of the pump shielding (but not including the motor) is 30 ft.   Typically, sodium pumps operate under very low system pressure, and there are no serious npsh problems with the low vapor pressure of sodium.   Most primary sodium pumps are of the pull-out type with integral shielding; the complete internal-pump assembly and rotating element can be removed as a unit for maintenance.   Pump-shaft length requires a bearing in the sodium.   The extremely low viscosity of sodium at high temperatures precludes conventional hydrodynamic sleeve bearings, and so special hydrostatic-pressurized bearings were developed.   They can be described more accurately as shaft-centering devices with no metal-to-metal contact at normal pump speeds.

Sodium pumps of this type are for free-surface operation (as opposed to the wholly contained electromagnetic pumps previously discussed).   Therefore, a cover blanket of low-pressure inert gas isolates the sodium from the atmosphere.   This cover gas is contained where the shaft comes out of the pump by a face-type gas seal.

Pump Selection   *Reactor Systems:* For today's large light-water-reactor systems, controlled-leakage pumps are almost universally used.   For the sodium-cooked breeder-reactors of the future, vertical centrifugal pumps for free-surface operation seem to be most popular.   *Head and*

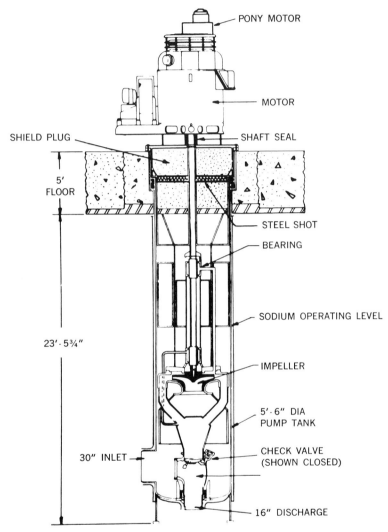

Fig. 12-10   A primary-loop sodium pump for the Enrico Fermi breeder reactor rated at 1,000 hp, 12,000 gpm, and 1000 F. (*Byron-Jackson Pump Division, Borg-Warner Corp.*)

Capacity: These vary so widely that it is impossible to give average ranges at the present stage of reactor development. *Materials:* Stainless or low-carbon steels are suitable for parts contacting the liquid. Surface-hardened stainless-steel shafts on stellite or Colmonoy bearings, or stellite on stellite (different grades) are generally suitable for liquid-sodium pumps. Pumping and flow characteristics of water and liquid sodium are fairly similar. Stainless steels, types 304, 316, 346, etc., are suitable for pump parts contacting reactor coolants. *Drives:* Electric motors are the sole type of drive used today. To supply power to a-c electromagnetic pumps, a unipolar-type generator with a NaK hydro-type brush collector is popular. *NPSH:* Usually not a factor. *Control:* Centrifugal pumps can use (1) infinite-step control with (*a*) a wound-rotor motor and liquid or drum rheostat, (*b*) eddy-current coupling, (*c*) fluid coupling, (*d*) magnetic coupling; or (2) stepped-speed control with stepped transformer control for 100, 80, and 60 per cent speed steps. For a-c electromagnetic pumps, both voltage and frequency, separately or together, can be used to control speed over a wide range. *Packing:* See discussion in this chapter. *Number of Pumps:* In large power-reactor systems, a single reactor usually serves two or more steam generators with one circulating pump operating in each steam-generator loop. Spares are not provided since, in the event of pump failures, the remaining loops can adequately cool the reactor.

Emergency System: Large power reactors are protected against overheating during abnormal conditions by a series of redundant cooling systems. These provide auxiliary supplies of water to carry away heat in the event that the main coolant-circulating systems fails. Pumps for this service are usually high-capacity centrifugals with multiple spares and complex interconnections, so that coolant flow can be assured under any possible emergency condition.

Feed System: With the exception of the nuclear steam system, other nuclear-power-plant pumping services, such as boiler-feed, heater-drain, condensate, and condenser-cooling, are similar to fossil-fueled plants (Chap. 11). Size-for-size, however, feed-water flows are likely to be higher and heads lower because most present-day nuclear-power plants operate at a lower turbine-throttle pressure than their fossil-fueled counterparts. In the future, however, the need for higher thermal efficiencies in nuclear-power plants may cause turbine-throttle pressures and temperatures to increase, approaching those of modern, large fossil-fueled plants.

CHAPTER THIRTEEN

Petroleum Industry

FOR CLASSIFICATION PURPOSES, pumps used in the petroleum industry can be listed under eight categories—drilling, production, transportation, refining, fracturing, offshore-rig, portable, and proportioning pumps. Petroleum is the second most common liquid handled in pumps, water being the first. This being so, it is surprising that the number of types of pumps used in the petroleum industry is relatively small, compared with other industries. A high degree of pump standardization is characteristic of this industry because of the petroleum industry's tendency to move pumps from one location and service to another when process improvements require installation alteration. Although, to a certain degree, the petroleum industry has begun to accept the American Voluntary Standard (AVS) type of pump, the level of acceptance here is not as extensive as in the chemical and petrochemical industries (see Chap. 14).

Drilling Slush or mud pumps (Figs. 3-2 and 13-1) are almost invariably reciprocating units—either direct-acting steam-driven (Fig. 3-2) or horizontal duplex or triplex power pumps (Fig. 13-1). To permit easy inspection and maintenance of liquid-end valves they are generally the pot type. Figures 3-36 and 3-37 show two typical liquid-end valves for slush pumps.

For most drilling operations the slush pump must develop fairly high pressures—up to about 3,000 psi at moderate flow rates. Modern drilling practices require both series and parallel operation of power pumps. Series operation can increase the available hydraulic horsepower of a rig 25 to 50 per cent over single or parallel hookups. When specifying slush pumps it is common practice to refer to the pump in terms of hydraulic horsepower instead of capacity. Many slush pumps are rated in the 500- to 1,000-gpm class.

The slush or drilling mud handled by these pumps weighs 11 to 17 lb per gal. Precautions must be taken during pump design to keep mud and water from entering the power end. Direct-acting steam pumps are often fitted with telltales to indicate mud leakage past the outside-type packing used in the liquid end. Some power pumps have water-flushed piston rods to remove any mud leaking from the liquid-end packing boxes.

To reduce the magnitude of discharge-pressure surges, slush pumps are sometimes equipped with a surge chamber. These protect the pump, discharge line, and rotary hose from surge or oscillating pressures. But in some designs the variation in discharge pressure is less than 5 per cent. With this slight variation there is little chance of breakage of fluid-end studs, piston-rod failure, etc. Pump and driver life are prolonged.

Power-type slush pumps are usually belt-driven by an internal-combustion engine. In recent years torque converters between the drive and pump have become popular. Horizontal reciprocating power pumps are also used for jet drilling in slim-hole activities, core drilling, cementing service, geophysical and seismograph work, oil-line and gath-

Fig. 13-1 Power-type slush pump. (*Worthington Corp.*)

ering installations, fracturing, and water flooding. Smaller units are used for mud mixing, transfer, and jetting. They find extensive service in portable rigs for depths from 3,000 to 12,000 ft and unitized rigs for 12,000 to 20,000 ft or more. See Chap. 3 for additional details on reciprocating pumps.

Production Three types of mechanical pumping systems are used to lift oil from producing wells and deliver it to the surface of the earth—sucker-rod, hydraulic, and submersible systems. Pneumatic *gas-lift* systems are also used but they are not so common as mechanical methods. Flowing wells need no pumping apparatus so long as they produce at a satisfactory rate.

Sucker-rod Systems Pumps for these systems are classified by the American Petroleum Institute (API) as: (1) rod, stationary, heavy-wall barrel, top anchor (Fig. 13-2a); (2) rod, stationary, thin-wall barrel, top anchor (Fig. 13-2b); (3) rod, stationary, heavy-wall barrel, bottom anchor (Fig. 13-2c); (4) rod, stationary, thin-wall barrel, bottom anchor (Fig. 13-2d); (5) rod, traveling, heavy-wall barrel, bottom anchor (Fig. 13-2e); (6) rod, traveling, thin-wall barrel, bottom anchor (Fig. 13-2f); (7) tubing, heavy-wall barrel (Fig. 13-2g); (8) tubing, heavy-wall barrel, soft-packed plunger (Fig. 13-2h). Additional standard variations include,

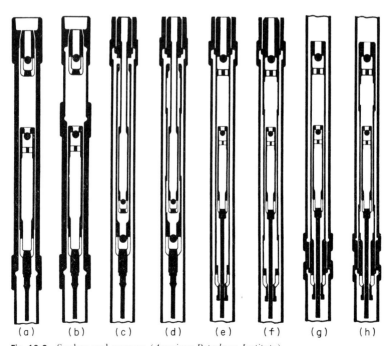

(a)   (b)   (c)   (d)   (e)   (f)   (g)   (h)

Fig. 13-2 Sucker-rod pumps. (*American Petroleum Institute.*)

Fig. 13-3 Walking-beam pumping unit for engine or motor drive.

for example, liner barrels in types one through five. See API Standard RP 11AR for complete specifications for these pump assemblies. This Standard covers inspection, transportation and handling, operation, and assembly and disassembly.

Tubing pumps generally have a maximum bore greater than rod or insert pumps. Also, with a tubing pump the barrel is attached to the lower end of the tubing string and the plunger is run from the sucker-rod string. A rod pump is operated as a complete unit on only the rods. Both types are normally fitted with ball suction and discharge valves. Metallic or nonmetallic seals are used.

To provide the reciprocating motion to the pump plunger, a walking-beam-type pumping unit (Fig. 13-3) is commonly used. This may be engine- or motor-driven, depending on the well location and the economics of the installation. Crank-counterweighted pumping units are rated in terms of API Standards.

Choice of a sucker-rod pump for a given well is influenced by many factors, including production rate desired, well depth, allowable pumping speed, and pump efficiency. One procedure is to select arbitrarily several plunger diameters, speeds, and stroke lengths for the production rate required. With these data the actual displacement and pumping-unit loads can be computed. Rod and tubing stretch and sucker-rod stress, as well as pump efficiency, must be carefully considered. Multiple sucker-rod installations have a single prime mover driving two or more pumping units. The equipment is arranged so the downstroke load of one well counterbalances the upstroke load of the other well. Pump speeds used for single and multiple wells commonly range from about 6 to 24 strokes per minute with stroke lengths ranging from 20 to 168 in. or more. Capacities range from 25 to 1,220 bbl per 24 hr, depending on plunger diameter, stroke length, and speed.

Hydraulic Pumping   Also called rodless systems, these consist of a reciprocating triplex pump (Fig. 3-4) mounted on the surface, a subsurface pumping unit, connecting tubing, and hydraulic fluid. The surface pump discharges power oil to the subsurface unit, producing a reciprocating action of a piston which results in a pumping motion. Rated speeds range from about 5 to 100 strokes per minute at strokes from 1 to 30 ft. Capacities vary with pump speed, plunger diameter, stroke length, and depth of setting. Typical capacities range from 72 to 4,500 bbl per day at 15,000- and 2,700-ft depths, respectively. Over-all efficiency of a hydraulic pumping system is high compared with sucker rods. It is advisable to use the services of a manufacturer when choosing this type of system because pump choice and application are highly specialized jobs.

Submersible Pumps   Motor-driven submersible pumps (Chap. 17) are used in some oil wells. Capacities of these units range from 20 to over 15,000 bbl per day, depending on pump size and lift. Power consumption is about 0.5 kwhr per bbl per 1,000 ft of lift. Pump efficiency in actual use usually averages over 50 per cent.

Portable Pumps   Portable self-priming single- and multistage centrifugal pumps are used in drilling test holes, jetting shot holes, and for fire protection. These resemble the units described in Chaps. 20 and 21.

Transportation   The United States today has hundreds of thousands of miles of oil pipelines. This total is made up of three classes of lines—crude-oil, gathering, and products pipelines. Since pipelines can readily compete with other transportation methods—tankers, trucks, etc.—continued expansion of pipe facilities is expected.

A number of important trends are noticeable in pipelines today. These are: (1) Use of larger diameters for pipelines—36 in. and larger being used in a few lines, compared with pre-World War II sizes of 12 in., or less. A 30-in. pipe requires three times as much steel as a 12-in. line but it handles over eight times as much oil per year. (2) Outdoor pumping stations, with no permanent covering for the pumps and their drives, predominate in new installations. (3) Automatic operation of main and booster stations is now almost universal.

Pipeline Pumps   Horizontal multistage split-case volute-type pumps (Fig. 11-3) and single-stage volute-type pumps (Fig. 13-7) are used in a large number of pipelines today. For this service they are often fitted with mechanical seals and are motor-, turbine-, or engine-driven (Chap. 8). Some horizontal and vertical multicylinder reciprocating pumps are used in main and gathering lines. But centrifugal pumps are more common and more popular for main-line service. In booster-line service the vertical turbine pump has found extensive application in recent years handling crude oil, gasoline, diesel fuels, etc. Rotary pumps for oil transportation are confined to small and medium capacities, except in tankers and bulk-loading stations where large units are used. See Chap. 8 for a number of details on pipeline pump drives.

Bulk-loading Stations Some pipelines terminate at bulk-loading stations for tankers, others at transfer and storage stations for road and rail transport. Horizontal single- and multistage volute-type pumps, vertical turbine pumps, and various special designs of vertical pumps (Fig. 13-4) are used in bulk stations, oil barges, underground tanks, etc. Combined turbine-propeller pumps (Fig. 13-4a) are also used where npsh is a critical factor.

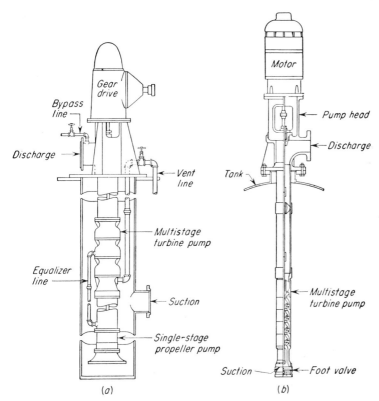

Fig. 13-4 (a) Combination turbine-propeller stripping pump. (b) Dispensing-service pump (*Johnston Pump Co.*)

Aircraft Fueling The end point of some fuel deliveries is an airport underground fuel-storage system. Pumps used to deliver fuel to airplanes resemble those for bulk loading. Jet fuels are particularly hazardous and must be carefully handled because static electricity arcing across the liquid surface in tanks filled at rapid rates can ignite the fuel.

Static charges are formed during petroleum pumping. By reducing the pumping rate, smaller charges are built up. One firm fills JP-4 storage tanks at 50 bbl per hr for the first 3 in. of oil in the tank, raising the flow rate to 100 bbl per hr for the next 2¾ ft. The tank is filled through its water-drawoff pipe until the depth in it is 3 ft. Specifications for fueling pumps are published by the Civil Aeronautics Authority, Army, Navy, and Marine Corps.

Pumping Stations While there have been many opinions expressed for and against outdoor pipeline pumping stations, their use seems to predominate in modern installations. Some pumping stations use electric-motor drives, especially in locations where dependable power supplies are conveniently available. Elsewhere, diesel engines and, lately, integral-gear gas turbines are common drivers (Fig. 13-5), and the trend toward the latter is growing. Both of these self-contained drive units frequently use the product pumped as their fuel. Injection stations often use reciprocating piston or plunger pumps because they must develop a pressure higher than that in the main line to force oil into it at an intermediate point in its run. Microwave communication and control systems are used in many modern pipelines between stations, crews, and other points. Newer pipelines are now using plastic pipe for some of their lines.

Fig. 13-5 Integral-gear, gas-turbine drive unit on pipeline service. (*Solar Division, International Harvester Co.*)

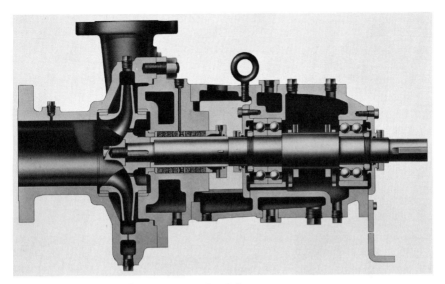

Fig. 13-6 Single-stage refinery pump. (*Allis-Chalmers*)

Refining API Standard gives specifications for centrifugal pumps for general refinery service, mechanical shaft seals, and vertical process pumps. Refinery pumps (Fig. 13-6) are required to have their casing split at right angles to the shaft if (1) the design temperature is 350 F or higher; (2) the pump handles flammable or toxic liquids with a vapor pressure greater than 14 psia at pumping temperature, or at 100 F, whichever is higher; (3) the pump handles flammable or toxic liquids with specific gravities less than 0.825 at pumping temperature, together with a suction pressure exceeding 150 psig; or (4) the suction pressure exceeds 250 psig. The casing of these pumps should be made of cast, forged, carbon, or alloy steel.

Refinery pumps handle a variety of hot and cold liquids at a wide range of pressures. The unit in Fig. 13-6 is built in a range of sizes to handle up to 1,300 gpm at heads to 600 ft, temperatures to 800 F. Two-stage diffuser refinery pumps (Fig. 13-6) can develop higher heads. It is fitted with a smothering gland to flush away flammable liquids leaking past the packing. It also cools the shaft and sleeve. The shaft packing is cooled by water circulated through cavity. Mechanical seals may be used instead of packing.

Process-type pumps (Fig. 13-7) are widely used for medium-duty services in modern refineries. Capacities of these units vary, but a typical line handles 20 to 550 gpm at pressures to 300 psi, temperatures to 500 F. Most designs are fitted with smothering glands for refinery service. Either packing or a mechanical seal is used. Maintenance is

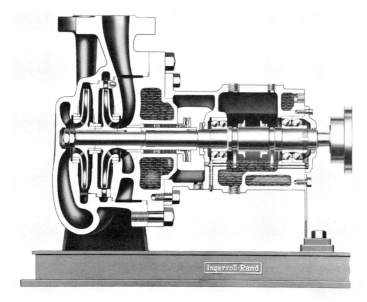

Fig. 13-7 Two-stage opposed-impeller refinery pump. (*Ingersoll-Rand Co.*)

simple. Spacer-type couplings, often used in refinery and process pumps, permit disassembly from the drive end without disturbing piping connections.

Horizontal and vertical barrel-type pumps find many uses for hot and cold refinery charge services. Because of the high pressures and temperatures, and the corrosive action of crude oils, hot-oil charge pumps present many design problems. Figure 11-5 shows a high-pressure barrel-type pump suitable for this service.

Many other types of centrifugal pumps are used in refineries, petrochemical, gasoline, and similar petroleum plants. These include close-coupled, self-priming portable, vertical turbine, regenerative turbine, chemical, submersible, sump, and sewage pumps. These are described in other chapters in this book.

A large number of rotary and reciprocating pumps are also used in refining. Rotary pumps handle crude and refined products and may be steam-jacketed when pumping extremely viscous crudes. High-pressure reciprocating pumps are used to recycle lean oil and to handle natural gasoline and many other products, depending on the process. Other reciprocating pumps feed chemicals, gum inhibitors, metal deactivators, desalt crudes, odorize natural gas, etc.

Special Pumps A number of these are used in refining. For residuum, flash-tower bottoms and black-oil recycle where hard coke of irregular size and shape is carried, a coke-crusher-type pump is used. These contain a crusher in the suction chamber located in front of the impeller eye. Any solid coke in the pumpage must pass through the crusher before entering the impeller. The pulverized coke particles are small enough to pass freely through the pump and all equipment in the system.

Alkylation units often use propeller-type circulating pumps built into the reactor or contactor. Horizontal propeller pumps are also used. Double oil-flushed mechanical seals are used for the shafts of these pumps. The usual capacity of the main pump is 10,000 gpm or higher at heads of 50 ft or less. The pump is often driven by a steam turbine through a reducing gear at 500 to 1,200 rpm. Pumps handling acids or hydrocarbons should have provisions for injecting oil into the stuffing box or mechanical seal, if there is any danger of corrosion or leakage. Use the Hydraulic Institute correction factors when choosing a pump to handle viscous liquids or hydrocarbons; see Chap. 6. In rich- and lean-oil service, two or more pumps on a single shaft (Fig. 15-5) are sometimes used. To solve packing problems today many refineries and pipelines are standardizing on mechanical seals for almost every pump (Fig. 13-10); other refineries and pipelines are making wide use of stuffing-box-less pumps (Fig. 13-11).

Water Flooding This and other similar operations—salt-water disposal, hydraulic pumping, oil-well acidizing, underground storage, cementing service, reactor charge, bottom-hole pump, liquid hydrocarbons, formation fracturing, etc.—use horizontal or vertical multicylinder

Fig. 13-8 Single-stage process-type pump. (*Deming Division, Crane Co.*)

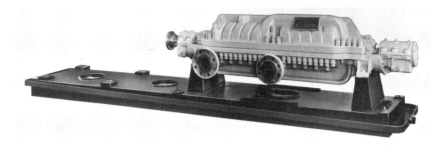

Fig. 13-9 Eight-stage high-pressure barrel pump. (*Byron-Jackson Pump Division, Borg-Warner Corp.*)

reciprocating pumps (Figs. 3-5 and 11-4). Pressures developed range up to 12,000 psi; capacities are moderate, usually being under 500 gpm. Most of these pumps are engine-driven, often being mounted on trailers for easy movement from one location to another. Alloy-steel cylinders and pistons are used. Skid mounts are provided, if desired.

Offshore Rigs These units, becoming increasingly popular, use the same drilling techniques and equipment as land-based rigs. The same is true of drilling barges. Electric-motor drives are standard for most pumps in offshore rigs and barges because motors require less space and are lighter than engines of comparable power. Circulating and bulk mud are stored aboard the rig. Fire protection is extremely important on these rigs and barges and special fire pumps are used. Production is by means of submersible or hydraulic pumps.

Portable Pumps Several applications for this type of pump are given earlier in this chapter. Others include removal of basic sludge and water (BS&W) from tank bottoms to slush pits and cleaning of these pits. Reciprocating plunger or diaphragm pumps are popular for this. Self-

(a)   (b)

Fig. 13-10 Mechanical seals for refinery service. (*a*) A single seal for moderate pressures. (*b*) A double seal for higher operating pressures. (*Byron-Jackson Pump Division, Borg-Warner Corp.*)

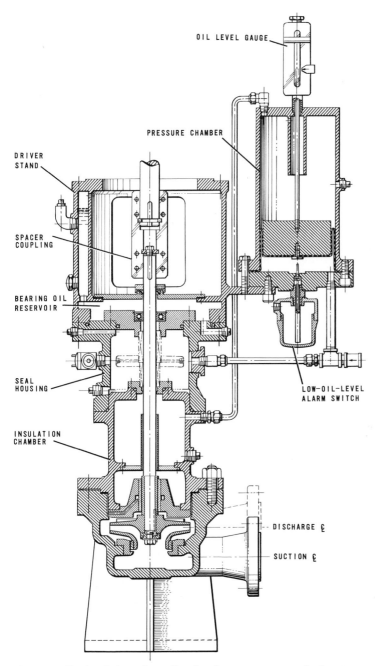

OIL LEVEL GAUGE

PRESSURE CHAMBER

DRIVER
STAND

SPACER
COUPLING

BEARING OIL
RESERVOIR

SEAL
HOUSING

LOW-OIL-LEVEL
ALARM SWITCH

INSULATION
CHAMBER

DISCHARGE ₵

SUCTION ₵

Fig. 13-11   Sectional view of a stuffing-box-less process pump for low-temperature service. (*Bingham Pump Co.*)

priming centrifugal pumps are used for transferring salt water, circulating tank bottoms, and field gathering systems. In refineries they circulate cooling water, pump out sumps, load and unload tank cars and transports, and are used in oil separation and fire fighting.

Metering and proportioning pumps (Chap. 3) are used in refining, water flooding, transportation, and many other operations. Some applications are given earlier in this chapter. For a number of other pumps used in petroleum operations, see Chap. 14.

Stuffing-box-less Pumps These handle a variety of liquids in petroleum operations, including hot oil at temperatures to 750 F and hydrocarbons at −150 F. Special pumps can be built for higher or lower temperatures. Units of the type in Fig. 13-11 are leakproof and explosionproof. They are used for a number of other services—synthetic rubber and plastics, natural gasoline, hydrogenation of edible oils (food processing), oil transfer and bulk loading, unwatering and drainage of construction pits, sumps, cofferdams, vehicular underpasses, and in marine salvage operations.

CHAPTER FOURTEEN

Chemical Industries

THE CHEMICAL INDUSTRIES are major users of all types of process pumps, particularly centrifugal pumps. Centrifugal pumps are used so extensively and for such a wide variety of services that the need for standardization of dimensions and operating characteristics has long been evident. Initially, individual manufacturers found it desirable to design standard lines of chemical-type centrifugal pumps with great interchangeability of parts.

In 1949, the Manufacturing Chemists Association requested that the American Standards Association form a committee to study the desirability and possibility of developing a standard centrifugal pump for the chemical industry. In 1955, ASA formed the B-73 sectional committee co-sponsored by the Manufacturing Chemists Association and the Hydraulic Institute. Although ultimately specifications developed by the committee were not approved as such, the manufacturing Chemists Association published them as a proposed American Voluntary Standard (AVS). Early in 1967, the Hydraulic Institute published "Tentative Standard Centrifugal Pumps for Process Use," based on AVS specifications. Now, well over a dozen manufacturers produce the so-called AVS pump.

CENTRIFUGAL PUMPS

Chemical Pumps Large numbers of vertically split casing *chemical pumps* are used today for a wide variety of routine chemical-process applications. Manufactured by a number of major pump builders, they are designed to have the wildest possible application in process service without the necessity for extensive changes in materials, packing, drive, etc. Materials used in construction of the major parts of these pumps include bronze, iron, carbon or alloy steels, glass, plastic, graphite, hard rubber, stoneware, stainless steel, and a number of other corrosion- and erosion-resistant metals and synthetics. The entire aim in the design of these pumps is complete corrosion-erosion resistance when handling acids, alkalies, and other liquids; easy installation, operation, and maintenance; and dependability.

Figure 14-1 shows two common designs of chemical pumps. Usually motor-driven, modern chemical pumps are end-suction volute types having either a semiopen or a closed impeller, heavy-duty ball bearings for the shaft, and a conventional stuffing box with provisions in some designs for water cooling the packing when the temperature of the liquid pumped exceeds about 300 F. A mechanical seal can be substituted for the packing, if desired. Some manufacturers will provide an adapter which can be bolted to the casing in the field to permit changing from packing to a mechanical seal, or vice versa. The vertically split casing permits use of a one-piece circular gasket. Though the two units in Fig. 14-1 are single-stage, similar standard designs are also available with two stages. Figure 14-2 shows a typical abrasive-resistant mechani-

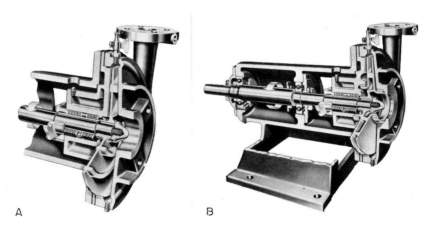

A   B

Fig. 14-1 Chemical pumps. (*A*) Semiopen impeller. (*B*) Closed impeller. (*Peerless Pump Division, FMC Corp.*)

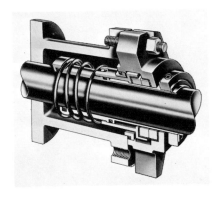

Fig. 14-2 Mechanical seal for use with abrasive liquids.

cal seal used with this type of pump. The pumps in Fig. 14-1 are built in sizes from 1 to 6 in. with capacities up to 1,200 gpm at heads to 231 ft, temperatures to 600 F.

Typical liquids handled include acids, bases, salts, acetates, hydrocarbons, chlorides, starches, oils, etc. Many pumps are cradle-mounted, others, intended for handling higher-temperature liquids, have center-line mounting. Grease lubrication is standard for liquid temperatures up to 400 F. Above this, oil is used.

AVS Pumps As previously mentioned, the American Voluntary Standard (AVS) pump has become widely available in many material combinations. Figures 14-3 and 14-4 show the operating range and standard dimensions of these AVS chemical pumps. To simplify piping connections, AVS pumps have center-line suction and discharge nozzles. Impellers can be either fully open or semiopen types. A wide range of

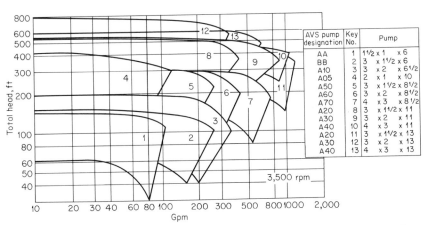

AVS pump designation	Key No.	Pump
AA	1	1½ x 1 x 6
BB	2	3 x 1½ x 6
A10	3	3 x 2 x 6½
A05	4	2 x 1 x 10
A50	5	3 x 1½ x 8½
A60	6	3 x 2 x 8½
A70	7	4 x 3 x 8½
A20	8	3 x 1½ x 11
A30	9	3 x 2 x 11
A40	10	4 x 3 x 11
A20	11	3 x 1½ x 13
A30	12	3 x 2 x 13
A40	13	4 x 3 x 13

Fig. 14-3 AVS pump operating range at 3,500-rpm covers 1- to 3-in. pumps up to about 1,000 gpm. (*Allis-Chalmers.*)

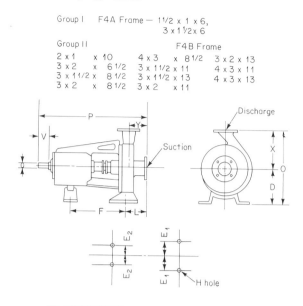

Group I   F4A Frame — 1 1/2 x 1 x 6,
                     3 x 1 1/2 x 6

Group II   F4B Frame

2 x 1   x 10   4 x 3   x 8 1/2   3 x 2 x 13
3 x 2   x 6 1/2   3 x 1 1/2 x 11   4 x 3 x 11
3 x 1 1/2 x 8 1/2   3 x 1 1/2 x 13   4 x 3 x 13
3 x 2   x 8 1/2   3 x 2   x 11

Dimensions	F4A frame	F4B frame
CP —	17 1/2	23 1/2
E_1 —	3	4 7/8
E_2 —	0	3 5/8
F —	7 1/4	12 1/2
H —	5/8	5/8
L —	4	4
U —	7/8 — 3/16 x 3/32	1 1/8 — 1/4 x 1/8
V —	2	2 5/8
Y —	4	4

Fig. 14-4 AVS pump standard dimensions cover two frame sizes for 1- to 3- in. pumps. (*Allis-Chalmers.*)

stuffing-box constructions, packed or sealed, cooled or uncooled, meet the majority of chemical industry needs.

Figure 14-5 shows a recent addition to many manufacturers' line of pumps for the chemical industry. It is a vertical in-line type where the pump and motor are completely supported by the piping system. While there are many dimensional requirements for previously discussed horizontal pumps, there is only one important dimension for vertical in-line pumps—the distance between flanges. This type of pump simplifies piping layouts and saves valuable floor space. Although they do not yet have a standard, at least one manufacturer has developed a line of vertical in-line pumps for the chemical industry which takes advantage of part-interchangability with correspondingly rated horizontal pumps.

Process Pumps This type, while resembling the chemical pump in many respects, is generally designed for somewhat higher working temperatures. The same design thinking is, however, behind them—

Fig. 14-5 Vertical in-line chemical pump is supported by the piping system. (*Crane Co.*)

simplicity, dependability, interchangeability. With some manufacturers there is a close resemblance between the design of their chemical and process pumps; with others the two types are widely divergent.

Figure 14-6 shows a typical modern process pump. The pump is supported at its horizontal center line to assure correct alignment throughout the entire operating temperature range of the unit. Packing or a mechanical seal can be used. The casing is generally made extra thick, to provide a liberal allowance for corrosion and erosion, with a high factor of safety. Figure 14-7 shows the casing-corrosion allowance used in one design of process pump.

One manufacturer estimates that with but six sets of liquid ends, suitable for mounting on a motor, turbine, or bearing frame, 111 different sizes and arrangements of his chemical and process pumps are available.

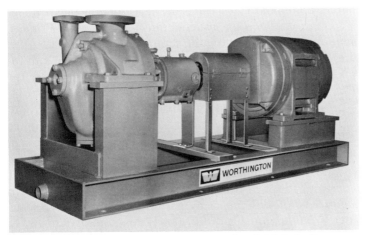

Fig. 14-6 Process pump is center-line supported to hold shaft alignment regardless of temperature. (*Worthington Corp.*)

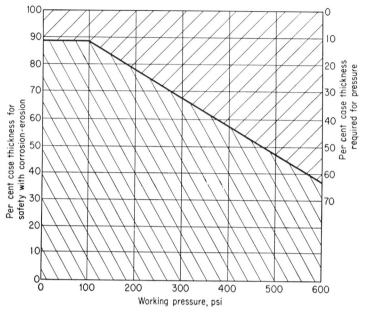

Fig. 14-7 Typical corrosion-erosion allowance in a process-pump casing.

Combining these sizes with the different materials available, and standard design modifications, there are over 60,000 available pumps. Yet all these units are built with standard parts. Some newer pumps today have special collectors fitted to their packing boxes to collect and drain away any leakage from the box. Double mechanical seals are used for extremely corrosive or high-temperature liquids.

Though there are no fixed rules, the head developed by most chemical and process pumps today is low to moderate—20 to 800 ft of liquid. Capacities range up to about 5,000 gpm, temperatures to 800 F.

High-pressure Services Horizontally split multistage volute pumps (Fig. 14-8) are popular in modern chemical and process plants for medium- and high-pressure hydraulic services. The unit shown is available for pressures of 300 to 1,000 psi at capacities to 700 gpm. Where much higher pressures are required, barrel-type pumps (Fig. 11-5) are often used. Note how the impellers of the unit in Fig. 14-8 are opposed, to balance axial thrust within the pump. Ball or sleeve bearings are used. Where only limited npsh is available, vertical single- or multistage pumps are often applied for chemical handling today. They are available in a variety of designs, with packing or mechanical seals. Capacities range to 5,000 gpm, heads to 1,000 ft, temperatures to 400 F, or more.

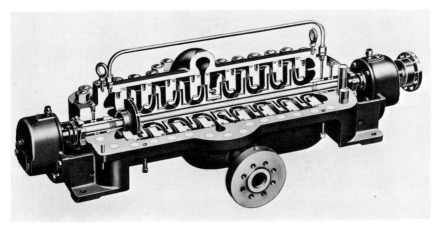

Fig. 14-8 Horizontally split multistage pump for general-process use. (*Ingersoll-Rand Co.*)

Zero-leakage Pumps Canned- and special-motor pumps of various designs (Fig. 1-17*b*) have found wide use for zero-leakage service. A number of other designs finding use in chemical plants today are discussed in Chap. 12. Many designs developed for nuclear-energy applications are also being applied in chemical plants. These include the canned-motor and electromagnetic types of pumps. Units like the one in Fig. 17*b* have capacities ranging to 300 gpm, heads to 195 ft. Two-stage canned-motor pumps are now available.

Acid and Slurry Pumps Units for these services may be individually designed by the pump manufacturer to meet a specific set of conditions. Since the service conditions for acid pumps are often severe (Fig. 14-9), careful selection and application are extremely important. Where special designs are not required, rubber-, Teflon-, or neoprene-base coverings are available for the casing, cover, and impeller. Figure 14-10 shows a typical Teflon-lined pump, Fig. 14-11 a pump for acids and slurries. All-lead, stainless-steel, solid plastic, and solid rubber horizontal and vertical pumps are also used for acid pumping. The unit in Fig. 14-11 can be fitted as shown with a variety of stuffing-box constructions —chemical-packing or mechanical-seal. The drip collector is made of chemically resistant material to protect the bearing from corrosion. The shaft is axially adjustable to take up impeller wear and restore operating conditions.

Molten-metal Pumps Besides the units discussed in Chap. 12, and earlier in this chapter, a number of other designs are available. A packingless design for handling molten metals is shown in Fig. 14-12. The pump in Fig. 14-12 has a heavy tapered shaft to obviate the need for

Fig. 14-9   Pump handling phosphoric acid. (*Worthington Corp.*)

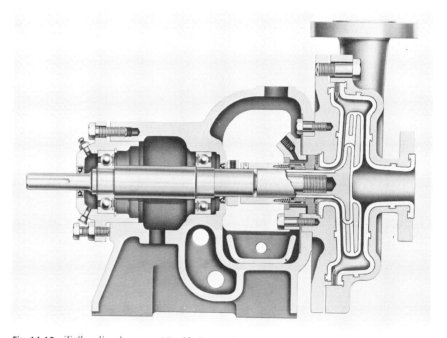

Fig. 14-10   Teflon-lined pump. (*Goulds Pumps, Inc.*)

a bearing below the coverplate. A packing box below the bearing is intended to confine only fumes or pressure, not liquid. Some vertical pumps of this general type have the liquid inlet at the top of the impeller, instead of from below, as in Fig. 14-12.

Packingless pumps are also available in horizontal designs, but the vertical type is more popular for handling metals. Some pumps for molten

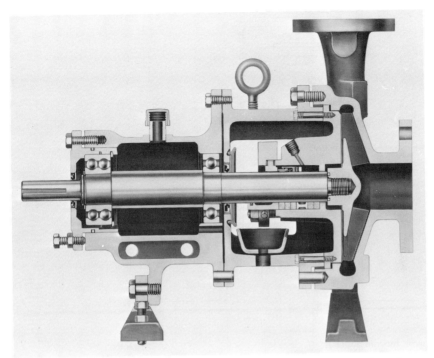

Fig. 14-11 Acid and slurry pump with two possible stuffing-box constructions shown. (*Goulds Pumps, Inc.*)

metals are arranged so that the suction well can be insulated, steam-jacketed, electrically heated, or cooled by air or water. Almost all units for this service today are vertical, single- or multistage. Their head range is often small, 5 to 200 ft, while the common capacity range is 10 to 20,000 gpm. Allowable temperatures range up to 1500 F. Pump elements must be kept at the melt temperature during pumping to prevent "freezing." Air circulation or a few rings of packing serve to contain fumes from the liquid handled. Table 14-1 lists a number of materials handled by centrifugal pumps in chemical and metallurgical services.

Agitator and Solids Pumps Horizontal and vertical axial-flow and propeller pumps are often used for agitating or mixing service, either by connection in a pipeline or by being fitted with a vertical cone through which the mixture is made to rise. Figure 14-13 shows typical agitator installations using horizontal and vertical axial-flow pumps.

Solids pumps (Fig. 14-14) are designed to handle solutions containing large percentages of suspended abrasive materials. The unit shown is suitable for solutions like gypsum, chemical pulps, zinc tailings, dolomite, bauxite, etc. It is a single-stage end-suction unit made of a special

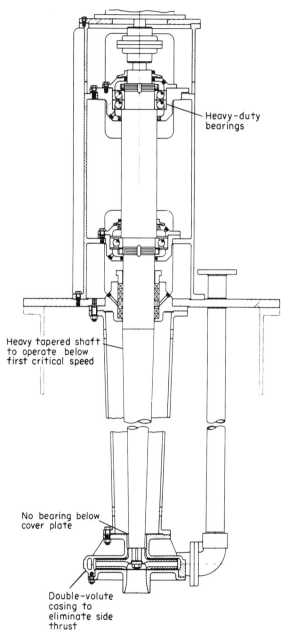

Heavy-duty
bearings

Heavy tapered shaft
to operate below
first critical speed

No bearing below
cover plate

Double-volute
casing to
eliminate side
thrust

Fig. 14-12 Pump for molten metals. (*Lawrence Pumps, Inc.*)

TABLE 14-1   Typical Operating Conditions for Molten-metal and Hot-service Pumps*

Temp, F	Specific gravity	Pumped liquid	Head, ft	Gpm	Rpm
1450	1.55	Molten magnesium	15	20	1,800
1250	1.43	Molten salt	48	50	1,800
1200	10.0	Molten bismuth	75	50	1,800
1000	1.69	Molten salt	36	240	1,500
950	1.98	Molten salt	25	600	1,200
850	1.75	Molten salt	45	17,000	720
850	1.75	Molten salt	55	9,000	900
850	1.75	Molten salt	45	4,600	1,200
850	2.14	Hot caustic	16	50	900
850	10.6	Molten lead	8.7	600	900
840	6.0	Molten alloy	19	50	1,800
800	10.6	Molten lead	65	50	1,800
800	10.6	Molten lead	60	50	1,800
800	10.0	Molten bismuth	22	50	1,200
750	10.6	Molten lead	22	85	1,200
750	10.6	Molten lead	60	50	1,800
720	6.0	Molten alloy	30	11	1,800
700	11.34	Molten lead	12	280	720
650	0.8	Clay oil	124	2,300	900
490	0.97	Dowtherm A	25	140	1,800
475	0.90 1.00	Dowtherm A	14	1,500	1,800
260	1.9	Sulfur	6	3,000	720

* Byron Jackson. Pump Division, Borg-Warner Corp.

cast alloy. A suction wear plate is held in place by bolts which extend through the casing wall. Capacities of this unit range from 175 to 10,000 gpm, heads to 140 ft.

Turbine Pumps   Single-stage end-mounted side-suction regenerative turbine pumps (Fig. 1-17a) find many applications in chemical plants for handling acids, acetates, salts, propane, butane, ethane, Freon, etc. Usual capacities range from 5 to 100 gpm at differential pressures to 250 psi. They are available in a wide variety of construction materials for different liquids. Either packing or a mechanical seal is used to prevent shaft leakage in these pumps.

ROTARY PUMPS

For small to moderate capacities up to 250 gpm and pressures to 600 psi, single-screw pumps (Fig. 14-15) are popular for a wide variety of caus-

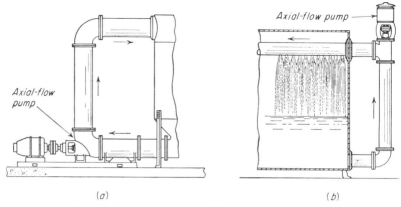

(a)   (b)

Fig. 14-13 (a) Typical agitator using an axial-flow pump. (b) Bottle-washer agitator or pasteurizer.

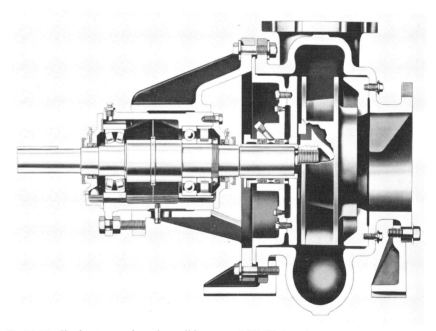

Fig. 14-14   Single-stage end-suction solids pump. (*Allis-Chalmers.*)

tics, acids, dyes, solvents, soaps, latex, resins, etc.   Two- and three-screw pumps also find some use in this service.   The single-screw design is preferred where semisolids are handled because it does not break them up.   When special materials and construction are used for this pump it can develop pressures up to 1,000 psi.

Fig. 14-15 Single-screw pump for chemical applications. (*Robbins &
Myers, Inc.*)

Gearless neoprene-impeller pumps (Fig. 14-16) are popular for lower
pressures and capacities. The maximum head developed is about 35
psi; capacity ranges up to 100 gpm at speeds up to 1,750 rpm.

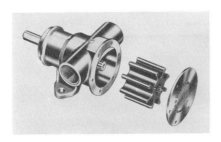

Fig. 14-16 Gearless neoprene-
impeller pump. (*Jabsco Pumps.*)

Other types of rotary pumps used in the chemical field include the
internal-gear, lobe, swinging-vane, flexible-tube, and roller-vane units.
See Chap. 2 for more complete data on rotary pumps.

RECIPROCATING PUMPS

Controlled-volume, metering, and proportioning pumps (Figs. 3-9 and
3-7 find use in chemical and metallurgical processes where small quantities of liquid are to be delivered. For larger flows, diaphragm pumps of
various designs are popular. Figures 3-14 through 3-16 show several
typical units.

Compressed-air-operated diaphragm pumps (Fig. 14-17) find wide
use in chemical and metallurgical plants for handling slurries, liquors,
acids, crystalline products, etc. Slow-speed units, operating at 15 to 30
strokes, they are built in capacities from 5 to 60 gpm at pressures to 100
psi. A timer I, solenoid valve O, and a diaphragm-and-spring-operated

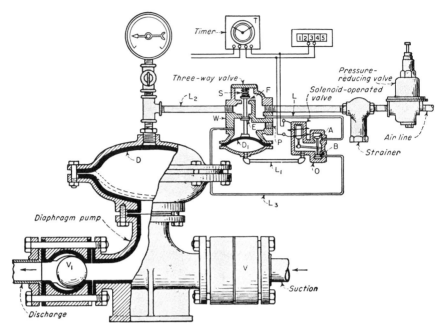

Fig. 14-17 Air-operated timer-controlled diaphragm pump runs at 15 to 30 strokes per minute for capacities of 5 to 60 gpm, pressure to 100 psi.

three-way valve W form the control for actuating diaphragm D. The air pressure is held at 5 to 10 psi above that required to operate the pump. Single-acting, this pump has an intermittent flow, which can be smoothed by an air chamber on the discharge. From 3- to 10-ft npsh is needed to fill the pump, depending on pipe size and liquid viscosity.

The diaphragm pump in Fig. 14-18 develops 300 psi when handling $1/8$ to $1/2$ gpm. The liquid pumped is on the left side of diaphragm D, while oil or hydraulic fluid which transmits the motion of piston P to the diaphragm is on the right. As a comparison, the pump in Fig. 3-16 handles 5 gpm at 80 psi at 200 rpm; 10 gpm at 430 rpm. At speeds less than 200 rpm, the pump can be operated at 125 psi.

Larger reciprocating pumps are often power-type units made of special alloys, stoneware, hard rubber, etc. Some metal pumps are fitted with porcelain acid-resisting cylinders and plungers. Figure 14-19 shows a typical high-pressure triplex power pump for use in chemical and process applications. Typical applications include handling of acids, paints, abrasives, etc.

Pump Selection *Capacity:* Determine the required capacity from the process conditions. Be certain that the exact capacity, temperature, and viscosity are known. *Head:* Since many chemical and process pumps

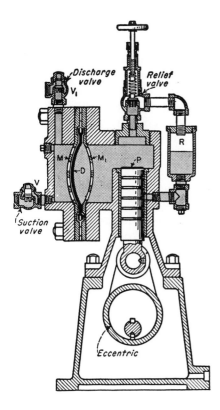

Fig. 14-18 High-pressure pump operates with two fluids: pumped and pumping.

Fig. 14-19 High-pressure triplex power pump. (*Kobe, Inc.*)

have a rather narrow head range, it is extremely important that the required head be accurately determined. See Chap. 4. *Materials:* Many chemical liquids are corrosive, abrasive, or highly alkaline. So the construction materials chosen are of utmost importance. See Chap. 6 and the present one for details. *Drives:* Electric motors are the most common drive, with steam turbines second. Choose the drive only after thorough investigation of the atmospheric conditions surrounding the pump. Do not overlook the fact that the liquid handled may be hazardous or that the pump may operate for long periods at or near its shutoff point. *NPSH:* Some chemical pumps require extremely high npsh values. So be sure to know the exact value required for the pump chosen and arrange the piping to ensure having a higher available npsh at all times. *Control:* Throttling of the pump discharge is perhaps the most common control method for chemical and process pumps because they are usually designed to run at constant speed. *Packing:* Stuffing boxes should be of generous depth, with provision for sealing the packing of hot-liquid pumps by means of cold water or another suitable liquid. Use a single or double mechanical seal for the shaft when positive leak prevention is required. Smothering-type glands and water or steam jackets may also be required. See Chap. 9. *Number of Pumps:* Few standby pumps are used in the usual chemical-process application because the availability of modern units is so high. Where pump failure would cause an expensive outage, use a standby pump. *Steam Jackets:* Many types of pumps for chemical use can be supplied with an integral steam jacket, if desired. Jackets are used with viscous liquids and when it is necessary to keep the liquid pumped at a certain temperature level.

Other Types   The chemical and process industries use large numbers of pumps for water supply, power generation, waste disposal, fire protection, refrigeration, etc. These pumps are discussed in other chapters in this book. Steam- or air-jet pumps, operating on the principle of an ejector, find some use where another type of pump is unsuitable or where a mixing and pumping action is desired.

CHAPTER FIFTEEN

Paper, Textiles, and Rubber

PULP AND PAPER MILLS use centrifugal pumps almost exclusively today for process operations. A few reciprocating pumps are applied for chemical-feed service and some rotary pumps for oil-hydraulic uses. However, their number is small compared with centrifugal pumps.

Pump Types Since centrifugal pumps in pulp and paper mills handle a wide variety of liquids—from clear water to acids, caustics, liquors, stock, and white water—there are a number of units built for specific jobs in these mills. Some resemble or are identical to the *chemical pumps* described in Chap. 14. They are end-suction single-stage bracket-mounted units with either vertically or diagonally split casings. Units handling stock with a consistency greater than 1 per cent by weight are usually called paper-stock pumps and have special impellers (Fig. 1-27E). Figures 15-1 and 15-2 show two views of paper-stock pumps. They are designed to handle sulfite, soda, kraft, and waste stock up to 6.5 per cent consistency. In special cases they can be used for higher consistencies. A steep head-capacity characteristic ensures approximately constant flow, regardless of change in suction head. Two other paper-stock pumps of modern design are shown in Chap. 6. Vertical pumps are often used to save floor space.

Other features of modern stock pumps include a minimum of casing

Fig. **15-1** End-suction paper-stock pump. (*Worthington Corp.*)

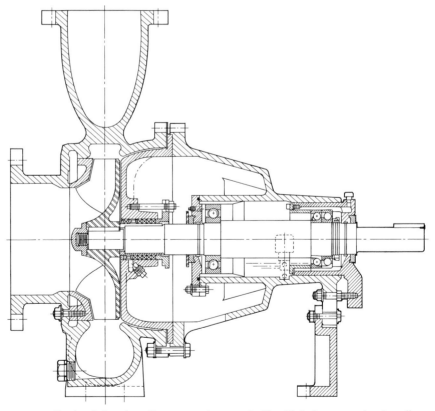

Fig. 15-2 Sectional drawing of paper-stock pump in Fig. 15-1 shows nonclog impeller. (*Worthington Corp.*)

joints to reduce the possibility of leakage, renewable metal liners for both sides of the impeller, a variety of impeller designs, use of a booster impeller to ensure good suction flow, and rubber lining of the casing and impeller to give longer life to the pump parts contacting the liquid handled.

Where the liquid handled is clear water, or other noncorrosive nonabrasive neutral liquids, standard general-purpose pumps, as described in Chap. 17, find wide use. Figure 15-3 shows a typical installation.

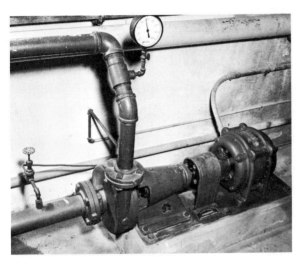

Fig. 15-3 Process-type general-purpose pump handling 100 gpm of make-up water in a paper mill.

Single-stage horizontal elbow-type propeller pumps are used in mills where large-volume low-head conditions prevail, as in the ammonium sulfate crystal process, and others. The pump serves in place of an elbow, reducing piping costs.

Screw feeder pumps (Fig. 15-4) are designed to introduce high-consistency pulp into the impeller eye of pulp pumps. They force pulp through the pulp-pump suction pipe, eliminating air binding or clogging. When used with the screw feeder pump, the standard pulp pump will handle stock to 8 per cent consistency, while special-duty pulp pumps will handle stocks of 8 to 10 per cent consistency. Vane-type feeders are also used for this service with stocks up to 8 per cent consistency.

For barking or grinder service, where relatively clear cool water is handled at medium to high pressures and capacities, standard horizontally split multistage volute-type pumps (Fig. 1-15) find use. For other

TABLE 15-1 Process Pumps for Paper-mill Service*

Service	Liquid handled	Type pump	Type impeller	How fitted	Remarks
Groundwood-pulp Process					
Grinder pressure supply	Water	1, 2	Closed	B	Flat head-capacity characteristics
Grinder shower	White water	1, 2	Closed	B	
Bull screen to flat screen	½–1% stock	3	Closed-open	C	Pump must handle min sphere size of 3 in.
Flat screen to decker	½–1% stock	3	Closed-open	C	
Flat-screen shower	White water	1, 2	Closed	B	
Decker chest to beaters	3–5% stock †	3	Closed-open	C	
Decker shower	Fresh water or white water	1, 2	Closed	B	
Decker chest to wet machine	1–3% stock	3	Closed-open	C	
Sulfite-pulp Process (Acid)					
Acid to digester	Acid	4, 5	Closed	S	
Digester circulating	Acid	5	Open-closed	S	Water-cooled stuffing box
Blow pit to knotters	1–3% stock	3	Closed-open	A	
Blow-pit shower	Water	1	Closed	B	
Knotter shower	White water	1, 2	Closed	B	
Knotter to fine screen	1–2% stock	3	Closed-open	C	
Fine-screen showers	Fresh water or white water	1, 2	Closed	B	
Screened stock to deckers	1–2% stock	3	Closed-open	C	
Decker showers	White water	1, 2	Closed	B	
Decker chest to beaters	3–5% stock †	3	Closed-open	C	
Decker stock to wet machine	2–3% stock	3	Closed-open	C	
Tailings from fine screens	1–3% stock	3	Closed-open	C	
Bleach liquor to bleachers	Caustic	3, 1, 5, 4	Closed	C	
Decker stock to bleachers	4–5% stock †	3	Closed-open	C	
Bleach stock to washers	2–4% stock	3	Closed-open	C	
Chlorinated stock to washers	2–3% stock	6	Open	R	
Sulfate-stock Process					
White liquor to digester	Caustic	4, 5	Closed	S	
Digester circulating	Caustic	5	Open-closed	S	Water-cooled stuffing box and sleeve
Diffuser stock to knotters	1–4% stock	3	Closed-open	C	Pump must handle min sphere of 3-in. size
Knotter showers	White water	1, 2	Closed	B	
Stock to fine screens	1–2% stock	3	Closed-open	C	
Fine-screen showers	White water	1, 2	Closed	B	
Screened stock to deckers	1–2% stock	3	Closed-open	C	
Decker showers	White water	1, 2	Closed	B	
Decker stock to beaters	3–5% stock †	3	Closed-open	C	
Decker stock to wet machine	2–3% stock	3	Closed-open	C	
Black liquor to storage	Black liquor	3, 4	Closed	C }	12 Bé (316 stainless steel if hot)
Black liquor to evaporators	Black liquor	3, 4	Closed	C }	
Concentrated black liquor	Black liquor	3, 4	Open	S	Water-cooled stuffing box for temp above 200 F
Green liquor	Green liquor	3, 4	Open	S	Water-cooled stuffing box for temp above 200 F
Paper Mill					
Beater stock to jordans	2–4% stock	3	Closed-open	C	
Jordan stock to Vor-trap	1–2% stock	3	Closed-open	C	
Machine stock to head box	3–4% stock	3	Open	B	Very steep head-capacity curve
Fan pump	0.4–0.9% stock	3	Closed-open	C	For capacities 2,000 gpm or less
Fan pump	0.4–0.9% stock	3, 1	Closed	C	For capacities over 2,000 gpm
Machine showers	White water	1, 2	Closed	B	
White-water removal	White water	1, 5	Closed	B	

1. Single-stage double-suction horizontally split volute pump; capacities 30 to 250,000 gpm or more; heads to 475 ft.

2. Single-stage end-suction vertically split volute-type close-coupled pump; capacities to 2,500 gpm; heads to 550 ft.

3. Single-stage end-suction vertically split warped-vane paper-stock pump, or vertical single-stage side-suction side-discharge paper-stock pump; capacities 175 to 10,000 gpm; heads to 270 ft.

Fig. 15-4 Screw feeder pump for delivering high-consistency stock to the stock pump. (*Byron Jackson Pump Division, Borg-Warner Corp.*)

4. Single-stage side-suction vertically split volute-type pedestal-mounted pump, or single-stage side-suction vertically split double-volute pedestal-mounted pump; capacities 20 to 1,300 gpm; heads to 260 ft.

5. Single-stage end-suction vertically split pedestal-mounted pump; capacities to 3,500 gpm; heads to 550 ft.

6. Single-stage end-suction pedestal-mounted volute-type rubber-lined pump; capacities to 2,400 gpm; heads to 100 ft.

A, all-bronze pump; B, bronze-fitted pump; C, all-iron pump; S, stainless-steel pump; R, rubber-lined pump.

* Adapted from data published by Allis-Chambers Mfg. Co.

† Consistencies as high as 8% may be handled if a standard pump is used in conjunction with a stock feeder.

services like boiler feed, general process, sewage, water supply, and fire protection the pumps described under these headings in other parts of this book are used. Many metering and proportioning pumps are employed in paper and pulp mills for pH control of stock, metering colors, resin addition, emulsifying rosin size, etc. See Chap. 3 for details of these pumps. Single-screw rotary pumps (Fig. 14-15) are often used to deliver coloring materials in paper mills.

Pump Selection   The choice of pumps for pulp and paper mills varies considerably from one process to the next and, to a lesser extent, from one engineer to another. Table 15-1 summarizes the recommendations of one pump manufacturer. It gives the service, liquid handled, pump and impeller type, construction materials, and other pertinent data for 45 different uses in paper and pulp mills. It can be safely used for most mills—where unusual conditions prevail it is best to consult the manufacturer. Other factors in selection are discussed below.

Drive: Motor drive is by far the most common but steam turbines find some use, especially where exhaust steam is available or usable in a process operation. *NPSH:* Provide as much npsh as possible, using liberally sized suction piping with easy bends and the minimum number of fittings possible. *Control:* Many pumps in this service are driven at constant speed with throttling control.

Packing: Either standard packing or mechanical seals are used. Mechanical seals are coming into greater favor for many services because they eliminate repacking. Where standard packing is used, provide an outside seal connection to keep stock out of the packing. *Number of Pumps:* Generally one pump per service is used, without a standby. On certain critical applications, a standby pump may be used but this practice is finding less favor in newer mills. Multiple-pump arrangements with four or more pumps for different services driven by a single motor are often used (Fig. 15-5). *Piping:* Wood-stave and rubberlined piping is used in some mills for lines 6 in. in size and larger handling water, white water, pulp stock, etc. Steel pipe, however, is also common. When computing friction losses, be sure to use the correct data. Table 15-2 gives friction losses for typical wood-stave piping. See Chap. 6 for steel, spiral-welded, and cement-asbestos pipe.

Lubrication: Vertical stock pumps are often fitted with water-lubricated lower bearings. A separate water line, fitted with a solenoid valve interconnected with the motor circuit and arranged to open when the pump starts, is used. A sight indicator on the water flow line should also be used. *Materials:* See Chap. 6 and Table 15-1. *Couplings:* Spacer-type couplings, which permit easy removal of the rotating cement without disturbing the pump piping, are incorporated in many designs of stock pumps. To minimize alignment difficulties, universal-joint spacer-type

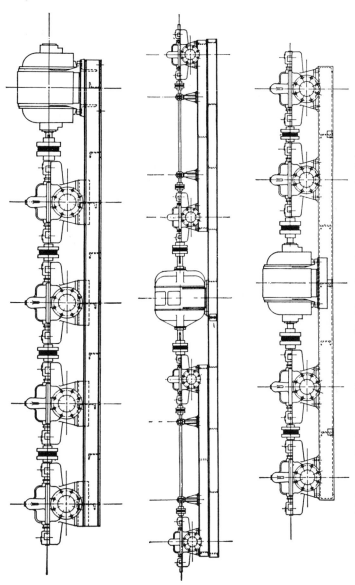

Fig. 15-5 Multiple-pump hookups used in paper mills. More than four pumps can be driven by a single motor. Belt drive also used.

TABLE 15-2   Pipe-friction Loss for Water

(Wood-stave pipe in good condition)

Diam, in.	Flow, gpm	Velocity, ft per sec	Velocity head, ft of water	Friction loss, ft of water per 100 ft of pipe
2	20	2.01	0.10	1.2
2	25	2.51	0.15	1.8
2	30	3.08	0.22	2.6
2	35	3.64	0.31	3.5
2	40	4.18	0.41	4.5
4	100	2.52	0.15	0.80
4	145	3.70	0.32	1.60
4	200	5.04	0.59	2.80
4	240	6.15	0.88	4.00
4	275	6.96	1.10	5.00
6	400	4.47	0.47	1.40
6	500	5.74	0.77	2.20
6	600	6.82	1.10	3.00
6	700	8.00	1.50	4.00
6	800	9.05	1.90	5.00
8	1,200	7.59	1.3	2.60
8	1,300	8.22	1.6	3.00
8	1,400	8.95	1.9	3.50
8	1,500	9.64	2.2	4.00
8	1,600	10.30	2.5	4.50
10	1,400	5.71	0.76	1.20
10	1,600	6.70	1.10	1.60
10	1,800	7.16	1.20	1.80
10	2,000	8.00	1.50	2.20
10	2,200	8.78	1.80	2.60
12	1,000	2.98	0.21	0.30
12	1,500	4.38	0.45	0.60
12	2,000	5.82	0.79	1.00
12	3,000	8.54	1.70	2.00
12	3,600	10.30	2.50	2.80

couplings are supplied with some designs of direct-driven stock pumps. *Caution:* Note that allowances must be made for reduction in pump head, capacity, and efficiency when handling stock. See Chap. 6. Use a large straight suction line. *Capacity Conversions:* To convert from tons of paper stock per 24 hr to gpm, use Table 15-3. Enter at the left at the capacity in tons and project to the right to the proper consistency column. Here read the equivalent flow in gpm. For example, 40 tons per 24 hr of 3.5 per cent air-dry-consistency stock is equivalent to 190 gpm.

TABLE 15-3   Paper-stock Capacity*

Tons per 24 hr	Air-dry consistency †							
	0.5	1.0	1.5	2.0	2.5	3.0	3.5	4.0
10	333	167	111	83.5	66.7	55.5	47.6	41.7
20	667	333	222	167	133	111	95.2	83.5
30	1,000	500	333	250	200	167	143	125
40	1,330	667	444	333	267	222	190	167
50	1,667	833	555	416	333	278	238	208
60	2,000	1,000	667	500	400	333	286	250
80	2,660	1,330	889	667	533	444	381	333
100	3,330	1,667	1,111	833	667	555	476	416
150	5,000	2,500	1,667	1,250	1,000	833	714	625
200	6,660	3,330	2,222	1,667	1,333	1,111	952	833
250	8,330	4,170	2,771	2,080	1,667	1,390	1,190	1,040

* Buffalo Pumps, Inc.
† Air-dry consistency is 10% greater than bone-dry consistency: 1% bone-dry stock = 1.1% air-dry stock.

$$\text{Gpm} = \frac{\text{tons per 24 hr}}{\text{air-dry consistency (as decim l)} \times 6}$$

TEXTILE MANUFACTURE

Pumps in textile mills handle a variety of common liquids. These include water, dyes, carbon bisulfide, acids, caustic soda, acetates, solvents, soda ash, alcohol, bleaches, hydrogen peroxide, salts, sizing, and butane. Some processes, for example, production of synthetic fibers, require extremely careful control of pumping conditions. In this particular application, traces of foreign matter might alter the fiber color. So complete freedom from corrosion or erosion products is essential. Choice of high-quality alloys for pump construction can therefore be justified in this service. All classes of pumps are used in modern textile mills.

Many metering and proportioning pumps are used in textile applications for handling bleach solutions, pH control of rayon wash water, color control in dyeing, wool carbonizing, etc. See Chap. 3 for details of various designs of this type.

Pump Applications Centrifugal pumps are widely applied in the processing and manufacturing departments of all types of textile mills. Where water and other clear cool noncorrosive liquids are handled, general-purpose pumps of various types (Chap. 17) are used. Process-type pumps (Chap. 13) are also used. For acids and bases, including such materials at hot size, chemical-type pumps (Chap. 14) are used. Some designs have the stuffing box under suction pressure and use a semiopen keyed impeller similar to that in Fig. 1-27B. For pumps in other than the processing section of the textile mill, see the various chapters in this book dealing with them. For process pumps, use the data given below. Some dye liquor pumps have an attached four-way valve, permitting flow reversal through the kier without stopping the pump. For handling latex, pumps having special provisions for quick disassembly are often used to allow easy cleaning. These units are similar or identical to sanitary-type centrifugal pumps (Chap. 16). Figure 15-6 shows a typical pump designed to handle latex.

Fig. 15-6 Available in a number of materials, this pump is designed to handle latex and many other liquids. (*Tri-Clover Division, Ladish Co.*)

Head and Capacity: Low to medium heads are used in textile mills, the required pressure seldom exceeding 200 psi. Capacities may be medium to large, depending on the application. *Materials:* Zeolite water softeners are popular in textile mills, giving a soft water which is easily handled in bronze-fitted pumps. For other liquids, see Chap. 6. *Drives:* Electric motors are most common, with some steam turbines being used in the larger plants. Small canned-motor pumps are growing in popularity in the textile field. They may someday find extensive use as kier-liquor-circulating pumps. *NPSH:* Provide sufficient npsh, especially with hot size and similar liquids. *Control:* Throttling of the discharge is a common type of pump control in this service. *Packing:* Teflon packing and mechanical seals are popular, depending on the liquid handled. *Number of Pumps:* Usually one per service, with relatively few standbys being used.

RUBBER PROCESSING AND MANUFACTURE

Pumps in the rubber industry handle a number of different liquids, including solvents, softener oils, caustic soda, pigments, liquors, latex, acids, butadiene, styrene, brine, catalysts, soap solution, modifiers, etc. Centrifugal pumps find many uses where small to large flows at low to high pressures are required. Reciprocating metering and proportioning pumps are applied for small flows at high pressures. Typical uses include metering fillers, lubricants, and other compounding ingredients; handling gels and oxides; controlling the flow of butadiene and styrene; etc. In the production of rubber accelerators a duplex adjustable-stroke reciprocating pump operating at 1,000 psi is used. It is fitted with a variable-speed drive. Rotary pumps also find use for a number of services.

Pumps for rubber processing and manufacture are chosen in much the same manner as for paper, textile, and chemical plants. See the pointers given in this and other chapters in this book. Metering and proportioning pumps handle butadiene, styrene, fluoride gels, zinc oxide, fillers, lubricants, and compounding ingredients in synthetic and natural-rubber production. Single-screw rotary pumps are used to handle rubber cement solvents.

CHAPTER SIXTEEN

Food Processing and Handling

SANITARY PUMPS are specifically designed to handle foods. They often have a number of special features not necessary in other types of service. To be successful in food applications a pump (1) must be highly corrosion-resistant, (2) must be easily taken apart for cleaning, (3) must not churn the food or cause it to foam, (4) must have an absolutely tight lubrication system, (5) must be free of any wear or rubbing of internal parts during operation, (6) must have packing that is positively sealed from the casing interior, and (7) must have internal passages of the casing smooth and free of sharp corners and other abrupt breaks in the surface.

A large number of centrifugal, rotary, and reciprocating sanitary pumps which successfully meet these requirements are available today. Many are fitted with sanitary connections on the suction and discharge nozzles so the impeller and casing can be easily removed for cleaning. The piping and fittings used with sanitary pumps are usually stainless steel, nickel alloys, hard rubber, glass, or plastic. The pump itself is made of stainless steel, monel, aluminum, iron, glass, porcelain, or some special alloy. Besides resisting the food, the construction materials must withstand the detergents, soaps, and germicidal agents used in cleaning and flushing the pump. In many food processes the pump must be opened and cleaned at least once a day.

Centrifugal Pumps Because multibladed impellers damage many fragile foods, bladeless impellers (Fig. 7-27) and single- and two-bladed screw impellers are often used in centrifugal pumps designed to handle a variety of foods.   Apples, oranges, corn, strawberries, lima beans, oysters, shrimp, eggs, olives, brussels sprouts, fruit juices, whole fish, and a variety of other foods are safely handled by bladeless impellers.

Process-type centrifugal pumps (Fig. 16-1 and Chap. 14) are available for many foods having a low solids content—cane wash, grape juice, tomato juice, tomato purée, sugar liquor, vegetable oil, etc.   They are built in capacities up to 2,000 gpm at heads to 200 ft by one manufacturer.   A single pump design covers this range, making selection simpler.   Most large manufacturers have a design of this type.   Chemical-type centrifugal pumps (Fig. 14-1) are also used for foods having a low solids content.   However, their use is generally confined to corrosive foods.

Figures 16-2 and 16-3 show a sanitary pump fitted with an open impeller designed to eliminate crevices and pockets.   Units of this design are furnished with a single inert-carbon balanced seal in close-coupled construction, a packing gland, a water-cooled double-rotary seal, or a single inert-carbon balanced seal.   They are built in capacities to 1,200 gpm at heads to 320 ft and handle milk, tomatoes, soups, etc.   The

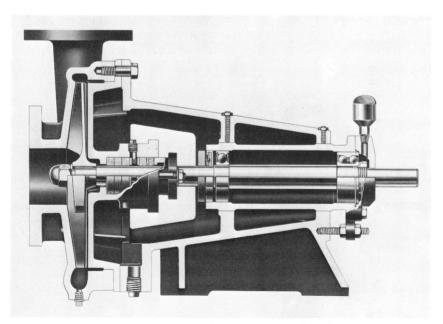

Fig. 16-1   Process-type centrifugal pump handles a variety of foods. (*De Laval Turbines, Inc.*)

Fig. 16-2 Typical close-coupled sanitary centrifugal pump. (*Tri-Clover Division, Ladish Co.*)

pumping head and all its parts are made of type-316 stainless steel.   Recommended drives include electric motors, turbines, and internal-combustion engines, either direct or belt-connected.

Other popular pump designs for food handling are shown in Figs. 6-12, 6-13, 15-1, and 15-2.   These units are used for mashes, pulps, solids, etc.   Portable centrifugal pumps also find some use in the food industries.   They must, of course, be of sanitary construction if they are

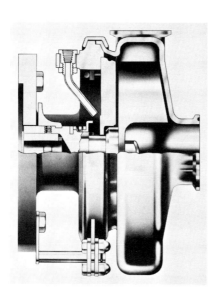

Fig. 16-3 Cutaway head of the pump in Fig. 16-2. (*Tri-Clover Division, Ladish Co.*)

used in the processing or manufacturing portions of the plant. Non-clogging open-impeller end-suction process-type pumps find many applications handling beet-pulp production. In sugar mills, vertical mixed-flow medium-head pumps are the principal type used for carbonation circulation. Large flows at moderate heads are required in this service. Stainless-steel impellers and welded-steel discharge columns are used. Milk-of-lime or kieselguhr pumps are generally single-stage single-suction open-impeller units designed to handle abrasives. The casing is bolted together for easy removal. It has a hardness of 600 brinell. Other process units for sugar mills include standard-liquor, pan-storage, saccharate-milk, kelly-excess, thin-juice, filter-feed, and high-melter pumps. All are generally single-stage process-type pumps.

Breweries and Distilleries Some breweries use sanitary-type pumps in all services in which the product is handled. Others employ conventional end-suction single-stage all-bronze or bronze-fitted units for mash, wort, beer, spent-grain, and filter-mass pumps. The yeast pump is often a rotary unit of sanitary construction. Spent-grain and filter-mass pumps should have a suction head of 10 to 12 ft of water. The filter mass usually has a consistency equal to about 5 per cent paper stock. Portable centrifugal pumps are sometimes used to transfer wort from settling to fermenting tanks and beer from fermenting to aging tanks. The suction and discharge lines are usually hoses. Cotton or flax packing, tallow-impregnated, is common in brewery service. Open impellers are used in wort, mash, yeast, and some beer pumps. Total head for these is usually 100 ft or more.

A large number of pumps are used in distilleries. End-suction single-stage all-bronze-fitted process-type pumps are common. The capacity required usually varies from about 15 to 150 gpm, heads from 25 to 100 ft. Typical units used are gin-filter, whisky-filter, treated-gin-and-spirits, carbon-treating, cistern-room, rye-whisky, spirits heads-and-tails, high-wine, thin-stillage, thick-stillage or whole-slop, rerun heads-and-tails, rerun-spirits, continuous-tails, badger-spirits, and badger-tails pumps. Where conventional centrifugal pumps are used in breweries and distilleries, the piping system is often cleaned by flushing with a 10 per cent solution of hot caustic soda. This is followed by a water wash, leaving the pumps and piping sterile.

Reciprocating Pumps The flexible-tube pump (Fig. 16-4) has a series of fingers A through L which compress the tube in sequence from left to right. This produces suction and discharge. At the instant shown in Fig. 16-4, finger A has closed the tube and trapped a quantity of liquid in it. Fingers J, K, and L are lifting to release the liquid, while the others compress the tube to force the liquid out. Discharge pressures up to 25 psi are practical at capacities from 0.1 to 0.5 gpm at a speed of 475 rpm.

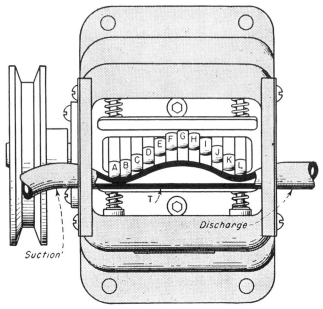

Fig. 16-4 Flexible rubber-tube pumping element is squeezed by fingers.

The tube is made of gum rubber, neoprene, or tygon. Its inside diameter ranges from $3/16$ to $1/2$ in.

Another flexible-tube pump is shown in Fig. 16-5. It has a loop of tubing which is compressed when the rollers surrounding the shaft are rotated, trapping liquid ahead of them and forcing it out the discharge.

The bellows pump in Fig. 16-6 is designed to handle 0.12 to 9 gpm at pressures up to 30 psi. By limiting the expansion and contraction of the bellows B, the capacity can be adjusted from zero to the rated value. The liquid being pumped passes through the stainless-steel suction and discharge assembly and the bellows B and spring expanders S, made of the same material. Cam C, acting on lever D, causes the bellows to expand and contract when the shaft to which C is attached rotates. Capacity is adjusted by raising or lowering sleeve F with nut N. Units like this can supply color or flavor extracts in canning, bottling, distilling, etc.

High-pressure horizontal and vertical reciprocating pumps (Fig. 3-4) are used to supply liquids for hog washing, cleaning of carcasses, and sanitary cleaning of the plant and its equipment. Centrifugal pumps are also used for hog washing and other sanitary services. Metering and proportioning pumps of various types (Figs. 3-6 and 3-7) are applied in breweries and distilleries for many jobs, including feeding of chill-

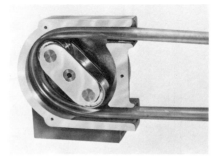

Fig. 16-5 Flexible-tube pump has rollers to compress tube. (*Little Giant Pump Co.*)

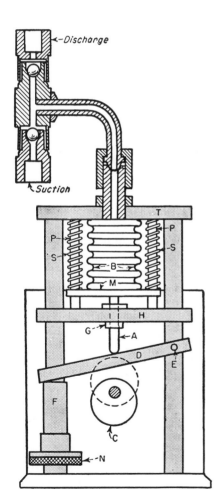

Fig. 16-6 Cam-actuated pump
having a stainless-steel bellows.

Fig. 16-7   Rotary sanitary pumps having solid crescent-shaped flexible rotors. (*Waukesha Foundry Co.*)

proofing compounds, handling filter-aid beer slurry, pumping acids or ammonia to the beer to adjust the pH, metering acid to wort prior to fermentation, etc.

Rotary Pumps   Several different types of rotary pumps are built for sanitary service. They are used for handling a variety of foods. Types available in sanitary constructions include internal-gear (Fig. 2-3), single-screw (Fig. 2-7), swinging-vane (Fig. 2-10) pumps, and a number of others. Some of these are gearmotor-driven at reduced speed in certain services to prevent agitation and churning of the product handled. See Chap. 2 for a discussion of the various types of rotary pumps. The single-screw pump (Fig. 14-15) is popular for a multitude of services including mustard, potato salad, chow mein, pastes, etc. It can be made of stainless steel in capacities to 250 gpm at pressures to 600 psi.

Figure 16-7 shows two types of rotary pumps, one fitted with crescent-shaped rotors and the other with flexible rotors. These units are built specifically for sanitary service and are widely used to handle products like soups, fricassees, and strawberries in the manufacture of jams, jellies, pie fillings, baby foods, apple sauce, candy, cookie batter, custards, and many others.

Both designs are positive-displacement units. They are self-priming and deliver a constant volume unaffected by speed or pressure. Available in nickel alloys, bronze, stainless steel, hastelloy, and other materials, these pumps may be driven directly, or through gears, belts, etc. They are mounted on regular or sanitary bases which are designed to permit easy, effective cleaning.

The pockets formed by the impellers during rotation are large enough to permit pumping of solids without damage or waste. The units in Fig. 16-7 can be fitted with O-ring rotary seals. Every part in contact with the food is designed for quick easy washing and sterilizing. All parts are smooth and flush, with no difficult-to-clean areas. The materials of construction do not taint or change the flavor of the food. Figure 16-8 shows a special food-handling pump that not only transfers the product (as a pump) but also, through shear action, serves to mix, emulsify, blend, or whip the product—all without aeration or with controlled aeration.

Pump Selection   *Capacity:* The required capacity of the pump depends on the tonnage of food being handled. Use caution when converting the process capacity in tons to pump capacity in gpm. In general, choose the pump to operate at its highest efficiency at normal flow, or at a point just to the right of this on the *HQ* curve. With solid foods, the ratio of water to solids varies. For best pumping conditions, maintain the highest ratio possible—keep velocities high enough to prevent settling yet low enough to minimize product damage. Typical ratios: string beans, 3 gal per lb; peas, 1 gal per lb. *Head:* The head to be developed by sanitary pumps seldom exceeds 100 ft. Use large pipes and easy bends to keep friction to a minimum. This reduces the total head in the system, permitting easier flow with less foaming, agitation, and damage of the food. Choose pump speed to give the desired system head.

Materials: Many factors other than simple corrosion must be considered. Pump parts contacting the food are the casing, impeller, shaft, packing, pistons, rods, or gears, depending on the class of pump used. Stainless steel is suitable for almost every food, except those containing salt brines. Many alloys containing copper, like monel metal, are suitable for salt brines but not for foods like peas, corn, or lima beans because with these the copper may discolor the skin. Do not use bronze pumps for salt brines intended for food canning because there is a chance of discoloring the food. Some acids, and most alkalies, cause aluminum to corrode. A number of states have regulations regarding sanitary pumps. Adherence is mandatory and the responsibility for

Fig. 16-8   Shear pump transfers food products and also performs processing operations at the same time. (*Waukesha Foundry Co.*)

determining and adhering to the regulations rests with the pump user, not the manufacturer. So be sure to choose pump materials to prevent discoloring and chemical, bacterial, or taste contamination or tainting.

Drives: All types used, but electric motors are most common. *NPSH:* Be sure to determine the required npsh for the pump chosen and install the pump so the required value is always available. *Control:* Variable-speed couplings and drives are often used. Some pumps have built-in capacity controls. *Packing:* Every effort should be made to keep the packing box free and clear of food. Mechanical, O-ring, rotary inert-carbon, and water-cooled seals are widely used to prevent contaminating the food handled by sanitary pumps.

Many pumps have outboard bearings to assure complete separation of the food and lubricant. *Number of Pumps:* One pump per service is almost always used, with hardly any standby units, except in certain critical services. *Piping:* Sanitary piping is made of stainless steel, nickel alloys, hard rubber, glass, or plastic. Pumps designed for food handling have male-threaded connections with sanitary acme threads. Consult the piping manufacturer for the exact friction loss in the piping to be used. The loss varies from one type of nonmetallic pipe to another.

A large number of pumps other than sanitary types are used in many food industries. For data on them, consult the proper chapters in this book. Where local laws or practices require daily disassembly and cleaning of pumps in food plants, note that these apply only to the sanitary pumps handling food or materials used in its manufacture. The other pumps in the plant ordinarily do not require such frequent cleaning. The U.S. Public Health Service has a number of recommended codes for food handling. They are extremely useful in system design.

CHAPTER SEVENTEEN

Water Supply

THE UNITED STATES TODAY draws more than 200 billion gal of water every 24 hr from its water resources. Pumps move almost every drop of this water. Of this total, an impressive 80 billion gal is said to be industry's share. The serious water shortages of recent years have focused attention on water supply and use. Correct pump selection and application are major factors in securing better results from existing and new supply facilities.

Ground-water Supplies Water pumped from wells below the surface of the earth supplies about 25 per cent of the total amount used daily in the United States. The balance comes from surface supplies. In industrial plants, about 13.5 per cent of the water used comes from wells, 19.8 per cent from public water supplies, and 66.7 per cent from surface supplies. Brackish water constitutes about 21.3 per cent of the industrial water intake. Especially in areas of acute water shortage, desalination processes are receiving increased attention. These are used for sea and harbor water as well as brackish ground-water supplies. In many cases, desalination is combined with steam-electric plants where advantage is taken of the large amount of waste heat available. It is interesting to note that 52.6 per cent of the total industrial water intake is recirculated for reuse. Since pumps of somewhat specialized design are used for water wells, ground-water supplies will be discussed first.

Deep Wells Multistage diffuser-type pumps (Figs. 17-1 and 1-23) are widely used for deepwell service. Units of this general design are commonly called *vertical turbine pumps.* They should not be confused with regenerative turbine pumps, discussed in Chap. 1.

Vertical turbine pumps may be oil- or water-lubricated. A shaft-enclosing tube is used with oil-lubricated bearings (Fig. 1-23). The liquid handled by the pump serves as the lubricant in water-lubricated pumps (Fig. 17-1). No shaft tube is used, leading to the term *open-line-shaft pump.* Vertical turbine pumps may be driven by electric motors, steam turbines, or gasoline or diesel engines (Fig. 19-4). Water-lubricated pumps are used where water absolutely free of oil is required or where certain regulatory bodies feel that there is a remote chance that oil used for lubrication might contaminate the water pumped. There are, however, official advocates of both types of construction.

Deepwell vertical turbine pumps are commonly manufactured for drilled wells 6 in. in diameter and larger. In many areas the most economic diameter for a drilled well is 12 in., but intermediate sizes are popular in industrial and small municipal installations. Pumps up to 30 in. in diameter have been built and are available. Besides water, these pumps can handle oil, volatile liquids, chemicals, etc.

Multistage pumps for deepwell service develop heads to over 1,500 ft and handle flows up to 30,000 gpm. The number of stages chosen depends on the head to be developed, the pressure rise being the same in each stage. High-head pumps may have 20 or more stages, but the majority of units used today have considerably fewer stages.

Fig. 17-1 Three-stage motor-driven water-lubricated vertical turbine for deep and shallow wells. (*Johnston Pump Co.*)

Impellers are usually closed or semiopen. The diffusers (Fig. 17-1) extend upward in the pump bowls. For average water conditions, the materials used for the impeller include bronze, close-grained gray iron, high-nickel iron, and porcelain-enameled iron. Porcelain-lined bowls are also available but are not too widely used because the impeller is subject to more cutting action than the bowls. Note that the well casing is not part of the pump.

Submersible-motor Pumps   In this design (Fig. 17-2) a vertical diffuser-type centrifugal pump is mounted directly above a small-diameter motor which operates submerged in the well water at all times.   The discharge pipe, also called the column or riser pipe, supports the weight of the pump and motor.   Motors used with pumps of this type are designed for long service without attention.   Should a motor failure occur, the entire pump must be lifted from the well.   This is a disadvantage in extremely deep wells.

Many submersible-motor pumps are built for heads up to 12,000 ft and capacities to 400 gpm at liquid temperatures to 270 F.   Some large pumps of this type have over 300 stages.   Many submersible-motor pumps of various designs are used today for both shallow and deep wells, especially where the well is crooked.   A lubrication connection between the motor and the surface, as well as a power cable, is necessary with all pumps of this type.

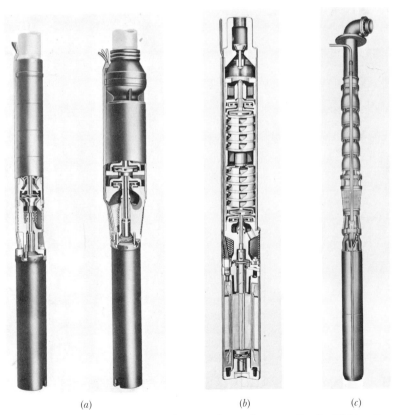

(a)   (b)   (c)

Fig. 17-2   (a) and (b) Small submersible-motor deepwell pumps. (*Deming Division, Crane Co.*) (c) Large submersible-motor pump. (*Byron Jackson Pump Division, Borg-Warner Corp.*)

Jet Pumps   These (Figs. 17-3 and 7-3) combine a single-stage centrifugal pump at the top of the well with a jet or ejector nozzle located at the suction screen in the well. A portion of the water discharged by the pump flows down and through the ejector, where it helps to improve flow into the pump and up the discharge tube.

For shallow wells with a lift of less than 25 ft, the jet is often located on

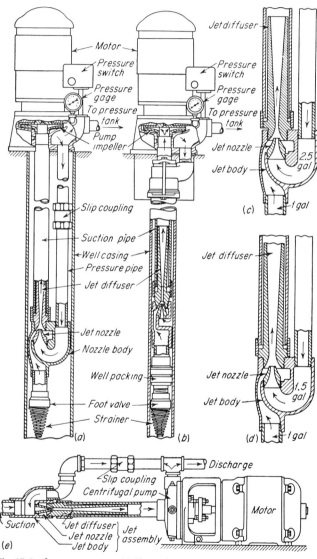

Fig. 17-3   Jet-pump types. (*a*) Two-pipe. (*b*) One-pipe. (*c*) High-lift low-capacity. (*d*) Low-lift high-capacity. (*e*) Connecting jet near the pump gives a steep *HQ* curve. (*Ingersoll-Rand Co.*)

the surface, in the pump casing, instead of in the well. This permits easier maintenance. For greater depths the jet is in the well, and the pump, which may be either horizontal or vertical, is on the surface.

Jet pumps are best suited for lifts of 25 ft or more with capacities up to 50 gpm net discharge (= pump capacity − quantity to jet). Lifts to 125 ft are common, with some pumps operating at lifts of 150 ft. In general, the efficiency of a jet pump on high lifts is low; often designs are better for high-lift service.

Helical-rotor Pumps   These resemble water-lubricated turbine pumps, except for the liquid end and its connection to the shaft. Instead of an impeller the pump is fitted with a helical rotor operating in a bihelical stator (Fig. 17-4). Water trapped in the stator depressions is positively displaced by the continuous upward moving contact of the rotor with the stator. A flexible drive tube above the rotor dampens the effects of rotor movement in the stator. Units of this type are designed for deep wells at capacities from 500 to 3,300 gph at lifts up to 1,000 ft. They are used in drilled wells having an inside diameter of 4 in. or more.

Reciprocating Pumps   Relatively few reciprocating pumps are used for industrial wells today because the various types of centrifugal units available are better suited for this service. Figure 17-5 shows the components of a modern reciprocating pump for water-supply service. The pumping head (Fig. 17-5*a*) may be used with a number of types of liquid ends, one of which is shown in Fig. 17-5*b*. This double-acting liquid end is located in the well, below the surface of the water. Reciprocating well pumps are built in capacities to about 300 gpm and heads to about 800 ft of water.

Surface-water Supplies   Surface waters supply over two-thirds of the water used in industry. *Close-coupled vertical turbine pumps* (Fig. 17-6) find many applications in this service today. They resemble the vertical turbine pumps described earlier in this chapter but are usually engineered

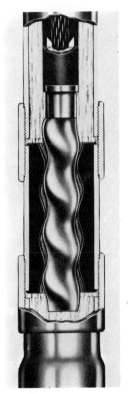

Fig. 17-4 Helical-rotor deepwell pump. (*Peerless Pump Division, FMC Corp.*)

for a much shorter setting. These units are used for pumping from lakes, rivers, ponds, pits, sumps, etc., where small to medium flow at high pressure is required. Capacities of one line range to 30,000 gpm, heads to 1,500 ft.

(a)   (b)

Fig. 17-5 (a) Pumping head. (b) Double-acting cylinder. (*Deming Division, Crane Co.*)

For medium to large flows at medium pressure, vertical mixed-flow pumps (Fig. 1-24) are often used. They are built to operate at speeds between about 400 and 1,750 rpm to deliver 500 to 100,000 gpm at heads from 20 to 100 ft. This type, usually having a specific speed between 4,200 and 9,000, is ideal for handling surface waters from rivers, lakes, and other sources. It is exceptionally well suited for applications requiring too large a capacity for the vertical turbine pump and too high a pressure for the propeller pump. It fills the gap between the vertical turbine and propeller pump.

Propeller pumps handle flows to over 200,000 gpm at heads from 1 to about 50 ft. Specific speed is above 9,000. See Chap. 7 for a discussion of pump intakes.

General-purpose Pumps These are often horizontal single-stage bronze-fitted volute-type centrifugal pumps (Fig. 17-7) designed to handle clear cool liquids at ambient or moderate temperatures. They

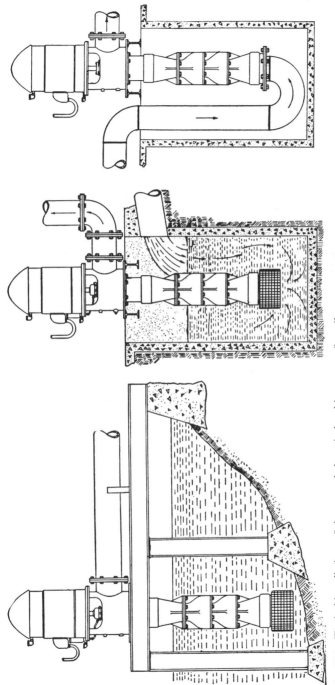

Fig. 17-6 Typical installations of close-coupled vertical turbine pumps for tailwater, open-sump, and booster services.

353

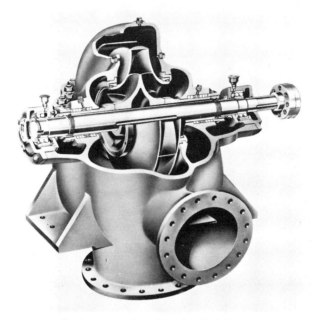

Fig. 17-7 Single-stage twin-volute double-inlet general-purpose pump. (*Worthington Corp.*)

find a large number of applications in water supply, particularly for surface waters.

Close-coupled end-mounted cradle-type centrifugal pumps for water supply and general-purpose duties are growing in popularity. This design (Fig. 17-8) permits complete separation of the liquid end of the pump from the bearings, allows easy maintenance without disturbing the piping, and uses only one packing box or mechanical seal. These units are usually the single-stage volute type, but two-stage volute-type units are also available. Capacities range up to 2,800 gpm, heads to 525 ft. Another popular design is shown in Fig. 19-1.

Municipal Water Supply The same types of pumps described above for industrial water supply are used in municipal installations. Since pumps on municipal water-supply service operate around the clock, the combined pump-motor efficiency is usually carefully evaluated at several operating capacities. These evaluations often assign cost values to efficiency which, when capitalized, are used to compare pump bids and arrive at the best possible pump-motor selection for the particular installation. As with pumps in industrial service, a wide variety of drives are employed. Electric motors are most common, with internal-combustion engines, particularly sewage-gas units, ranking second in munic-

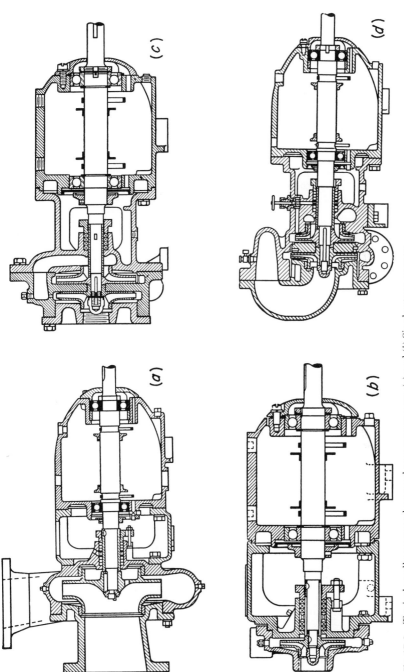

Fig. 17-8 Typical cradle-mounted general-purpose pumps. (a) and (b) Single-stage pumps. (c) and (d) Two-stage pumps. (*Ingersoll-Rand Co.*)

TABLE 17-1   Typical Water Needs of Cities*

Population	Domestic, gpm	Fire, gpm	Total,† gpm
1,000	65	1,000	1,065
2,000	140	1,500	1,640
4,000	300	2,000	2,300
6,000	500	2,500	3,000
10,000	900	3,000	3,900
20,000	1,900	4,500	6,400
40,000	4,200	6,000	10,200
60,000	7,000	7,000	14,000
100,000	12,000	9,000	21,000
150,000	19,000	11,000	30,000
200,000	27,000	12,000	39,000

* For cities and towns predominantly residential and commercial; where industries predominate, a greater flow will probably be required than the total given here.
† Includes water for domestic and fire use.

ipal installations. Steam turbines are also used. Table 17-1 gives the water requirements of typical cities and towns. When choosing pumps for municipal water supply, it is wise to size the piping and equipment for future growth.

Building Water Supply   Commercial, industrial, and residential buildings often require one or more pumps to boost the city or supply pressure to a suitable level. Centrifugal pumps are almost invariably used for this service. Three types of systems may be used within the building to supply the various fixtures: (1) overhead tanks, (2) pneumatic systems, and (3) tankless systems. An overhead tank is often used where the required flow is greater than 100 gpm. Tank capacity is usually 20 to 50 times the gpm capacity of the pump. Pneumatic systems are used where an overhead tank is unsuitable or where the required capacity is less than 100 gpm. Tankless systems are most common in small buildings where the water demand is small or where the building water pressure is excessively low during peak demands.

Pumps used for building and municipal water supply booster service are often bronze-fitted horizontally split volute or diffuser units, but vertically split volute or diffuser units are also used, as well as vertical turbine and horizontal regenerative turbine pumps. Single-stage pumps are common in buildings up to 16 stories, with multistage units being used for higher buildings, or where higher pressures are required for process or other special uses. General-purpose standard-fitted pumps are the most common choice for building water supply. They may be either horizontal or vertical.

Water consumption, gpm per fixture, varies from one class of building to another. Table 17-2 shows the typical requirements quoted by one pump manufacturer. To determine pump capacity, multiply the total number of fixtures in the building by the correct factor and apply any of the applicable notes in the table. To determine the head against which the pump must operate, find the sum of: (1) height from the pump center line to the highest fixture; (2) pressure required at highest fixture, expressed in feet, and (3) friction head in pump discharge line. This is the total head on the pump, unless there is a pressure head available on the pump suction. If there is, its value expressed in feet of water is subtracted from the above sum. The pressure required at the highest fixture is usually 15 psi (or 35 ft of water). If there is a suction lift, add it to the discharge head.

example: What are the required capacity and head of a pump to serve an office building having a total of 150 fixtures if the height from the pump to the highest fixtures is 125 ft, pressure needed at the top fixture is 15 psi, the friction loss in the discharge pipe is 20 ft, and the pressure head on the suction of the pump is 10 psi?

solution: Capacity required, using data from Table 17-2, is $(150)(0.7) = 105$ gpm. Total head on the pump is $125 + 15 (2.31) + 20 - 10 (2.31) = 157$ ft of water $= 68$ psi.

Swimming Pools General-purpose bronze-fitted horizontally split double-suction single-stage volute pumps are commonly used for refiltration and backwashing in swimming pools. Table 17-3 shows the pump capacity required for pools of various sizes with different refiltra-

TABLE 17-2   Water Requirements of Buildings*

Type of Building	Water Consumption, Gpm per Fixture †
Hotel	0.8
Apartment house	0.3
Hospital	0.4
Office building	0.7
Mercantile	0.6

* The Deming Co.

† For less than 50 fixtures, reduce pump capacity 50%. For more than 150 fixtures, increase pump capacity 15 to 25%. Increase pump capacity 25% if the majority of building occupants are women. Where the consumption has been measured by a meter, pump capacity should be three times the measured value. Pneumatic tanks in systems with uniform demand should have a capacity about 30 times the gpm capacity of the pump.

tion periods. Check with local authorities on the required refiltration period; many require 8 hr; others permit 10, 12, or 16 hr.

Pump head is composed of the friction in the piping system supplying the pool and connecting it to the filters, plus filter-bed, strainer, and inlet resistances. Total head generally ranges from 40 to 60 ft, depending on pool size. Head loss for spray nozzles supplying water to a pool is about 10 psi.

For filter backwashing, a flow about four times that normally required must be furnished. With three or more filters and backwashing of one at a time, the refiltration pump will have sufficient capacity. If only one filter is used, a separate backwash pump with a capacity four times that of the refiltration pump (see Table 17-3) should be provided. Refiltration pumps are also called pool circulating pumps. Reciprocating metering and proportioning pumps (Chap. 2) are used to feed chlorine and other liquid chemicals to the pool water.

Golf Courses, Parks, Airports   Lawn sprinkling for these and similar installations—cemeteries, race tracks, etc.—requires 27,150 gal per acre per week when there is no rain or dew. Pump capacity must be based on the occurrence of drought conditions. Healthy turf requires about 1 in. of water per week. With lawn sprinkling done at night for 7 hr, 7 days per week, rated pump capacity should be 10 gpm per acre of grass.

Single-stage double-suction horizontally split volute general-purpose bronze-fitted pumps are common for flat areas where the head required is about 100 psi. For hilly areas, multistage pumps of the same type are used. Head required is usually about 200 psi. Water at atmospheric

TABLE 17-3   Required Capacity of Pumps for Swimming Pools

(Bathing capacity per day assuming 24-hr operation on basis of refiltration in times given)

Pool capacity, gal	8 hr, 400 gal per bather		10 hr, 625 gal per bather		12 hr, 900 gal per bather		16 hr, 1,600 gal per bather	
	Bathers	Pump capacity, gpm	Bathers	Pump capacity, gpm	Bathers	Pump capacity, gpm	Bathers	Pump capacity, gpm
55,500	418	116	214	93	124	72	53	58
80,800	606	168	311	135	180	112	76	84
120,000	900	250	461	200	267	167	113	125
155,600	1,170	324	597	260	346	216	146	162
207,600	1,555	432	796	346	461	288	195	216
254,000	1,905	530	975	423	565	353	238	264
306,000	2,300	638	1,177	510	681	425	288	318
422,400	3,170	880	1,623	705	950	586	397	440
558,000	4,180	1,160	2,145	930	1,242	775	524	581

temperature is used. Since piping runs are often extremely long, correct sizing of the pipe is important. Motor or internal-combustion-engine drives are common in this service.

Fire Protection Centrifugal fire pumps are built in six standard sizes—500, 750, 1,000, 1,500, 2,000, and 2,500 gpm. Figure 1-20 shows a typical unit with the equipment required by most fire underwriters. This includes a relief valve, overflow cone, hose-valve manifold, air and starting valve, and pressure gages. These are furnished by the pump manufacturer. Table 17-1 gives the fire flow required for cities of various sizes.

When a fire pump is driven by a constant-speed motor and a wide range in pressure is desired over the rated capacity range, choose a unit having a steep HQ curve. This will give a higher pressure at small flows and is desirable where long hoses may be used. Choose a pump with a flat HQ curve when the unit has a variable-speed driver or a constant pressure is desired at all flow rates. The average fire pump usually has an HQ curve midway between the steep and flat.

At 65 per cent of its rated pressure, the pump should deliver not less than 150 per cent of rated capacity. With a steep HQ curve, the shutoff should not be greater than 130 per cent of the rated pressure; with an average HQ, 120 per cent; with a flat HQ, 110 per cent of rated pressure. Pump efficiency at rated discharge pressure must be between 55 and 75 per cent, depending on the rated capacity and net discharge pressure. Water should be provided to the pump suction under a positive head, if possible. Where a suction lift is necessary, it should not exceed 15 ft. For specific requirements related to fire-pump approval, installation, and operation, see the NFPA "Handbook of Fire Protection," published by the National Fire Protection Association, Boston 10, Mass.

Drives approved for centrifugal fire pumps include electric motors, steam turbines, and gasoline and diesel engines. There are a number of requirements that must be met by each type of drive if the installation is to meet underwriters' approval. See the above handbook for complete details.

Either horizontal or vertical centrifugal pumps may be used for fire protection. Single-stage units are generally horizontally split double-suction volute-type bronze-fitted pumps. All-bronze construction is also used. Capacities range from 500 to 2,500 gpm, heads from 40 to 150 psi. For higher pressures, two-stage units are used. They often have opposed impellers with an external loop for connection between the stages. Bronze-fitted, they have the same capacity range as single-stage units; heads range to 200 psi.

Jockey pumps are used to maintain a certain pressure on a sprinkler system at all times. They are designed to start and stop automatically.

Types used include units like those shown in Figs. 1-14, 1-15, and 17-8. They are usually bronze-fitted when handling water and are rated at 250, 500, 750, or 1,000 gpm.

Close-coupled vertical turbine pumps are also used for fire protection. Figure 17-6 shows typical units for an isolated industrial fire system. Vertical turbine fire pumps are built in capacities from 500 to 2,500 gpm, heads from 240 to 285 ft or higher. They are driven by 1,760-rpm motors; steam turbines or gasoline or diesel engines may also be used. The number of stages depends on the head required. They may be either oil- or water-lubricated.

Rotary pumps are also used for fire-protection service. As with centrifugal pumps, they are rated at 500, 750, 1,000, 1,500, 2,000, and 2,500 gpm at 40 to 100 psi, or higher.

Portable Pumps   Many centrifugal, rotary, and reciprocating pumps of the portable type have been developed in recent years. Discussed elsewhere in this book, they find a number of uses in industrial and municipal installations for fire protection, emergency water supply, swimming-pool cleaning, etc. Note that all fire pumps, whether portable or stationary, should be approved by an insurance laboratory—otherwise they may not be accepted by the insurance underwriter. Use of approved equipment in the recommended manner can measureably reduce insurance rates.

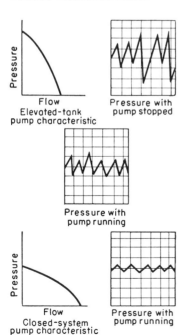

Fig. 17-9   Pressure characteristics for water-supply systems.

Water Treatment   Chemicals used for treating water for drinking and process uses in industrial and municipal systems include copper sulfate, activated carbon, chlorine, alum, iron sulfates, ferric chlorride, etc. They are fed to the water stream by a number of methods.   One is use of a metering or proportioning pump. Several designs are shown in Chap. 3.

Water Distribution and Storage   The same general types of water distribution and storage systems are used for industrial and municipal water-supply systems.   The most common industrial system uses an overhead storage tank fed by a pumping installation.   In plants where the required storage capacity is too large for economical overhead storage, ground basins are used.   Pneumatic systems are found in some smaller industrial installations and where pressure or flow boosting is required.   Tankless systems are relatively rare, except for plants near large bodies of fresh water.   These comments apply equally well to most small- and medium-sized municipal systems, except that pneumatic tanks are sometimes also used where it is desired to avoid an overhead tank for aesthetic reasons.   Figure 17-9 shows the pressure characteristics obtained with elevated-tank and closed systems.

CHAPTER EIGHTEEN

Sewage and
Sump Services

PUMPS FOR HANDLING SEWAGE, SUMP, BILGE, and other waste waters are almost always centrifugal units today because they can deliver solids without difficulty, have high efficiency, and are readily installed in pits, sumps, and other localities.

Large Pumps Where large quantities of sewage are handled, the vertical or horizontal mixed- or axial-flow-type pump is common. These develop low to moderate heads at large flow rates. Volute-casing mixed-flow pumps (Fig. 1-5) are recommended for services where solids or trash are contained in the liquid and the unit operates at a high load factor. Pumps chosen for sewage service often have a closed three-vane-type impeller. Typical axial-flow impellers are shown in Fig. 1-27. While some horizontal axial- and mixed-flow pumps are used in sewage service, the majority of new pumps of this type installed today are vertical units. Speeds of these large pumps are generally low—in the 200- to 1,200-rpm range.

Nonclog Pumps Raw sewage may contain a variety of solids—sticks, rags, rocks, hair, etc. These can clog the pump and damage rotating or stationary parts, reducing pump efficiency or causing complete stoppage of the unit. To prevent this, a number of clogless or nonclogging pumps have been developed. Though the design details differ from one manufacturer to another, most pumps of this type have impellers with at most two or three vanes (Fig. 18-1) or none at all (Fig. 7-27). The impeller may be either open or closed, but the closed type seems to be more popular at present. Usually, the clearance between the vanes is large enough to allow any solid entering the pump to pass out through the discharge. In large pumps, wearing rings are sometimes water-flushed to keep out grit and other materials that would accelerate wear. In some designs the suction pipe is 25 per cent larger than the discharge; in others both are the same size. The smallest discharge size is generally 3 in., though some 2- and 1½-in. pumps are also built. The smaller sizes are used for handling thin sludge or liquids which carry no suspended matter. It is common practice to state the maximum-diameter solid a pump of this type can handle without clogging. Thus 4-in.-diameter solids pass through the usual 8-in. pump.

Nonclogging trash or sewage pumps are built as either horizontal or vertical units. Figure 18-2 shows a typical vertical sewage pump with the motor mounted directly on the pump frame, while Fig. 18-3 shows a vertical extended-shaft type. Both have enclosed-type two-vane impellers. Present trends in sewage-system design indicate a decided preference for vertical pumps in almost all types of installations. The advantages of vertical installation include the need for less floor area, simpler piping connections, the avoidance of gas-accumulation problems in the pump suction, and the possibility of using extended shafts to isolate the motor from its pump. Figure 20-6 shows a typical installation of vertical-shaft sewage pumps handling storm water. Close-coupled pumps of this design are also available.

Dry and Wet Pits In a *dry pit* (Fig. 18-4) the pump, which may be vertical or horizontal, takes its suction through a pipe from an adjacent sump or wet well. The exterior of the pump is dry at all times, permitting easy inspection and maintenance. Also, there is less chance of cor-

Fig. 18-1 Two-vane nonclog sewage-type closed impeller. (*Worthington Corp.*)

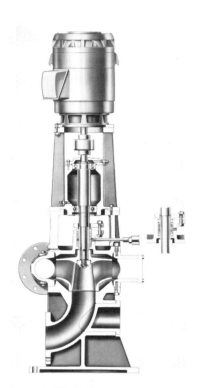

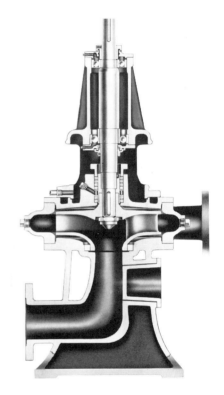

Fig. 18-2 Vertical sewage pump with motor mounted directly on pump. (*Worthington Corp.*)

Fig. 18-3 Vertical extended-shaft dry-pit nonclog sewage pump. (*The Weinman Pump Mfg. Co.*)

rosion of the pump casing, shaft, bearings, and other parts. The unit in Fig. 18-4 is float-controlled. In some pumps a patented strainer is fitted on the discharge side. Sewage enters through it while the pump is stopped and any solids present are trapped, while the water flows through the pump and into the wet pit. When the pump starts, after the water in the wet pit reaches a predetermined level, a check valve above the strainer closes and the discharging water flushes the solid matter in the strainer out the discharge line.

In a *wet pit* the pump is immersed in the liquid handled. Figure 1-22*b* shows a typical pump of this type. It can be installed in a round, square, or rectangular metal or concrete sump. Where one pump does not have sufficient capacity, two or more pumps may be used in a single sump. Single and double installations are popular for handling sewage, sump, seepage, and drain water in buildings, industrial plants, power plants, etc. Most pumps of this type are oil- or grease-lubricated and are fitted

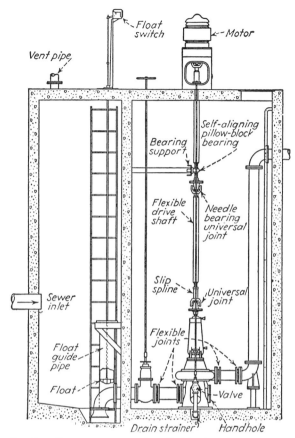

Fig. 18-4 Deep-sump pump installed in dry pit.

with a suction strainer having an inlet area five times that of the impeller eye or inlet. In sumps deeper than 6 ft, an intermediate bearing for the pump drive shaft is usually provided.

Sump Pumps These are called by a number of different names—wet-pit pumps, ejectors, bilge pumps, utility pumps, etc. Most pumps for this service are vertical (Figs. 1-22b and 18-5) and may be installed singly or in duplicate. They handle sewage, seepage, and drainage from buildings, but in many installations they handle only seepage and sewage. In others they handle only sewage from fixtures below the sewer line. Open nonclogging-type impellers are common. The pump manufacturer will, in many instances, supply the sump tank with the pump, if requested. In others, only the pump is supplied, while in some cases the pump and tank lid are furnished. Pumps of this type are almost always single-stage units because the solids in the liquid interfere

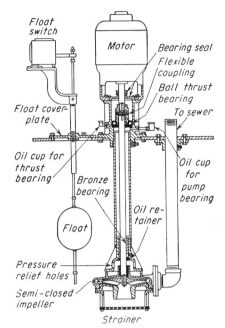

Float switch

Motor

Bearing seal

Flexible coupling

Ball thrust bearing

Float cover plate

To sewer

Oil cup for thrust bearing

Bronze bearing

Oil cup for pump bearing

Oil retainer

Float

Pressure relief holes

Semi-closed impeller

Strainer

Fig. 18-5   Wet-pit sump pump.

with operation of multistage pumps. However some 2-stage units (Fig. 20-2) are used. The sump or pump inlet should be fitted with a screen to remove solids and strings.

Cylindrical sumps are probably the most common in industrial plants and commercial buildings. In marine work, rectangular tanks are used to conserve floor space.

Portable Sump Pumps   A number of units of this type have come on the market in recent years. Figure 18-6 shows one popular pump of this type. It is completely submersible and needs only two connections—one to a discharge pipe and the other to the motor. The pump is fitted with a handle for easy movement from one spot to another. Pumps of this type are handy in industrial plants, commercial buildings, and similar

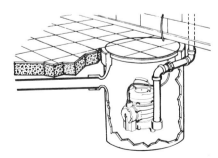

Fig. 18-6   Portable sump pump. (*Kenco Pump Division, The American Crucible Products Co.*)

installations.   A strainer is fitted to the intake to prevent solids from entering.   Short legs hold the pump off the floor.   Capacities and heads are moderate.

Underground Stations   These resemble dry pits and are generally fitted with two vertical close-coupled nonclog sewage pumps.   They are used extensively to extend sewer lines to new areas.   The pumps, which run alternately, draw raw sewage from a low line and discharge to a higher one, a force main, or a discharge manhole.   Completely equipped automatically controlled stations of this type are available from some manufacturers.   An air-bubbler system controls pump operation, the bubbler pipe extending into the suction manhole.   The welded-steel pump chamber and entrance tube, pumps, controls, starting equipment, etc., are furnished in these stations.   A small centrifugal sump pump is used to keep the chamber dry at all times.

Mill-scale Recovery   Figure 18-7a shows two vertical turbine pumps handling settled water in a steel-mill scale-recovery system.   These units are float-controlled and handle only clear water.   Liquids containing abrasives are not ordinarily handled by vertical turbine pumps.   Cutless rubber bearings (Fig. 18-7b) are used in vertical turbine pumps when the liquid is or may be slightly abrasive.

Sludge Pumps   Reciprocating diaphragm pumps (Fig. 3-8) are good for handling sludge from settling tanks and other sources.   Capacities range to 300 gpm and higher, depending on the diameter and number of diaphragms.   Plunger pumps with ball-type valves also find much use in sludge handling.   They resemble the diaphragm pump (Fig. 3-8) and have a walking-beam, V-belt, eccentric, or crankshaft drive.   On many, the stroke length is adjustable.   Capacities range up to about 9,000 gph at heads to 70 ft.

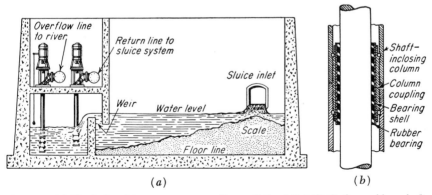

(a) (b)

Fig. 18-7 (a) Vertical-turbine sump pumps. (*Ingersoll-Rand Co.*) (b) Cutless rubber shaft bearing for vertical turbine pump.

Several different types of centrifugal pumps are built to handle sludge. In one, a double-flight screw and cutters on the pump casing and impeller cut up any solids or stringy substances entering the pump. Small clogless sewage pumps (Fig. 18-3) are also popular for sludge handling.

Comminuting-type pumps are widely used to handle liquids containing sticks, rags, strings, etc. These solids are common in sewage sludge. The comminuting pump has a single-vane semiopen impeller provided with self-sharpening stellited cutting edges which cut the solids into small pieces so they can be easily handled by the pump. A stationary shear ring in the suction cover aids in the cutting action. Capacities range up to 250 gpm at heads to 50 ft.

Pneumatic Ejectors   These (Fig. 18-8) handle sewage and sludge in a number of installations—sewage-lift stations, industrial waste-treatment plants, and hotels and office buildings where the basement is below the sewer line. With the basement below the sewer line the sewage must be lifted from the lowest outlets and discharged to the sewer. Sewage enters the hermetically sealed receiver and remains there until a predetermined level is reached. Compressed air is then automatically admit-

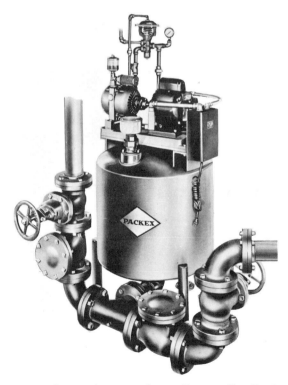

Fig. 18-8   Pneumatic sewage ejector. (*Yoemans, Clow Corp.*)

ted to the receiver and the sewage is discharged through a check and gate valve to the line. This type of unit comes complete with an air-cooled sliding-vane air compressor, controls, receiver, etc. Capacities of the usual pneumatic ejectors range from about 20 to 2,000 gpm, depending on size. Heads range up to about 50 ft at air pressures up to 50 psi. Water-operated ejectors are built for suction and discharge sizes of 1½ to 4 in. They are not so widely used as the pneumatic type.

Recirculation Pumps   These handle effluent from filters or secondary tanks, pumping it to a primary settling tank, distributor, or dosing tank. The liquid is relatively clear, having only small particles in suspension. Therefore there is no likelihood of clogging and a standard single-stage volute-type or regenerative-turbine pump can be used. Where the flow is large, about 0.5 mgd, either a vertical or horizontal propeller pump may be used. Head for this service is low—about 10 ft.

Metering and Proportioning Pumps   These (Chap. 3) are used in large and small sewage plants, such as those for villages, institutions, industrial plants, and camps. Chlorination for odor control and disinfection is accomplished by feeding hypochlorite with some type of metering pump. In large sewage-disposal plants, metering and proportioning pumps are used to handle ferrous or ferric chlorides in vacuum filtration. Figure 18-9 shows a typical proportioning pump for feeding

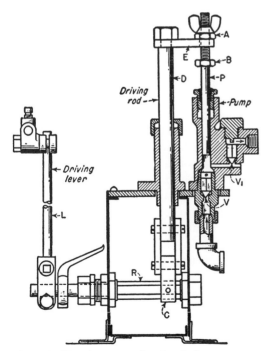

Fig. 18-9   Metering pump for chemicals.

chemicals in sewage and similar systems. All its pumping parts are outside the case enclosing the operating mechanism. It can be mounted directly on a reciprocating direct-acting steam pump. The stroke of the feeder is adjusted by changing the lost motion between nuts A and B. Other uses for metering pumps include pH control and sewage sampling. Capacities of the unit shown, with $\frac{1}{4}$- to 1-in. plungers, range from 1 pint to 17.5 gal per hr at pressures to 5,000 psi.

Pump Selection *Capacity:* With large sewage pumps the capacity is computed by finding the total amount of sewage entering the main from all the laterals connected to it. The same is true of smaller pumps, except that fewer plumbing fixtures are involved. The flow from various plumbing fixtures, in gpm per fixture, is: water closets, 6; urinals, lavatories, bathtubs, $\frac{1}{2}$-in. hose connections, and floor drains, 5 each; shower baths, 8; slop sinks, 15. The capacity to handle seepage water in basements and similar areas varies somewhat, but 5.0 gpm per 1,000 sq ft of floor space with clay soils and 7.5 gpm per 1,000 sq ft with sandy soils are common. These, however, must be carefully checked at the building site. Seepage varies with the nearness of the site to a large body of water, a river or lake. Where the rate of infiltration indicates the possibility of flooding the basement, the drive motors and controls should be located above the highest water level. See Chap. 20 for a discussion of storm-water drainage pumps.

Head: This varies with the installation. Large sewage pumps generally operate at low or moderate heads up to about 50 ft. In sump-pump installations, the distance from the low water level in the sump to the highest point of discharge, plus pipe and fitting friction, comprises the head. *Materials:* Bronze-fitted, all-bronze, and all-iron construction are common. Special materials are used for extremely corrosive sewage. *Drives:* Electric motors and internal-combustion engines (using sewage-gas fuel) are widely used. Steam-turbine drives, while used, are not so common as the others. A few water turbines are also used.

Control: Float actuation of pump controls is common, the pump being started and stopped, as necessary. See Chap. 20. *Number of Pumps:* For sewage and sump service, several pumps are often installed because loads vary widely. Using two or more pumps permits economical part-load operation and gives added safety because the failure of one pump will not cause complete stoppage of the installation. *Lubrication:* Oil or grease lubrication is generally used for pumps handling gritty liquids, to prevent excessive bearing damage and wear. *Piping:* A number of different materials are used for sewer piping—steel, cast iron, concrete, plastic, clay, etc. So when computing the total head on the pump, be sure to use the head loss for the actual piping material chosen.

example: An industrial plant has 10 water closets, 5 urinals, 5 shower heads, 3 slop sinks, and 20 floor drains. What pumping capacity is required to handle this flow if half is discharged to a sewer and half goes to a sump pit before being discharged to an overhead sewer?

solution: Total flow, using values given above, is $10(6) + 5(5) + 5(8) + 3(15) + 20(5) = 270$ gpm. Since only half this quantity flows to the sump pit, the minimum pump capacity required is $270/2 = 135$ gpm. To this must be added the capacity to handle any seepage or drainage in the building.

example: The sump in the above plant serves a floor area of 10,000 sq ft. What pumping capacity is required if the plant is built on clay soil?

solution: Pumping capacity for seepage in clay soils is 5 gpm per 1,000 sq ft. Hence, capacity $= (10,000/1,000)(5) = 50$ gpm. This flow is added to the sewage capacity, or $135 + 50 = 185$ gpm. A pump with a capacity of 200 gpm would probably be chosen to give a small margin of safety.

Sump Size: A rule for planning industrial sumps recommends a size having a storage capacity equal to at least twice the pump capacity in gallons per minute. In the above example this would be $2(200) = 400$ gal. This capacity should be obtainable between the high- and low-water levels in the sump. If a 6-ft-diameter sump is chosen for this installation, its capacity is 210 gal per ft of depth. So a depth of about 2 ft is needed to store the 400 gal between the high- and low-water levels in the sump. Where drains leading to the sump are run under the floor, as is usual in industrial and commercial buildings, the distance between the floor level and the bottom of the pipe must be added to the storage depth to obtain the depth of the low-water level below the floor line. Drains are usually run 2 ft under the floor. In addition, a clearance of at least 1 ft must be left between the bottom of the pump suction and the floor of the pit. Adding these dimensions gives $2 + 2 + 1 = 5$ ft, the minimum allowable depth for the sump in this plant. *Sump Location:* Usual practice is to locate a sump close to an exterior wall so the basement floor can be pitched toward the sump, assuring positive drainage of seepage and wash water. If possible, locate the sump in an accessible corner where the pump is protected but readily available for inspection and maintenance.

CHAPTER NINETEEN

Air Conditioning and Heating

THE RAPID STRIDES IN AIR CONDITIONING in recent years have led to an enormous growth in the number of installations in all types of buildings throughout the United States. Coupled with water shortages in various areas, the rise in the number of air-conditioning units has led to extensive use of water-conservation devices like cooling towers, evaporative condensers, etc. These units generally employ one or more pumps to circulate water over their cooling surfaces.

Though not so well publicized, there has also been a steady improvement in heating systems of all types for industrial and commercial buildings. Much of the improvement taking place can be traced to better design and construction of the pumps used in the various heating systems.

AIR CONDITIONING

Water Supply City, well, or surface water supplies are used in air conditioning for cooling the air and condensing refrigerants and steam. Well water is often the most desirable because it is usually available at lower temperatures than city or surface water in Northern areas of the

United States. Typical well-water temperatures at depths to 200 ft are as follows: southern Florida, 77 F; Charleston, S.C., 67 F; Norfolk, Va., 62 F; Baltimore, Md., 57 F; New York, N.Y., 52 F; Milwaukee, Wis., 47 F; Los Angeles, Calif., 67 F; San Francisco, Calif., 62 F; Seattle, Wash., 52 F. Below a depth of 200 ft, the temperature of the water increases about 1 F for each 64 ft of depth.

Pump Uses Four services are commonly met in air-conditioning installations in industrial, commercial, and institutional buildings. These are (1) water-supply, (2) air-washer circulation, (3) chilled-water, and (4) condensing-water pumps. The greatest variation exists in the first service because water may be available from one or more of three sources—wells, surface supplies, or city mains. Chapter 17 contains a comprehensive discussion of pumps for general water supply. These pumps may supply the air-conditioning system in addition to their other loads, or separate pumps of the same types as described may be used. Clear cool water, in sufficient quantities, is the only requirement of the usual air-conditioning system, from the standpoint of supply. Bronze-fitted centrifugal pumps are therefore common, with relatively few rotary and reciprocating pumps being used.

Air-washer Circulation Single-stage end-suction volute-type close-coupled and pedestal-mounted motor-driven horizontal and vertical centrifugal pumps are widely used on small and medium-sized spray-type air washers. In large air washers, closed-impeller horizontally split single-stage single- or double-suction motor-driven volute pumps (Fig. 19-1) are popular. Some vertical pumps are also used. It is general practice today to fit each air washer with an individual pump. The liquid circulated is a mixture of chilled and recirculated water, the amount being controlled by a three-way valve, in the pump suction line. Free of solids and relatively noncorrosive, the water is readily handled by standard bronze-fitted pumps. If the air-conditioning system is used for winter heating, the same pumps are employed, except that they circulate warm water instead of chilled water. Capacity and head requirements are almost the same as for summer air conditioning. Horizontal pumps are common for large air washers; vertical pumps are often used with small air washers to conserve space.

The head to be developed by air-washer circulating pumps depends on the pressure needed at the spray nozzles and the losses in the piping and fittings. Usual spray nozzles operate at inlet pressures of 15 to 40 psi. A strainer is often used in the pump discharge pipe to prevent solids from plugging the spray nozzles. Head loss in this strainer must not be neglected when computing the total head on the pump. Where the air-washer pump receives liquid from the water-supply or chilled-water pump under a positive head, be sure to subtract this head from the

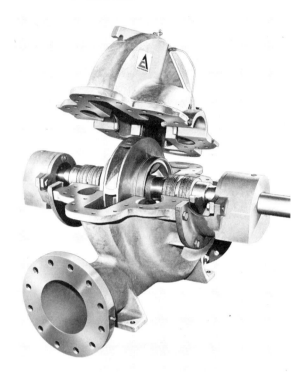

Fig. 19-1 Single-stage double-suction volute-type centrifugal pump for general water services. (*Allis-Chalmers.*)

total head to be developed. Usual air washers are designed to supply 5 to 6 gpm per min per 1,000 cu ft of air. Spray nozzles are rated at about 0.4 to 2.5 gpm per nozzle, depending on the nozzle size and its inlet pressure. Flooding nozzles for the washer eliminator plates are arranged to operate at 3 to 5 psi and are installed to deliver 3 to 6 gpm of

Fig. 19-2 Double-suction single-stage closed impeller with its shaft, lantern rings, and ball bearings. (*Weinman Pump Mfg. Co.*)

water per ft of washer width. Some authorities recommend 3 gpm of spray water per bank of plates, per square foot of washer pump either on or alongside the unit it serves. Figure 19-2 shows a typical impeller used in a horizontal single-stage pump for air-conditioning jobs.

Chilled-water Pumps These circulate water from the chiller to air-cooling coils in the air-conditioning unit. Pumps chosen for this service are the same general type as used for air-washer circulation, except that the head and capacity may be greater. Bronze-fitted construction is almost always suitable. The total head developed includes the loss through the chiller, chilled-water coils, and the static suction and discharge heads. The latter may be quite large, especially in tall buildings. With a water temperature rise of 4 to 12 F in the coil, 2 to 6 gpm per ton (= 12,000 Btu per min) is circulated by the chilled-water pump. Where a single pump supplies a number of coils, compute the total head on the pump by using the coil farthest from the pump, unless conditions cause a greater head loss during flow to and through some other coil. Chiller and chilled-water coil head losses are usually in the 20- to 30-ft range. Another type of centrifugal pump used for chilled-water circulation is the vertically split case single- or multistage unit. It is bronze-fitted for this service.

Condensing-water Pumps These circulate water from the city main, surface or well supply main, or from the cooling tower, to the condenser serving the refrigeration unit used in conjunction with the air-conditioning system. Bronze-fitted pumps, of the same types as discussed for air washers, are often used in condensing-water systems. However, the required head and capacity may be larger than for air-washer service. Cooling-tower circuits are usually sized for 30 gal deg per min per ton. With a condensing-water flow rate of 3 gpm per ton of refrigeration, the temperature rise through the condenser would be $30/3 = 10$ F. Many air-conditioning installations are designed for a condensing-water flow rate in the 3-gpm range.

Head to be developed includes the losses in the condenser, cooling tower, fittings, pipe, and any static lift. Where possible, it is wise to design the system so the chilled-water and condensing-water pumps can be served by a single standby unit. To increase the head against which the standby pump operates, when there is a marked difference between the heads of the two systems, an orifice plate or throttling globe valve is used in the lower-head system.

In recent years, regenerative turbine-type pumps have become popular for all types of air-conditioning service. As with other pumps in this service, they are often fitted with mechanical seals instead of standard packing. This reduces the amount of maintenance required. When the chilled water in an air-conditioning system requires treatment, reciprocating metering or proportioning pumps (Chap. 3) are often used to feed chemicals and control pH.

REFRIGERATION

Evaporative Condensers These have grown rapidly in numbers since the introduction of new regulations to reduce the water shortages in many cities. Single-stage end-suction close-coupled motor-driven bronze-fitted volute-type pumps, mounted horizontally or vertically, are popular for this service. They are often supplied as original equipment with the evaporative condenser and are designed for discharge pressures in the 20- to 30-psi range at capacities from 25 to 150 gpm.

Brine Pumps Two types of salt brine are in common use—calcium chloride and sodium chloride. Nonelectrolytic brines include the glycols, alcohol, acetone, varsol, etc. Brine pumps resemble the chilled-water units described earlier, except that all-iron construction is used for calcium chloride brine, and all-bronze or bronze-fitted construction for sodium chloride brine. When computing the head to be developed by brine pumps, note that the specific gravity of salt brines is greater than water, varying from 1.04 to as high as 1.29. Pumps handling low-temperature brine are usually insulated with granulated cork. A brine velocity of 5 to 7 ft per sec in the piping is considered reasonable. Head developed must be sufficient to overcome losses in the brine cooler and the cooling coils served by the pump. Regenerative turbine pumps are also used for brine circulation at temperatures above and below freezing. Brine-mixing pumps resemble brine-circulation pumps.

Refrigerant Pumps In steam-jet systems, water is the refrigerant. The chilled-water pump is the same as described above, but the condensate pump generally resembles the units for this service described in Chap. 11. Often bronze-fitted, the condensate pump handles water at about 100 F. Its npsh must be carefully computed because it takes suction from the condenser hotwell.

Other refrigerants, like the freons, ammonia, ethane, and propane, must also be pumped in certain refrigeration installations and in various manufacturing operations. Several rotary pumps, especially the external-gear type (Chap. 2), are popular for this service. Regenerative turbine pumps (Figs. 1-4 and 1-17a) and horizontal power pumps (Fig. 3-4) are also widely used because their characteristics are well suited to refrigerant handling.

Figure 19-3 shows a gearless double-impeller-type rotary pump for handling refrigerants, brine, oil, and many other liquids. Fitted with an impeller made of a resilient synthetic material running in a bronze

Fig. 19-3 Gearless double-impeller rotary pump. (*Eco Engineering Co.*)

casing, the pump needs no lubrication. It will operate in either direction, reversal of rotation causing the suction to become the discharge. Where high pressure differentials are required in refrigeration systems, two external-gear pumps may be connected in series to develop up to 300 psi.

Pumps handling refrigerants must be carefully checked for npsh. Provide a suction head of at least 8 ft of liquid. Suction piping should be designed so there is little or no possibility of vapor accumulation. Mechanical seals are often used instead of packing. Excessive heat generation in the pump should be avoided. Process and refinery-type centrifugal pumps are also used to handle refrigerants. For extremely low temperatures, the mechanical seal used in these and other pumps must be made of special materials to prevent embrittlement at low temperatures.

COOLING TOWERS AND PONDS

Small Towers These commonly use either horizontal or vertical single-stage end-suction volute-type close-coupled motor-driven bronze-fitted pumps, similar to evaporative condensers. One pump per tower is commonly chosen, with the pump usually being supplied as original equipment with the tower. Capacity of the pump is often in the 100-gpm range and the head is about 100 ft.

Large Towers The recent water shortages have stimulated the use of large towers throughout the United States for many services—refrigeration, steam power plants, internal-combustion engine plants, etc. With these towers the pumping installation is usually extensive, involving two or more pumps per tower; in some installations a dozen high-capacity high-head pumps may be used.

Vertical close-coupled bronze-fitted multistage turbine pumps (Fig. 17-6 and Chap. 17) are extremely popular today for large cooling towers because they require little space, are always primed, and are easily ac-

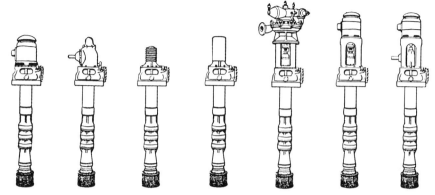

Fig. 19-4 Drive arrangements for vertical turbine pumps: motor, right-angle gear, V-belt, flat-belt, steam-turbine, solid-shaft motor, combined motor or turbine and gear. (*Peerless Pump Division, FMC Corp.*)

commodated in the usual tower basin. Water-lubricated units are generally recommended but the oil-lubricated type may also be used. Figure 19-4 shows seven different types of drive arrangements used. Horizontally split single-stage double-suction motor-driven volute-type pumps (Fig. 1-14) are also applied for cooling-tower water circulation. Reciprocating metering or proportioning pumps (Chap. 3) are employed for chemical feed in treating the tower water.

Tower circulation rates vary with the process served. Standard mechanical-compression refrigerating units circulate 3 gpm per ton of refrigeration; steam-jet units, 9 gpm per ton; diesel engines, 0.5 gpm per hp; steam power plants, 0.6 gpm per kw. When figuring the pump head, note that water must be delivered to the top of the tower.

Cooling Ponds Both horizontal and vertical pumps are used for cooling ponds. But because little lift is required for ponds, the horizontal general-purpose pump (Fig. 1-13) is used more often than a vertical pump. The usual spray nozzles employed in cooling ponds require an inlet pressure of 5 to 10 psi. Nozzles are spaced at 6- to 12-ft intervals; each nozzle discharges 25 to 50 gpm, depending on its size.

BUILDING HEATING

Two types of pumps are commonly used in building heating systems—hot-water circulating pumps and condensate-return pumps. The first type is invariably a centrifugal pump of a design similar to that in Fig. 1-17a or 1-22a. Bronze-fitted, these pumps are often equipped with a mechanical seal to reduce maintenance to a minimum. When oil-lubricated, they have a large oil reservoir to permit operation for long periods without attention. Single-stage side-suction pumps, they can be installed in the pipeline without use of bends or elbows. Pumps like Fig. 1-14 are often used to boost the water pressure in buildings.

Condensation Sets These (Fig. 19-5) are built in a number of different designs, but the centrifugal type is the most popular today. For gravity heating systems, the condensation set generally consists of one or two bronze-fitted single-stage centrifugal pumps and a cast-iron receiver with an automatic float switch to start and stop the pump. Some rotary, regenerative-turbine, and reciprocating pumps are also built for this service. It is usual practice to size the receiver so it has a storage capacity about 1.5 times the gpm quantity of condensate returning from the system during normal shutdown periods of the pump or pumps. The pump has a discharge capacity three to four times the condensate return rate, in gpm. Many standard pumps for this service have a capacity of 1.5 gpm per 1,000 sq ft of equivalent direct radiation (edr).

Condensate-return pumps should be used whenever the existing gravity head cannot return the water to the boiler. Vertical condensation pumps and receivers and float control are available for installations where the return lines are below the plant floor level or too low for horizontal pumps. They resemble the sump-pump installation in Fig. 18-5. Many condensate-return pumps are motor-driven, but turbine drive is also used. Standard horizontal and vertical single and duplex condensation sets are built in capacities up to 150,000 sq ft edr. Regenerative turbine pumps are also used for condensate-return service in sets similar to those described above.

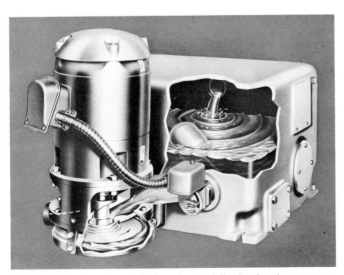

Fig. 19-5 Condensate set designed especially for heating systems. (*Hoffman Specialty Mfg. Co.*)

Fig. 19-6   Vacuum heating set has separate pumps for air and condensate. (*Nash Engineering Co.*)

Vacuum Heating Pump   These (Figs. 19-6 and 1-21) are used in vacuum heating systems to remove air and condensate from the piping. *Low-vacuum* pumps maintain a vacuum of 5.5 in. Hg or less on the system. *High-vacuum* pumps maintain a vacuum greater than 5.5 in. Hg. With condensate at 160 F, the pump should handle a minimum of 0.3 cu ft per min of air and 0.5 gpm per 1,000 sq ft edr at 5.5 in. Hg vacuum. But for usual jobs one manufacturer recommends the following capacities per 1,000 sq ft edr: single pumps, 1.0 to 1.25 gpm and 1.0 cu ft per min; duplex pumps, 1.0 gpm and 0.75 cu ft per min. Both the above are at 10 in. Hg vacuum. For subatmospheric systems at 20 in. Hg vacuum, the air-handling capacity should be 2.0 cu ft per min for single pumps and 1.0 to 2.0 cu ft per min for duplex pumps.

The unit in Fig. 19-6 is fitted with one or two air pumps, a receiver, one or two condensate pumps, and automatic controls for each pump or set of pumps. The air and condensate pumps are independently controlled, the air pump being a liquid-piston type (Fig. 23-7). The condensate pump is a single-stage type. They are built single or duplex with capacities to 150 gpm. Instead of a separate air pump, the unit in Fig. 1-21 has a venturi tube, through which the liquid pump discharges, creating a vacuum. This vacuum draws air and condensate from the heating system. Either one or two pumps may be used, depending on the capacity desired. They are built in capacities to 100,000 sq ft edr. In recent years many vacuum heating pumps have been built to discharge at pressures up to about 100 psi, to eliminate the need of a separate boiler-feed pump between the vacuum pump and the boiler.

Reciprocating vacuum heating pumps are used in some plants. They are either single or duplex and have an air-separating tank on their discharge. Pump displacement should be six to ten times the volume flow of condensate from the system. The pump may be steam- or motor-driven. All duplex heating pumps, reciprocating or centrifugal, can be run singly or together, depending on the load.

High-temperature Water   Widely used in Europe for many years, high-temperature high-pressure hot-water heating systems have gained popularity in the United States. Pumps used for this service are generally motor-driven   single-stage   single-inlet   1,750-rpm   bronze-fitted 150-ft-head units with water-cooled bearings and glands. Mechanical seals are becoming popular for these pumps. Two pumps are often used in space-heating installations, three if the load fluctuates widely. Where there is danger of a power failure, one pump is turbine-driven.

Note that pumps in this service handle much larger volumes of water than those in low-temperature work because the density of the water decreases as the temperature increases. A manually controlled connection between the pump discharge and suction is provided to prevent flashing in the suction line. Pump head is computed using the resistance of the *index circuit,* which is the longest supply and return run of the system. To this must be added the resistance of the boiler, equipment heated, fittings, etc.

CHAPTER TWENTY

Irrigation and Flood Control

THE SERVICES are, in general, characterized by two requirements—large water flows at relatively low or moderate heads. Today irrigation, land drainage, and flood control are based on accurate scientific data. As a result, it is possible to predict with greater accuracy the amount of water needed and from this to choose the class and type of pump best suited for the existing requirements.

Types of Irrigation Four common types of irrigation systems are used today: (1) basin, (2) borders or checks, (3) furrows, and (4) sprinkling. From the standpoint of the pump, the first three systems require the least amount of developed head, once the water is drawn from the ground or surface supply. The last system, sprinkling, may require fairly large heads because of the loss in the piping between the pump and sprinklers, and the pressure head required at the sprinkler inlets. Minimum nozzle pressure required for sprinklers varies from 20 to 100 psi. Large nozzles use the higher pressures. The head to be developed by a pump in irrigation service may be high if the water is secured from a deep well. But with surface supplies or shallow wells, the head required is generally fairly low.

Pumps A number of different types of centrifugal pumps are used for irrigation. These vary from small portable tractor-driven units to large propeller or mixed-flow pumps. Much depends on the type of irrigation being performed, source of water supply, weather conditions, etc. Reciprocating and rotary pumps find relatively little use in irrigation today because they do not have the characteristics required for this service.

Portable self-powered pumps for irrigation resemble the unit shown in Fig. 21-4. They are usually self-priming. Similar pumps, without an engine, are built for temporary attachment to a tractor drive shaft. Capacities of portable pumps for irrigation range up to about 6,000 gpm. For sprinkler irrigation, 2-stage volute-type engine-driven pumps are often used.

Vertical turbine pumps find widespread use in irrigation service in many parts of the United States. They are perhaps most commonly applied to deepwell service, their discharge being conducted by suitable means to the fields being irrigated. Close-coupled vertical turbine pumps are often used as boosters for land irrigation when the water reaches the field at too low a pressure. In areas beyond the reach of power lines, these pumps are often driven by gasoline or diesel engines.

Vertical turbine pumps are generally chosen where the flow required is low or moderate and the head to be developed is high. This is a most common characteristic of deep wells.

Close-coupled motor-driven pumps are finding some use for irrigation, either as permanently mounted units (Fig. 20-1) or as temporary

Fig. 20-1 Close-coupled motor-driven irrigation pump being primed.

units mounted on skids sloping from the bank into the river or lake. With the second arrangement a length of flexible discharge pipe is used between the pump and the pipe to the field. This permits the pump to be raised or lowered on the skids with change in water level.

Submersible pumps (Chap. 17) finds some use for irrigation water supply. Submersible propeller pumps, popular for graving-dock unwatering, are also used in some larger irrigation projects. A typical 24-in. submersible propeller pump handles 40,000 gpm at 40-ft head, while a 16-in. pump of the same design handles 8,000 gpm at a 50-ft head.

Standard single-stage and multistage propeller pumps (Fig. 20-2) and axial-flow pumps are extremely popular for low to moderate lifts at large capacities. As shown in these illustrations, pumps of this type are usually arranged to be supported on feet below the motor, the column extending downward into the water. The distance between the discharge elbow and the motor can be varied, if desired. This permits locating the motor above the high-water level. Oil- or water-lubricated designs may be used. Ordinary, bronze-fitted pumps are suitable for many irrigation services. Most pumps of this type are easily supported on relatively light structural-steel frames.

Ditch pumps are low-priced standardized-design propeller pumps built for heads up to about 20 ft and capacities to about 10,000 gpm. Levee-incline pumps are propeller-type units designed to be supported by the sloping face of a levee. They are usually installed inside a protective tube—for example, a 20-in. pump is inserted in a 34-in. tube; a

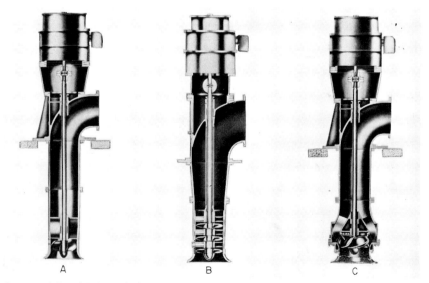

A   B   C

Fig. 20-2   (*A*) and (*B*) Vertical propeller pumps. (*C*) Vertical mixed-flow pump.

16-in. in a 28-in. tube; and a 14-in. in a 24-in. tube. The tube may be mounted on a simple wooden sled to permit easy and safe movement of the pump into or out of the water. A low-cost shed is often built over the upper end of the pump to protect the motor. The levee-incline pump has an extremely low installation cost.

Tailwater pumps are single-stage vertical turbine units designed for reclaiming irrigation runoff water. One manufacturer builds these units in four standardized oil-lubricated designs: 2, 3, 5, and 7.5 hp. Each is available in three lengths—6, 8, and 12 ft. The pump is designed to operate in a sump from which the runoff water is pumped back to the high side of the field for reuse, conserving the water supply and allowing more thorough irrigation. Capacities range up to about 1,000 gpm, heads from 15 to 30 ft. For lower heads in this capacity range a ditch pump is used.

Pump Selection *Capacity:* This varies with the crop, amount of moisture in the soil, weather conditions, etc. Arid regions require 11,000 to 1,600,000 gal per acre per year. To find the quantity of water for irrigation, subtract from the total amount of water needed for the crop, the soil moisture, and the rainfall. The basic quantity unit in irrigation is the acre-inch = 27,154 gal = 112.2 gpm per 12-hr pumping day = 0.25 sec-ft = 0.2479 acre-ft. All other multiples are easily derived from this unit. Crops like sugar beets, potatoes, alfalfa, corn, wheat, oats, barley, peas, and beans require from about 1.0 to 3.5 acre-ft per acre per year. Exact requirements are available in irrigation texts. Water losses in sprinkler irrigation run 20 and 40 per cent in cool and hot dry climates, respectively. This loss must be replaced by supplying additional water to the field. Multiply the number of sprinkler heads by the flow per head to find total flow required. Add about 2 per cent of this for leakage. Sprinklers discharge 1 to 25 gpm, depending on size, inlet pressure, etc.

example: How much pumping capacity is required to apply 3 in. of water to a 10-acre field in 15 hr? What is the flow per sprinkler if 200 are used?

solution: Capacity, gpm = (450)(water depth, in.)(area, acres)/time, hr. In this equation, the value 450 is a constant. Or, capacity, gpm = 450(3)(10)/15 = 900 gpm. Then, flow per sprinkler = (total flow)/(number of sprinklers) = 900/200 = 4.5 gpm per sprinkler.

Head: In horizontal irrigation where the slope of the terrain causes the water flow, the pump need develop only enough head to raise the water to the point of distribution. This may be a central area in small installation. On large farms the water is piped to an outlet head, from where it flows to the crops. Piping and fitting losses, head loss at the outlet, and suction lift must then be computed, using the longest run as the index circuit. Sprinklers require an inlet pressure of 25 to 100 psi, the larger nozzles requiring the higher pressures because their discharge is greater. The pump must develop sufficient head to overcome the suction lift, and friction in the piping, fittings, and laterals (about 20 psi) and deliver the water to the nozzles at the required inlet pressure. In existing sprinkler systems a booster pump may be required if the present pump does not develop sufficient pressure. This is often a close-coupled vertical turbine pump. See Chap. 4 for head calculations.

Materials: Bronze-fitted pumps are suitable for most clear waters, though all-bronze and all-iron construction is also used. See Chap. 6. *Drives:* Motor, gasoline or diesel engines, and steam turbines are popular, with the first two seeing greatest use. *NPSH:* Try to provide as much npsh as possible, particularly with horizontal pumps. *Control:* Start-stop control is probably most common because extreme accuracy of flow regulation is not necessary. *Piping:* With portable pumps, field ditches must be deep enough to submerge the suction strainer and foot valve. A turtleback or cylindrical strainer should be fitted to the pump suction pipe. For sprinklers, size the main line for a 10-psi, or less, loss. Use a larger pipe size when the sum of the standby charge, hourly pumping cost, and cost of the pump exceeds the extra cost of the next pipe size. The Johnston Pump Company recommends 3-in. pipe for 50 to 100 gpm, 4-in. for 100 to 200 gpm, 5-in. for 200 to 350 gpm, and 6-in. for 350 to 600 gpm. Portable supply lines are not suitable for flows exceeding 600 gpm. *Pumping Cost:* The total cost runs between about $3 and $12 per acre-foot per season for motor-driven pumps.

DRAINAGE AND FLOOD CONTROL

Drainage Pumps   Recovery of land by planned drainage is an important phase of providing more acreage for farming. Drainage service resembles irrigation in many respects—the lifts are low to moderate and the required capacity is usually large. Vertical propeller, mixed- and axial-flow and nonclog pumps, as well as horizontal units, are popular for this service. In recent years vertical pumps appear to have become somewhat more popular in some areas because they require less space, have high efficiency, and are relatively easy to install. Mixed-flow pumps are often used when the lift is 30 to 60 ft.

Figure 20-3 shows a typical vertical propeller pump mounted on the bank of a drainage ditch. The motor has grease-packed ball bearings and the entire installation resembles the levee-incline pump described earlier, except that an enclosing tube is not used for the pump. Two other typical drainage-pump installations are shown in Figs. 7-38 and 7-39. As these illustrations show, the lift required is generally low or moderate. A number of large drainage stations are operated by the United States government to reclaim and protect various land areas.

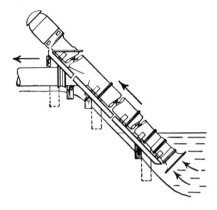

Fig. 20-3 Vertical propeller pump on inclined mount. (*Johnston Pump Co.*)

Figure 20-4 shows a typical station of this type in Florida. The interior of this station is shown in Fig. 8-14.

Flood Control The functions of drainage and flood control are often combined in a single station, where both these needs exist. This is the purpose of the stations in Figs. 20-4, 20-5, and 8-14. They also supply water for irrigation purposes. The pumps in the station in Fig. 20-4 are horizontal axial-flow propeller units while those in Fig. 20-5 are vertical propeller units. Standardized pumps are used for small and medium-sized stations, but in large stations the pumps are often specially designed for the particular site, station, and flow conditions. Extended-shaft pumps (Chap. 18) find many applications in smaller drainage, flood-control, and irrigation stations.

Pump Selection *Capacity:* With tile subsurface drainage systems the water flow the pump must handle is about 7 gpm per acre; with ditch or tile surface drainage, and field crops, 10 gpm per acre; and with truck crops, 15 gpm per acre. Capacity in any drainage or flood-control installation should include the runoff from gravity, plus seepage. Records and data from the Geological Survey, Weather Bureau, Soil Conservation Service, and the U.S. Army Corps of Engineers are invaluable in determining the pump capacity required. In general, the pump manufacturer must be supplied with all data from which the pump capacity was computed so a check can be made. These data should include the high and low flood stages, area runoff, drainage coefficient, seepage, soil type, and the pumping head at various flood stages.

Fig. 20-4   Suction side of large low-lift flood-control and irrigation station. (*Fairbank, Morse Pump Division, Colt Industries.*)

Head: This varies with the pump installation (Figs. 20-3 and 20-5). However, the majority of drainage and flood-control pumps operate at low heads—up to about 60 ft.   Be sure to compute the head for the minimum-water-level conditions.   Siphons (Chap. 4) can be used to advantage.   *Materials:* Bronze-fitted pumps are satisfactory except for

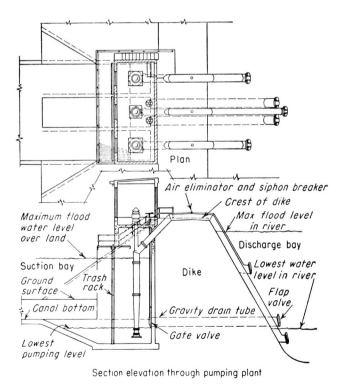

Section elevation through pumping plant

Fig. 20-5   Typical land-drainage pumping station. (*Johnston Pump Co.*)

extremely abrasive liquids where special materials should be used. *Drives:* Electric motors, internal-combustion engines, or steam turbines may be used. *NPSH:* Provide at least the minimum required. *Control:* Start-stop float and variable-speed control are common. A few pumps use adjustable impeller vanes to vary either the head or capacity. This gives high efficiency over a wide range. *Packing:* Simple rubber gaskets are popular in low-head pumps.

Number of Pumps: Where the variation in lift is slight, say not more than 10 ft, a single pump is satisfactory, unless it does not have enough capacity. Then duplicate units should be used to provide the desired capacity. But where variation in lift is greater, say 25 ft or more, two or more pumps may be used. Choose one for the high-lift condition, one for the medium-lift, and possibly a third for the low-lift condition. The high-lift unit is often a mixed-flow pump, while the medium- and low-lift pumps are the propeller type. This permits operating each pump in the head range of its maximum efficiency. An alternate scheme is to use a single pump with a variable-speed drive. This is popular with many engineers. *Piping:* Provide erosion protection for the suction and discharge bays. Fit trash racks in the suction bay. Gates and spillways can be used to provide gravity flow at times of high-water level, allowing the pumps to be stopped.

STORM-WATER DRAINAGE

Flood control is a matter for state or Federal authorities because a relatively large area, possibly covering one or more states, is usually involved. But street and road drainage in cities and towns is a matter

Fig. 20-6   Mixed-flow pumps for storm-water service. (*Worthington Corp.*)

for local authorities, especially in areas where heavy rainstorms are likely. Gravity flow provides the means to drain rainwater runoff in many areas. It is successful if the main and lateral sewers are large enough to handle the flow. Also, they must be enlarged at regular intervals to keep up with the growth of the city.

But some cities and towns are in areas where the rainwater cannot be drained by gravity. This occurs when the land being drained is below the water level of nearby rivers, lakes, or bays. Two cities in which this occurs are New Orleans, La., and Detroit, Mich.

Vertical mixed-flow pumps (Fig. 20-6) or axial-flow diffuser-type

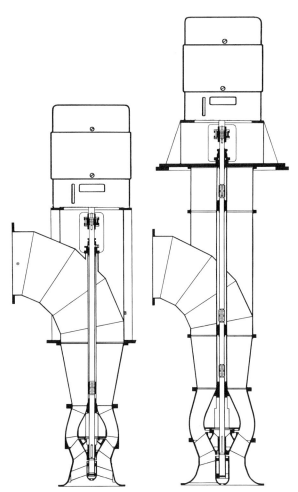

Fig. 20-7   Large vertical axial-flow diffuser pumps for flood-control service. (*Ingersoll-Rand Co.*)

propeller pumps (Fig. 20-7) are often used for this service. To provide maximum safety, the rainwater intake sewer is led to a large deep sump (often called a well) from which the pumps take their suction. They are piped to discharge to a nearby river, lake, or bay, or to a sewer leading to such a disposal point. Head is moderate, seldom exceeding 50 ft, but the required capacity may be large, 500 cu ft per sec per pump being fairly common. Mixed-flow pumps in this service turn at fairly low rotative speeds, 200 to 225 rpm being popular.

CHAPTER TWENTY-ONE

Mining and Construction

PUMPS FOR MINING AND QUARRYING OPERATIONS handle a variety of liquids, many of which contain abrasive solids or are acidic, or both. Applications in coal mining include dewatering, filter feed, handling heavy media, precipitate transfer, sludge handling, sump pumping, thickener feed, thickener underflow, waste disposal, coal washing, etc. Most pumps in this service are centrifugal, but reciprocating units also find some use.

In metal and nonmetallic mining and preparation, pumps are used in beneficiation process steps, dredging, filter feed, jig feed, drainage, tailings disposal and recovery, thickener feed, precipitate transfer, spillage handling, floor flushing, handling heavy media, etc. As in coal mining and preparation, centrifugal pumps are popular, with some reciprocating units being used.

Gravel pits and quarries use pumps to handle cement slurry, classifier water supply, corrosive mixtures, dredging, lime putty, drainage, sand transfer and wasting, etc. As with mining operations, centrifugal pumps are popular, with some reciprocating units being used. Rotary pumps find relatively little use in any of these fields except as original equipment supplied with many of the processing units.

Centrifugal Pumps Often called *mine pumps*, these units have heavy casing walls, giving ample corrosion allowance. Provisions are made to keep corrosive, often very low pH(acid), liquids from the pump shaft. Easy renewal of wearing parts is an important feature of all modern pumps of this type. Materials contacting the liquid handled should be chosen for maximum resistance to corrosion and erosion.

Figure 21-1 shows a rubber-lined pump for handling a wide variety of mining materials mixed with water or other liquids. Typical solids it will pump include ores, coal, cement, sand, gravel, filter aids, ground metallics, tailings, refuse, etc. Normal heads range up to 100 ft at capacities approaching 4,000 gpm. The impeller (Fig. 21-2) is a recessed type located so it is out of the main flow path. Design of the pump inlet reduces the possibility that solids will contact the impeller.

Rubber-lined pumps (Fig. 21-1) can make sizable savings in the re-placement costs for pumps handling abrasives. Special rubber linings of various compounds are available to handle a variety of slurries. Some rubber-lined pumps show an operating life ten to fifty times that of alloy pumps. One line of this type handles 10 to 3,000 gpm at heads through 140 ft. They pump materials from $1/8$ in. to 325 mesh. Open impellers are recommended for solids from 325 mesh to $1/16$ in., while closed impellers are used for solids up to $1/8$ in. in size.

Solids-handling pumps (Chap. 11) resemble mine pumps but are

Fig. 21-1   Rubber-lined pump for handling abrasive materials con-taining solids. (*Allis-Chalmers.*)

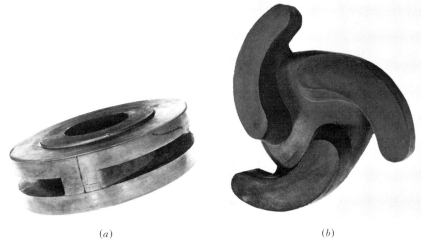

(a)   (b)

Fig. 21-2   (*a*) Closed impeller for pump in Fig. 21-1. (*b*) Open impeller for pump in Fig. 21-1. (*Allis-Chalmers.*)

often built to meet a specific demand in handling a certain type of solid. Casing and impeller are ruggedly built of special materials. Capacities of some units range up to 10,000 gpm, heads to 300 ft.

Self-priming rail-truck-mounted centrifugal pumps find much use in mines because they are readily moved to the proper site, require little headroom, and are easily operated once they are connected to the suction and discharge. Plastic, rubber, and other special types of piping are often used with mine pumps to keep wear to a minimum. Chemical-type centrifugal pumps are used to handle acid in copper-leaching

Fig. 21-3   Heavy-duty slow-speed slurry pump is easily disassembled (*Morris Machine Works.*)

plants. These pumps are generally permanently installed, instead of being portable.

Slurry Pumps Units of this type handle abrasive slurries, sand, chemical sludges, plant wastes, and similar products. Figure 21-3 shows a typical low-speed continuous-duty slurry pump built in sizes from 2 through 6 in. Its stuffing box is subject to only suction pressure, wearing parts can be reached without disturbing the piping, and the rotating element is adjustable. It handles tailings, slag, ores, etc.

Slurry pumps ordinarily handle solids in a range from 400 to 8 mesh, approximately, while sand pumps handle solids in the range of 35 to 4 mesh. Sealing water, introduced at the back of the impeller, may or may not be used, depending on the pump design. Dredge-type pumps are used for materials 3 mesh and larger. Many sand and slurry pumps are rubber-lined. Dredge pumps may or may not be lined with rubber. Some dredge pumps are specially built for a specific application and use one or more hard alloys to reduce wear.

Sumps used with sand and slurry pumps should be cylindrical, if possible, and should have sufficient capacity for the normal variations in the sump feed without overflow. The sump height should be great enough to allow for sufficient depth of liquid above the center line of the pump suction, plus surge capacity. The shape of the sump bottom should provide at least a 45-deg slope to the outlet.

Vertical Turbine Pumps This type is finding ever wider use in mines today because it is compact, highly efficient, and easily installed. Multistage units are commonly used because the head to be developed is moderate to high. For example, in a Pennsylvania coal mine, the 20-in. vertical turbine pump used delivers 5,000 gpm against a head of 480 ft. Pumps for this service can be designed to hang suspended from the surface of mine shafts, or they can be installed permanently on a foundation above the high-water level. Special mine dewatering pumps have been designed with two pumps on a common rig and lowered into flooded mines as a unit with their driving motor. Near the ground surface, where total heads are low to moderate, the pumps operate in parallel. When the mine has been dewatered to a depth where the total head exceeds the capability of the pumps operating in parallel, they are automatically switched to series-operation and develop enough total head in this mode to complete the dewatering job.

When corrosive acids or chemicals are handled, special metal or rubber linings are used in the pump discharge, column, and the exterior of the oil-enclosing tube. A mine disconnect device, permitting easy removal of the bowl from the column and shafting, may also be used. This makes pump maintenance easier. Vertical self-priming pumps are also popular in mine sinking and industrial construction.

Horizontal Pumps   This type is also widely used in mines and quarries. For example, in a Michigan copper mine, six horizontally split 2-stage 1,500-gpm 940-ft-head pumps, each driven by a 500-hp motor, are used as boosters at a depth of 1,077 ft below the surface. They receive water from six submersible-motor pumps installed at a lower level. By means of this arrangement, the large amount of explosive gas present in the water of this mine is kept in solution. The continuous airtight line from the lowest pumping depth to the surface ensures that no gas will leave the water and enter the mine workings. All parts of the pumps contacting the water are made of monel or bronze. The interior of the water piping is neoprene-coated.

Other Vertical Pumps   Submersible-motor pumps find some use in mines, as mentioned above. See Chap. 17 for a discussion of this and the other types of vertical and horizontal pumps mentioned here.

Vertical packingless pumps and units without lower bearings are used to handle relatively clear mine water. Pneumatically driven sponge pumps find many applications in mine shafts and winzes, sumps, etc. They are built in single- and 2-stage models developing heads to 300 ft at capacities from 25 to 300 gpm. The air pressure required for operation varies from 70 to 100 psi. Air flow is 105 or 125 cu ft per min, or higher, depending on pump head and capacity.

Coal Washing   In many mines the coal must be washed and cleaned when it reaches the surface. This is often done by drawing water from a settling pond near the tipple and discharging it to an elevated cone where the coal is washed and cleaned. In one installation 2,000 tons of soft coal are washed per day by a 1,000-gpm 150-ft-head single-stage 4-in. centrifugal pump. Since the settling-pond water becomes highly acidic—pH is sometimes as low as 2—the pump must be made of suitable acid-resisting materials. By providing a make-up-water pump and using no outlet on the settling pond, stream pollution in the area can be prevented. In many mines the pump for handling wash water is mounted on a float so it can be easily moved about the pond from one point to another.

Rotary Pumps   Gear-type rotary pumps, made of bronze or stainless steel and fitted with a variable-speed drive, are used to meter stock solution and diluent in separan feed systems. The pump should be protected with 16-mesh wire cloth when used in this service. Separan is a synthetic flocculant used with a wide variety of ores, uranium slimes, copper and zinc concentrates, etc.

Other types of rotary pumps find some uses in mining in hydraulic systems, lubricant transfer, etc. They do not, however, find as much use as centrifugal pumps.

Reciprocating Pumps   Direct-acting, diaphragm, and power-type reciprocating pumps are used in coal and metal mining. These units are usually horizontal because they may be operated in areas where the headroom is extremely low. Typical uses include dewatering, handling gold-bearing sludges, pressure grouting, and seepage disposal. Many direct-acting pumps designed for mine installation can be steam- or air-driven, depending on the requirements of the job.

Metering and proportioning pumps (Chap. 3) are used for control of pH in conditioners ahead of ore flotation cells, to feed depressing agents in differential ore flotation, in treating industrial wastes, and for a number of similar services in mining and quarrying.

Pump Selection *Capacity:* To provide positive control of mine drainage, it is common practice to split the required capacity amongst two or more pumps, with one or more standbys. This gives better flexibility in pump operation and maintenance. *Head:* When abrasive liquids are handled, the multistage centrifugal pump is unsuitable because it has close tolerances and absolute sealing is required between stages. Single-stage pumps are connected in series when the required head exceeds that which can be developed by a single pump. Power-type reciprocating pumps are often used for exceptionally high lifts in mines and quarries. *Materials:* These vary considerably from other types of industrial pumps. Nickel irons, manganese steel, white iron, good-quality gray iron, and natural and synthetic rubber perform well in abrasive services. Care must be exercised when handling sand because it will damage some materials suitable for handling large abrasive particles.

Drives: Motors are common, with internal-combustion engines, steam turbines, and air power also being popular. *Controls:* Many different types are used, including d-c variable-speed motors, adjustable-speed a-c motors, fluid and magnetic couplings, and gear and belt drives. *Speed:* Most solids-handling pumps operate at lower speeds than other industrial pumps, necessitating a speed reducer of some type when the driver is a motor. A pump speed range of 700 to 900 rpm is common. *Packing:* Liquids containing abrasives are usually poor lubricants. Clear water from an outside source may be used. Or a specially built centrifugal seal, available on some types of pumps, is used instead of a water seal. *Liquid Velocity:* Most tailings are pumped at a minimum velocity of 4 ft per sec, if they include slimes. Some authorities recommend 5 to 7 ft per sec for slime tailings, 7 to 9 ft per sec for slime-free tailings. Sand and gravel are pumped at velocities as high as 18 ft per sec, but the usual velocity for sand is in the range of 12 ft per sec.

Power Input: Multiply the clear-water horsepower by the specific gravity of the mixture handled to determine the power input required with slurries, sand, etc. *Froths:* Provide at least 6 ft npsh and use as large a pump as possible, to permit it to operate at low speed. Reduce the size of the discharge pipe so the liquid velocity is high enough to keep the solids in suspension.

Piping: Pumping solids either horizontally or vertically, or both, is easier than at angles between 15 and 75 deg with the horizontal. At these angles the solids tend to drop out of suspension, roll back down the line, and increase flow resistance. With usual solids the Williams and Hazen coefficient is 140, compared with 100 for clear water in 15-year-old pipe. However, this coefficient can vary widely, higher and lower, depending upon specific operating conditions such as per cent solids by weight, screen analysis, and velocity. Therefore, the friction losses in solids-handling piping must be carefully computed. Piping materials are discussed earlier in this chapter.

CONSTRUCTION

Large numbers of centrifugal and reciprocating pumps are used on construction projects of all types. Typical applications include drainage, jetting, and pressure grouting. Most units of this type are wheel-, truck-, skid-, or wheelbarrow-mounted to allow easy movement from one site to another. While some are motor-driven, most centrifugal pumps for this service are engine-driven (Fig. 21-4). Either gasoline or diesel engines may be used.

Centrifugal Pumps   These are single-stage for all ordinary jobs and handle up to 4,000 gpm. Heads range up to 110 ft. Capacity standards are established by the Contractors' Pump Bureau of the Associated General Contractors of America, Inc. It is best to purchase pumps meeting AGC standards. Two-stage self-priming pumps are also available for this service.

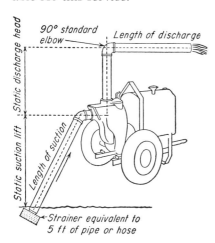

Fig. 21-4 Typical piping for a pump used in construction work.

Engine-driven units are furnished on a base, steel wheels, or pneumatic tires. The discharge elbow (Fig. 21-4), nipple, and suction strainer are furnished as standard equipment on gasoline-engine-driven pumps. The same is true for most diesel-driven portable pumps. Almost all centrifugal pumps built for contractors' service today are self-priming units.

Caisson Pumps   These are designed to be lowered into caissons, pits, mines, and other areas where pumping is required. They are often close-coupled motor-driven single-stage centrifugal pumps mounted vertically and fitted with a suitable metal sling for lowering into the pit in which they are used. Some direct-acting vertical reciprocating simplex and duplex pumps are used for this service too but they are being replaced by centrifugal pumps.

Jetting Pumps   These are generally single- or multistage centrifugal pumps used for gravel washing, rotary drilling, jetting, fire protection, and road construction. Portable, they are usually engine-driven. Jetting applications often use a 250-gpm 150 psi pump. In well drilling, the pump must handle drilling mud, cement slurry, and gravel up to $1\frac{1}{4}$-in. size. For gravel washing, a 225-gpm 85-psi pump is usually satisfactory, but larger sizes are used on big jobs. Figure 21-4 shows the vari-

Fig. 21-5   Portable pressure-grouting rig. (*Gardner-Denver Co.*)

ables which must be considered when computing the total head on pumps used in construction work.

Reciprocating Pumps   Diaphragm pumps are popular today for handling a wide variety of construction liquids.   They usually operate at relatively low speeds and use abrasion-resistant materials like rubber, cast iron, and manganese steel.

Steam- and air-driven direct-acting pumps are also used on many construction projects.   Horizontal duplex portable units (Fig. 3-1) are the most common type used today.

To fill voids or fissures in rock formations, foundations, roadbeds, and mine shafts, pressure-grouting techniques are often used.   Here, grout mixtures varying from $4\frac{1}{2}$ to 45 gal of water to a sack of cement are pumped into the voids using steam- or air-driven reciprocating pumps. Figure 21-5 shows a portable rig which includes the pump itself, together with tanks and mixers.

CHAPTER TWENTY-TWO

Marine Services

PUMPS AND ASSOCIATED AUXILIARIES are among the most vital parts of a modern ship. Careful choice is extremely important because service at sea, in either naval or merchant vessels, can be severe. Also, the pumps must conform to the rules or specifications of one or more regulatory bodies—the American Bureau of Shipping, U.S. Coast Guard Marine Inspection Service, Navy Department Bureau of Ships, Bureau of Yards and Docks, Army Transportation Corps, U.S. Engineers, etc. It is up to the engineer choosing a pump for marine service to determine which rules apply. The manufacturer ordinarily does not assume the responsibility for this but usually wants to know which rules apply, so he can supply a suitable pump. One of the characteristics of marine service is the wide variety of pumps used for different services. Many are described below.

Centrifugal Pumps These are used for auxiliary condenser condensate and cooling services, atmospheric drains, ballast, bilge, boiler-feed, booster, brine, Butterworth system, cargo, circulating, condensate, damage control, distiller, drainage, drinking-water, elevators, engine-cooling, evaporators, fire, flushing, fresh-water, gasoline, general-service, gun-cooling, heater-drains, hot-water-circulating, hotwell, ice-water, make-up-feed, potable-water, sanitary, and wash-water services.

As in stationary applications, the type of pump chosen depends on the head, capacity, liquid handled, etc. Many of these factors are discussed later in this chapter.

Rotary Pumps These are used for engine cooling; cargo loading; unloading and stripping; crude oil; elevator service; emergency bilge and steering gear; fuel-oil service and transfer; gasoline transfer; lube-oil circulation, service, and transfer; molasses transfer, stripper service; and transfer of oils, gasoline, sirup, etc. Rotary pumps for marine service are often mounted vertically to save space. Horizontal units are sometimes mounted on an intergral strainer, particularly for tanker and barge service.

Direct-acting Reciprocating Pumps These are used for air, ballast, bilge, boiler-feed, Butterworth system, cargo, circulating, combined air and circulating, donkey feed, drinking water, emergency feed, evaporator feed, fire, fog oil, fresh water, fuel oil, general service, hydraulic testing, ice water, lube oil, mate's pump, oil cargo, potable water, salt water, sanitary, transfer, vacuum, wash water, and waste-heat boiler feed. While the use of direct-acting pumps has declined in both marine and stationary plants, the decrease has not been quite so marked in ships. This is because the direct-acting pump has many advantages for marine services—including simplicity, reliability, satisfactory efficiency, etc.

Power-type Reciprocating Pumps These are used for ballast, bilge, boiler feed, drinking water, feed-water treatment injection, fire, fresh water, fuel oil, general pressure services, oily ballast, potable water, sanitary water, and wash water. Though this type finds somewhat less use than the direct-acting pump, it is well suited to certain marine services.

Pumps Used As in stationary practice, certain fairly standard choices are made for marine services. A number of these are outlined here to aid the engineer in his design work.

Boiler Feed: Low-pressure ships use 2-stage horizontally split volute-type bronze-fitted or all-bronze motor- or turbine-driven units. Medium- and high-pressure vessels use 4- or 6-stage horizontally split volute-type bronze-, marine-, or stainless-steel-fitted pumps. Multistage barrel-type diffuser pumps are also used in high-pressure ships. Another popular design for high pressures is shown in Fig. 22-1. It consists of a single-stage impeller at one end of the shaft and a velocity-staged steam turbine at the other end. Variations of this design have a 3-stage pump and are widely used in naval service. Close-coupled horizontal motor-driven single- and 2-stage pumps are often used for feeding donkey, waste-heat, and other small boilers. Horizontal and vertical simplex and duplex direct-acting bronze-fitted pumps are used for low-pressure boiler feed where liquid pressures up to 300 psi are suitable. Horizontal duplex and vertical duplex and triplex power

(a)

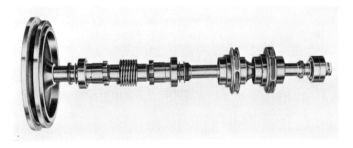

(b)

Fig. 22-1 (a) High-pressure turbine-driven boiler-feed pump uses a single shaft. (*Coffin Turbo Pump, FMC Corp.*)

pumps are popular for all classes of ships, whether high or low pressure. These units may be single- or double-acting. Figure 11-4 shows a typical unit for marine service. A high-speed centrifugal pump is shown in Fig. 22-6.

Ballast: Vertical single-stage double-suction volute-type vertically split motor- and turbine-driven pumps (Fig. 22-2) are popular. Horizontal single-stage double-suction volute-type pumps (Fig. 1-13) are also used. Where extremely large flows are to be handled, horizontal mixed-flow

Fig. 22-2 Single-stage double-suction vertically split volute-type marine pump. (*Worthington Corp.*)

pumps are often chosen.   Horizontal and vertical simplex and duplex direct-acting reciprocating pumps, as well as horizontal and vertical duplex and triplex power pumps, also handle ballast in many vessels. Usual ballast pumps are bronze-fitted or all-bronze and develop moderate head—about 100 psi.

Bilge: The same types of centrifugal and reciprocating pumps as used for ballast services are used for bilges, plus multistage vertical turbine, mixed-flow, horizontal close-coupled, regenerative-turbine, and single-suction horizontal volute pumps.   These units develop moderate heads —about 100 psi—and are bronze-fitted or all-bronze construction.   Submersible bilge pumps are vertical units fitted with motors capable of operating completely submerged in water.   This type is required on many vessels.

Butterworth Systems: Horizontal or vertical two-stage split-casing volute pumps for heads up to 1,000 ft, capacities to 3,200 gpm, are common for this service.   Horizontal or vertical duplex direct-acting 200-psi liquid presure 500-gpm reciprocating pumps may also be used.   Pumps for this service are bronze-fitted or all-bronze.

Brine: Centrifugal pumps are almost universal for brine circulation. Horizontal or vertical single-stage volute-type split-casing pumps for heads to 300 ft, capacities to 3,000 gpm, are the common choice. Close-coupled pumps developing about the same head and capacity are also used. All-iron construction is popular but bronze-fitted and all-bronze are also used.

Cargo: Vertical turbine pumps rated up to 12,000 gpm and 500-ft head handle oil, gasoline, etc. Horizontal and vertical rotary gear pumps handle a wide variety of oil cargoes. Figure 22-3 shows a popular unit for cargo handling. Some direct-acting steam pumps also handle cargo. They develop heads up to 250 psi for this service. Vertical close-coupled self-priming turbine pumps are popular for cargo unloading and close stripping, as are horizontal centrifugals.

Condensate: Single- and 2-stage horizontal and vertical volute-type split-casing centrifugal pumps are used for handling condensate. The construction of 2-stage units is identical to those described in Chap. 11. Single-stage vertical condensate pumps are the top-suction type. Condensate pumps for marine service are bronze-fitted or all-bronze. Reciprocating pumps are seldom used for condensate today.

Condensate Return: See Chap. 19. The same types of units described there are used aboard ship.

Condenser Circulating: Single-stage double-suction horizontal and vertical volute-type, as well as single-suction volute-type, mixed flow, and horizontal and vertical propeller pumps are used to circulate condensing water. Capacities range up to 50,000 gpm for the axial-flow and

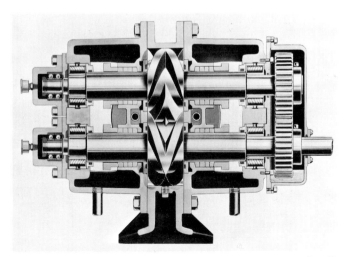

Fig. 22-3 Cross section through helical-rotor main cargo pump for oil, molasses, water, etc.

propeller pumps, but the head developed is relatively low. Both motor and steam-turbine drives are popular. The pumps are usually bronze-fitted or all-bronze construction. Reciprocating pumps are rarely used for main-condenser circulation. Some direct-acting simplex and duplex units are used for auxiliary-condenser circulation.

Damage Control: Horizontal or vertical single-stage double-suction volute-type split-casing pumps are used in this service. Head and capacity required are moderate.

Drinking Water: Two-stage volute-type, horizontal single- or 2-stage close-coupled, and single-stage regenerative-turbine pumps handle drinking water. Horizontal or vertical simplex and duplex reciprocating pumps still find some use in this service. Head required is moderate to high; capacity is moderate. Bronze-fitted or all-bronze construction is used.

Fire and Flushing: These resemble the units used in ballast, Butterworth, and brine systems. Underwriter-approved single- and 2-stage pumps are fitted with the equipment described in Chap. 17 and built in four standard capacities—500, 750, 1,000, and 1,500 gpm at 100 psi. The driving motor or steam turbine is rated at 75, 100, 125, and 200 hp, respectively, for these capacities.

Fuel Oil: Simplex and duplex horizontal and vertical direct-acting pumps are often used for fuel-oil service and transfer. Horizontal and vertical duplex power pumps are popular for fuel transfer. Gear-type rotary pumps, both vertical and horizontal, are probably the most widely used pumps for fuel transfer, loading, unloading, and lube-oil service. A typical unit is shown in Fig. 22-4.

General Service: See ballast, bilge, Butterworth, brine, and cargo pumps.

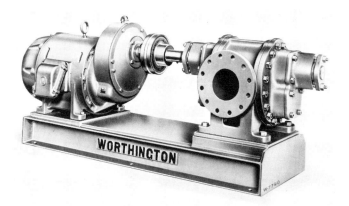

Fig. 22-4   Typical gear pump for marine service. (*Worthington Corp.*)

Gun Cooling: Single-stage close-coupled volute-type motor-driven bronze-fitted all-bronze or all-iron pumps are used for gun cooling on naval vessels. Heads range to 300 ft, capacities to 3,000 gpm.

Hot-water Circulation: See gun-cooling pumps.

Hotwell: See condensate pumps.

Ice Water: See gun-cooling and brine pumps.

Sanitary: See bilge, brine, fire, and general-service pumps.

Sewage: See Chap. 18.

Wash Water: See drinking-water pumps.

Dry Docks The most important requirement in this service is high capacity at relatively low heads. For dry-dock unwatering, single-stage double-suction horizontally split volute-type pumps are popular for capacities up to about 50,000 gpm at heads up to 300 ft. Horizontal and vertical mixed-flow pumps handling up to about 225,000 gpm are used in large dry docks. Submersible construction is used for these pumps when they are applied to floating dry docks. Vertical turbine pumps (Fig. 22-5) are also used in floating dry docks, as well as in cofferdams. Either bronze-fitted or all-bronze construction is used for dry-dock pumps. Submersible-motor propeller pumps are often used in graving docks where they handle large capacities —up to 100,000 gpm or more. See Chaps. 17 and 20.

Steering-gear Pumps Hydraulic steering is probably the most popular type today. Various types of hydraulic pumps (Chap. 23) are used.

Pump Selection *Capacity and Head:* A number of useful rules of thumb are helpful in choosing pump capacity and head, or in checking values found by other methods. They give pump capacity in terms of *gpm per* 1,000 *tons displacement of the vessel and are as follows:* Sanitary or flushing pumps, 1.6 gpm at 100 psi; ballast pumps, 35 gpm at 50 psi; bilge pumps, 35 gpm at 50 psi; fresh-water pumps for faucets, 3 gpm at 100 psi; ice-water-circulating pumps, 0.5 gpm at 35 psi. Other pumps, to which the above rule does not apply, are sized thus: Fire pump—use at least two 400-gpm 125-psi pumps, regardless of ship size. Large vessels should have more. Fuel-oil transfer pumps—use at least one 225-gpm

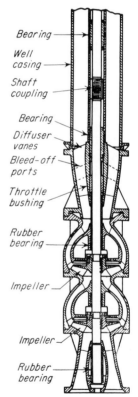

Fig. 22-5 Liquid end of 2-stage vertical turbine pump for dry-dock and sump service.

50-psi pump, regardless of ship size or power. Large vessels may require more. Refrigeration-condenser-circulating pumps—3 to 5 gpm per ton of refrigeration. Head should be at least 40 psi. Turbine lube-oil service pumps should deliver about 0.04 gpm at 50 psi, per shaft horsepower. Butterworth system—use at least one 450-gpm 200-psi pump per vessel, with two or more in large ships. Other pumps should be chosen in accordance with heat-balance calculations, underwriter requirements, or other governing factors.

Materials: Three common choices for marine pumps are bronze-fitted, all-bronze, and all-iron. However, many other types may be specified for extremely corrosive conditions. In naval service, cast irons are avoided because of their tendency to shatter under the shock of gun fire. See Chap. 6. *Drives:* Motors, generally d-c, drive most pumps on large vessels. Some a-c motors for pump drive are coming into use. Steam turbines are now often reserved for standby or auxiliary pumps, except in naval vessels, where the requirements differ from commercial service. *NPSH:* Be sure to provide sufficient npsh for all marine pumps. When pumps are above the water line of the vessel and they take their suction from the sea, fit a foot valve next to the valve on the sea connection. *Control:* Variable-speed motors are common; manual control is often used for direct-acting steam pumps.

Number of Pumps: Standby pumps are widely used aboard ships to prevent major breakdowns at sea. The usual practice is to power the standby pump from a source different from that for the main pumps. With the main pumps motor-driven, the standby units are often steam-driven. This is one reason why direct-acting reciprocating pumps are still so popular in marine service. Most pumps aboard ship should have at least one standby and often must be interconnected with a third pump. *Piping:* Fit sea suction connections with a cock or a valve and a strainer over the outboard opening. Locate the cock or valve close to the hull so the pump suction pipe can be removed. Discharge lines from pumps below the water line should be fitted with check valves. Be sure to design all piping systems in accordance with government and underwriters' requirements.

Special Pumps Because marine service differs so much from others, the pumping-system designer will find a number of special pumps available for use in various types of ships. These include reciprocating air pumps for steam condensers, combined air and circulating pumps, centrifugal gasoline pumps driven by water turbines, damage-control pumps close-coupled to steam turbines, submersible pumps driven through intermediate shafting by a motor located above a watertight deck, horizontal and vertical geared turbines driving vertical pumps, and 3-bearing 3-stage turbine-driven feed pumps. It is interesting to note that the recently developed *Texas towers* use a number of marine-type pumps, including two submersible-motor units (Chap. 17). Since marine pumps are so important to successful and safe ship operation, the designer should comply with the rules of the government and underwriting agencies having jurisdiction over the ship. Other chapters in this book contain a number of references to marine pumps. See the index for page numbers.

Dredge Pumps These handle a variety of materials—from silt and marl to coarse gravel and boulders. Pumps for dredges must handle large solids without clogging, withstand wear, and have a low power consumption and minimum maintenance cost.

Centrifugal pumps are the standard choice for modern dredges. They are generally end-suction single-stage volute pumps with large clearances to handle boulders, roots, and other large materials. Since abrasion and shock may be severe, special construction materials are used. For dredging clay or silt, wearing parts are made of hardened alloy steel or nickel irons; for coarse gravel, manganese steel is often used; for fine sharp abrasive sand, heat-treated alloy steel is common.

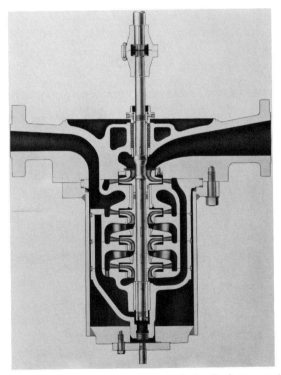

Fig. 22-6 Four-stage marine-type boiler-feed pump. (*Ingersoll-Rand Co.*)

There are only three wearing parts in these pumps—the casing, impeller, and suction disk.

Dredge pumps may be driven by electric motors, turbines, or internal-combustion engines. Standard pumps are built in a number of sizes. One typical line handles up to 195 cu yd per hr at 10 per cent solids and 12 ft per sec discharge velocity. At this rating it requires 31 bhp for each 10 ft of total discharge head.

When choosing a dredge pump, the manufacturer must be given complete information. This should include pump use, nature of material to be pumped—whether compact or loose, percentage of solids of different sizes, maximum size of solids, and the piping and drive arrangements. Dredge pumps are also used for sand and gravel production, abrasive solids, and land-filling operations.

New Power Sources Gas-turbine-propelled and nuclear-energy-powered vessels are beginning to see service at sea. Pumps for these vessels will not differ much from those described above and in Chap. 12.

Mechanical Seals These are being more widely used in marine pumps of all types because they offer a number of advantages. For example, in main cargo pumps for tankers they permit stripping cargo to almost 1 ft lower than with conventional packed pumps. Also, leakage of oil into tanker pump-room bilges is almost completely eliminated. By reducing shaft leakage to a minimum, many potential fire hazards are avoided. Also, maintenance requirements are reduced.

Yachts, Other Small Vessels Small vessels generally have 6-, 12-, or 32-volt electrical systems with only moderate generating and storage capacity. So water and other pumps for this service must be carefully designed. In pressure water systems small tank-mounted centrifugal pumps are popular. The pump discharges to the pressure tank, from where lines lead to the faucets supplied. Pumps without a pressure tank are also available. One unit for this service is rated at 5 gpm and 20 psi.

Where direct-reversing internal-combustion engines are used for driving a vessel, reversible centrifugal pumps may be used for the circulation of cooling water and lube oil. These pumps resemble the usual centrifugal but can be operated in either direction of rotation without damage. The head and capacity are identical in both directions.

Booster Feed Pumps Because space is often limited aboard ships, a booster feed pump may be used. This unit, generally of the centrifugal type, takes its suction from a direct-contact heater and discharges to the suction of the main feed pump. Booster feed pumps for marine service are generally multistage units, though a single-stage pump can be used where the required pressure is low.

CHAPTER TWENTY-THREE

Industrial Hydraulic and Vacuum Pumps

HYDRAULIC TRANSMISSION OF POWER by means of a pump and hydraulic motor is widely used in industry today. A few typical applications include machine tools of many types—presses, injection molders, extrusion presses, tube reducers, tube testers, welding machines, trimming machines, broaching, gear shaping, die casting, etc.—as well as a number of others throughout industry. These include such diverse uses as oil-well pumping, materials handling, elevators, cranes, winches, door opening, steel descaling, log debarking, hydraulic mining, and mechanical testing of materials for expansion, collapse, and shrinkage. Hydraulic systems use millions of pumps and all indications today point to even greater application of hydraulic power transmission in the future.

Pumps Three classes of pumps are used in hydraulic systems today: (1) reciprocating-piston, (2) rotary, and (3) centrifugal. Regardless of the type of pump used, it is arranged to discharge the hydraulic fluid to a motor, piston, or other power-conversion device. Often the motor is quite similar in design to the pump.

Reciprocating Pumps Figures 3-11 through 3-14 show four piston-type pumps built for hydraulic service. Today piston pumps are solving new problems in high-pressure high-temperature high-speed applica-

411

tions. Piston pumps with small clearances minimize volumetric leakage losses at higher operating pressures and speeds. Power capacity of piston pumps is proportional to the product of fluid pressure, pump speed, and displacement. Since weight and size are primarily functions of displacement, an increase in pressure and speed reduces the pump size for the same power output.

Small piston pumps can operate at speeds of over 6,000 rpm and are better suited to higher-speed applications than gear and vane pumps. A higher operating pressure also reduces the size of other units in the system, requiring less fluid volume. Failure of pressure-containing elements is not hazardous because a small discharge of fluid is enough to reduce the pressure to zero. Piston pumps have been used at pressures over 8,000 psi and at temperatures above 400 F. The discharge flow from high-speed multiple-piston pumps has only negligible pulsation. At higher speeds and pressures—over 1,500 rpm and 1,000 psi—the efficiency of the piston pump is greater than either the gear or vane pump.

Piston pumps are built for either constant or variable delivery, depending on the needs of the system they serve. Figure 23-1 shows the components of an axial-piston constant-delivery pump. It has nine major parts. Rotation of the cylinder barrel causes axial reciprocation of the pistons.

Four variable-delivery pumps are shown in Figs. 3-11 through 3-14. Delivery may be varied by one of several methods. In some units the piston stroke length is varied by changing the throw of a swash plate, cam, or eccentric. In others, a sleeve rides over piston port holes, with

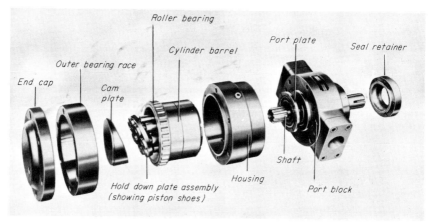

Fig. 23-1 Components of an axial-piston-type hydraulic pump before assembly. (*Denison Engineering Co.*)

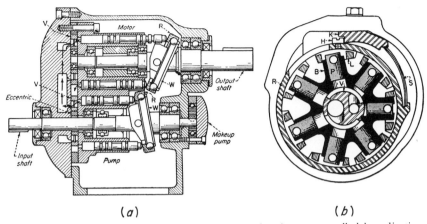

Fig. 23-2   (*a*) Pump and motor in this variable-speed unit are controlled by adjusting wobbler plates *W*. (*b*) Self-unloading radial-piston pump.

the sleeve control allowing a varying amount of liquid to escape to the inlet.

Figure 23-2*a* shows a variable-delivery pump and motor combined in a single housing. Output is changed by adjusting the wobbler plate *W*. A self-unloading radial-piston unit is shown in Fig. 23-2*b*. It unloads by swinging ring *R* to a neutral position if the discharge pressure rises above a fixed value. When the pressure returns to normal, spring *S* moves the ring back into position for liquid delivery.

Today piston pumps are built to handle capacities up to about 2,500 gpm. Pressures of 6,000 psi are fairly common, with 10,000 psi presenting no serious design problems. Fire-resistant hydraulic fluids permit operation at higher temperatures than oil's upper limit. The present maximum limit on viscosity for hydraulic fluids is 1,000 SSU.

Power Pumps   Modern horizontal and vertical power pumps (Figs. 3-4, 3-6, and 3-7) are used in many industrial hydraulic systems where the operating pressure is 2,500 psi or higher. They are generally considered as constant-speed constant-capacity units. Suction-valve unloaders and stroke transformers (Fig. 3-44) are the most common controls for variable-capacity service. Power pumps are confined almost exclusively to large-capacity high-pressure systems like those found in paper, steel, petroleum, and similar applications. They frequently use water containing a small amount of oil as their hydraulic fluid and are built for pressures up to about 10,000 psi.

Rotary Pumps   Gear-type rotary pumps (Figs. 2-2 and 2-3) as well as vane-type pumps (Fig. 2-11) are popular in hydraulic systems. Gear pumps develop up to 3,000 psi, vane pumps 2,000 psi. Two pumps can

(a)

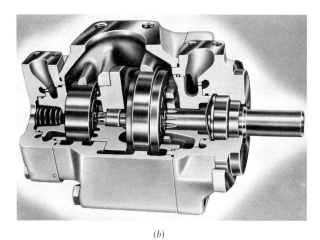

(b)

Fig. 23-3   (a) Dual-vane constant-delivery pump. (*Parker-Hannifin.*)
(b) Vane-type double pump provides hydraulic fluid to two circuits.
(*Denison Division, Abex Corp.*)

be arranged in series to develop twice these pressures if desired. Capacities vary from one manufacturer to another, but a range of 2 to 300 gpm is common for many standard makes.

Vane-type pumps (Fig. 23-3) are relatively small in size, operate at 400 to 2,000 rpm, and ordinarily handle mineral-base oils with viscosities from 70 to 300 SSU. They are usually built as constant-delivery units, but variable-delivery vane pumps are also available. The unit in Fig. 23-3 is a dual-vane type, each vane being composed of two parts sliding freely relative to each other and providing a connecting passage between the pocket underneath the vane and the space between vanes. Vane pumps are also built with two pumps in one casing. They provide two different flows from a common suction to such equipment as molding machines and machine tools using "hi-lo" circuits.

Gear pumps are normally constant-delivery units built in sizes to 200 hp for hydraulic service. They operate at speeds from 200 to 4,000 rpm, with one automotive installation operating at 9,000 rpm. Some pumps are *supercharged* by means of another gear or a centrifugal pump which discharges into the suction inlet of the main gear pump. Others have a venturi or orifice on the inlet of the pump to accelerate oil flow into the pump. By using a high-pressure oil seal on the pump drive shaft, the unit can be operated in either direction.

Gear pumps for hydraulic systems use spur, helical, or herringbone gears. They are low-cost units of simple design and construction. Dirt does not cause so much difficulty with them as with other types of pumps. Gear pumps, however, are noisier than other types.

Centrifugal Pumps Multistage split-case and barrel-type diffuser or volute pumps (Figs. 11-3 and 11-5) are used in large high-pressure hydraulic systems. Pure water or water containing a small amount of soluble oil is generally used as the hydraulic fluid with centrifugal pumps. This class of pump is particularly well suited for systems in which the liquid demand varies little from some average value. The pump must be guarded against overheating during periods of low or zero demand. This can be done by recirculating some of the liquid from the discharge to the suction side of the pump.

To safeguard centrifugal pumps from water hammer and seizure on the suction side, provide a tank containing a 10-sec water supply or a 15- to 20-ft static suction head. The first type of tank should have half its volume devoted to air space, while the second should have an automatic float-actuated make-up valve for water or the hydraulic fluid. Use orifice plates in the discharge line, installed 10 to 15 ft away from the pump, with a straight run of pipe between it and the orifice. This prevents some of the shock from being transmitted to the pump during sudden changes in liquid demand. When large cylinders are to be filled, check to see that the resulting head and capacity fall on the normal characteristic curve and that the pump does not operate on the break-off. A cushion tank with a 10-sec discharge capacity and an equal air capacity is often useful in absorbing shock on the discharge side of the pump. In steel-mill descaling, use a pump specially designed for this service. Several manufacturers have them available.

Air-driven Pumps To provide small amounts of hydraulic power at low, medium, or high pressures, air-driven pumps are often used. These units are ideal for testing tubing, valves, or pressure vessels and supplying power for die casting, plastic molding, or belt presses. Figure

Fig. 23-4   Air-driven pump delivers small amounts of hydraulic fluid at high pressures. (*Aldrich Division, Ingersoll-Rand Co.*)

23-4 shows one such pump.   Built with ⅝- to 3-in.-diameter plungers, it develops 103 psi hydraulic pressure with 10-psig air or 37,000 psi with 100-psig air pressure.   Special designs are available for higher pressures.   Each unit of this design is fitted with an air filter, regulator, lubricator, and hydraulic pressure gage.   They are built as either vertical or horizontal pumps.

Coolant Pumps   Many hydraulic systems and machine tools are fitted with independent pumps for circulation of the hydraulic fluid or a coolant through a heat exchanger.   Small single-stage side-suction close-coupled motor-driven centrifugal pumps (Fig. 23-5) are popular for circulation and coolant service.   Some gear pumps are also used, especially where high pressures are desired.   In large hydraulic systems an additional unit, called a replenishing pump, is often used.   This may be a centrifugal pump similar to that in Fig. 23-5 or a gear pump on the shaft of the main pump.   Complete coolant systems consisting of a pump, reservoir, flexible tubing, and connectors are also available.   The reservoir is rectangular or cylindrical and can be fitted with a filter to remove abrasives and other solids.

Excessive liquid temperatures reduce the viscosity of hydraulic oils, making them less viscous.   This can cause pump slippage, internal leakage, seizure of close-tolerance parts, poor functioning of precision-made controls, accelerated breakdown of vital properties, and the risk of varnish and sludge formation.   The importance of correct cooling of hydraulic fluids can therefore be readily seen.

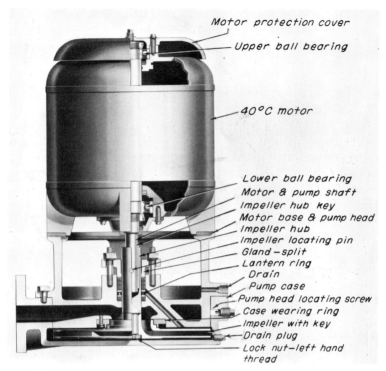

Motor protection cover

Upper ball bearing

40°C motor

Lower ball bearing
Motor & pump shaft
Impeller hub key
Motor base & pump head
Impeller hub
Impeller locating pin
Gland – split
Lantern ring
Drain
Pump case
Pump head locating screw
Case wearing ring
Impeller with key
Drain plug
Lock nut – left hand
thread

Fig. 23-5   Typical close-coupled motor-driven centrifugal coolant pump.

Hydraulic Fluids   Oil, water, a solution of water and oil, and a number of synthetic liquids are used in hydraulic systems today. When choosing a pump, check carefully to see that it can handle the hydraulic fluid to be used. This is important because at present the hydraulic fluid is often chosen more from a safety standpoint than from any other. Fire-resistant hydraulic fluids, made by mixing water and nonpetroleum synthetic materials, provide considerable added safety. Be careful to determine the exact viscosity of the hydraulic fluid at its working temperature because pump output and power consumption are directly related to liquid viscosity. Since the corrosive characteristics of different hydraulic fluids vary, the material of construction for the pump must be carefully chosen. If at all possible, the hydraulic fluid should have good lubricating qualities.

Accumulators   These are devices in which the hydraulic fluid can be stored during idle periods and held ready for instant use. They permit use of a smaller pump and electric motor, with both operating steadily. The result is a higher over-all efficiency and lower operating cost. The types of accumulators used today include weighted, air-piston, and gas-loaded. Accumulators must be carefully chosen for the particular system because incorrect applications can lead to operating difficulties. Be sure to provide all hydraulic systems with one or more relief valves capable of handling the total liquid capacity of the pumps in the system.

INDUSTRIAL VACUUM SYSTEMS

Liquid-piston Pump   A unit of this type is shown in Fig. 23-6. Single-stage pumps of this design maintain vacuums up to 27 in. Hg; standard 2-stage units go up to 29-in.-Hg vacuum. Capacities range from 3 to 9,600 cfm at 15-in.-Hg vacuum and 8 to 8,000 cfm at 25-in.-Hg vacuum. These pumps are used in a number of different industries and applications—paper, textile, vacuum extraction, suction filtration, priming centrifugal pumps, jet and surface condensers, vacuum pans, deodorizers in dairies, candy and sugar evaporation, solvent recovery, deaeration of latex, cellulose acetate, ceramic products, etc.

As shown in Fig. 23-6a, the pump consists of a round multiblade rotor, positioned eccentrically and revolving freely in a cylindrical casing partially filled with liquid. When the rotor revolves, a solid ring of liquid turns with it and follows the shape of the casing. Liquid between any two vanes enters and recedes from this space during pump operation, drawing in air or gas through the inlet port and discharging it through a discharge port. These pumps are sometimes used in series with air ejectors to achieve vacuums up to 29.5 in. Hg.

Clinical vacuum pumps are employed in hospitals to remove fluids from incisions during operations and for postoperative drainage, and in laboratories for filtration, instrument cleaning, and transfer of liquids and gases. Units like Fig. 23-7 are often applied in hospitals for vacuum service. They are located at a central point with a receiver, from which piping runs to service outlets.

Steam-jet Vacuum Pumps   These are used for many of the same services as liquid-piston pumps, as well as a number of others. They are built as single- or multistage units (2, 3, and 4 stages) and produce vacuums of 29.5 in. Hg or greater when using steam between 65 and 100 psig. These pumps utilize the kinetic energy of steam to entrain, remove, or compress gas and vapors. They may be built as condensing or noncondensing units. Standard materials of construction are used for noncorrosive gases; with corrosive gases, the pump interior is lined with rubber, plastic, stoneware, haveg, etc. Typical installations are discussed in Chap. 11. Other uses include many types of filtration processes, distillation, impregnation, absorption, drying, mixing, vacuum transfer, and packaging.

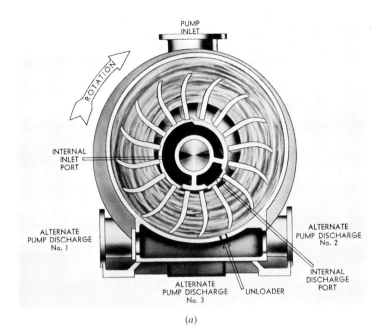

(a)

(b)

Fig. 23-6 (a) End view of liquid-piston impeller. (b) Exploded view of liquid-ring vacuum pump. (*Nash Engineering Co.*)

Pump Selection *Capacity:* This may be expressed in terms of horsepower output or fluid pumped, cubic feet per minute, quarts per hour, or gpm. The pump should be chosen to meet the displacement requirements of the motor, cylinder, or other units it serves. Generally, the load determines the type or types of motors required. This, in turn, fixes pump capacity. *Head:* Pumps may be built to deliver at two or

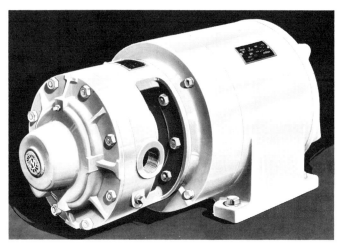

Fig. 23-7   Typical liquid-piston-type vacuum pump. (*Nash Engineering Co.*)

more pressures, depending upon the capacity required. So choose the head on the basis of the capacity required for the types and sizes of devices used in the system. *Materials:* Check to see that pump is suitable for the hydraulic fluid chosen. *Drives:* Motor drives are almost universal. *Number of Pumps:* Only one pump per use, without a standby, is chosen. This keeps costs down. *Piping:* Special pipe or tubing may be required for the highest pressures used today. Check with the pump manufacturer.

CHAPTER TWENTY-FOUR

Iron, Steel, and Other Industries

STEEL MILLS IN THE UNITED STATES use over 6 million gpm of water. Blast furnaces, rolling mills, and steelworks used a total of 3,108 billion gal of water in 1953. About 55 per cent of this water was recirculated back to surface sources, often purer than when it was taken into the plant. Over 90 per cent of the water for blast furnaces and 80 per cent for steelworks and rolling mills was taken from surface-water supplies. Under 5 per cent came from wells, with the remainder from public supply systems.

Pumps used in iron and steel production are often general-purpose units of various types, depending on the head and capacity requirements. For details on the selection and application of these and other pumps in steelmaking, see the earlier chapters in this book.

In descaling operations a pressure of 1,000 psi and over is often used. Water discharged at this pressure onto the metal produces rapid chilling. Also, the impact of the jet is sufficient to remove scale from the metal surface. A 1,000-psi jet of water easily cuts a hole in a 1-in.-thick pine plank. Vertical power pumps (Chap. 3) are widely used for descaling in steel mills, cleaning castings in foundries, and cleaning heat-exchanger tubes in refineries and chemical plants.

Natural-gas Processing A number of pumping operations are performed in the production, storage, and transportation of natural gas.

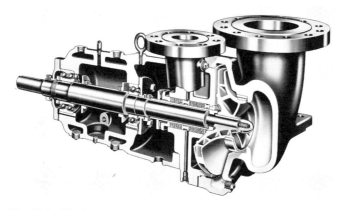

Fig. 24-1 Single-stage top-suction process pump. (*Peerless Pump Division, FMC Corp.*)

These include desulfurization of natural gas, storage of liquefied petroleum gases in underground basins, and handling ethylene glycol in natural-gas dehydration operations. Multicylinder power pumps are popular. Regenerative-turbine pumps are also widely used. They handle a number of liquefied gases, including propane, butane, and isobutane, and are fitted with explosionproof motors and switches in stationary service. For trucks transporting these gases, pumps driven by a power take-off are available. Regenerative-turbine pumps handling liquefied petroleum gases must have carefully sized suction pipes. Sizes usually recommended are 2 in. for 20 to 30 gpm, 3 in. for 45 to 60 gpm, and 4 in. for 100 gpm.

Split-case multistage volute and diffuser pumps are often used for high-pressure high-capacity services in natural-gas production. Single-stage end- and side-suction process pumps are also popular in natural-gas operations. Figure 24-1 shows a typical unit. It handles up to 1,000 gpm at heads to 300 psi and temperatures to 800 F.

Plastics Manufacture and production of plastics involves the use of a number of pumps. These include general-purpose, oil-hydraulic, and high-temperature water pumps, in addition to a number of other types for power production. See the earlier chapters in this book for details.

Newspapers Ink pumps often present problems in newspaper-printing plants because the ink, as received, is viscous and requires careful handling to prevent spillage and contamination. Rotary gear pumps are a common choice for pumping ink. In many plants a separate pump is used for each color of ink handled. A single standby serves all colors. Other pumps used in newspaper plants are covered in earlier chapters.

Paint Manufacture   Pumping paints, and the components used in their manufacture, is tricky.   Leakproof service is required and rotary pumps fitted with mechanical seals (Fig. 11-10 and Chap. 2) are popular.   They effectively handle paints, oils, pigments, etc.

Cosmetics   Small-capacity centrifugal and gear pumps find use in cosmetics production of all types.   No special problems are met in their application.   The same is true of pharmaceuticals manufacture.

Automotive   Plants producing cars and trucks are generally large and use many pumps.   For the most part these are standard units built to meet customary specifications.   General-purpose end-suction process pumps for this and a number of other services throughout industry often have eductor vanes (Fig. 24-2).   These reduce the pressure on the packing box so that at rated head of the pump the packing-box pressure is zero, or nearly so (Fig. 24-3).   Longer packing life is assured at lower maintenance costs.

Fig. 24-2   Rear, end, and front views of a single-stage impeller with eductor vanes.

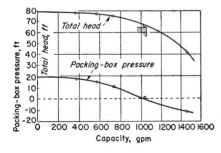

Fig. 24-3   Packing-box pressure obtained with eductor vanes.

Agriculture   A novel, water-wheel-driven pump (Fig. 24-4) feeds liquid fertilizers into irrigation water.   Since the speed and hence capacity of the fertilizer pump are a direct function of the irrigation water flow, a constant fertilizer-water ratio is produced at nearly all water

Fig. 24-4 Fertilizer-injection pump is driven by irrigation water flow. (*Moyno Pump Division, Robbins & Myers, Inc.*)

flows. Fertilizer capacities up to 50 gpm at pressures up to 600 psi can be handled.

Electronics Large quantities of pure water are used in the manufacture of television tubes. Since extreme purity is an important requirement, the pumps chosen are best fitted with mechanical seals. Other services in the electronics industry are standard. For television-tube production, general-purpose volute-type centrifugal pumps are common. The water must be filtered and purified before use.

Wherever large flows are involved, or high pressures, adjustable-speed operation should be considered. Figure 24-5 shows an interesting comparison of three methods of operation. Curve A shows the power input required with two centrifugal pumps operating in series and driven at constant speed with their output adjusted by throttling. When the discharge flow is regulated by adjusting the speed of both pumps, curve B is obtained. When the speed of only one pump is adjusted, curve C is obtained. Note, also, the flat HQ curve obtained with adjustable-speed drive.

Metal Working Coolant pumps (Chap. 23) and oil and water hydraulic pumps, covered in the same chapter, are the most common types used in metalworking. Careful choice is important because these units protect expensive equipment. Figure 24-6 shows a typical close-coupled pump for this and other services.

In certain industries the hydraucone-type centrifugal pump is used.

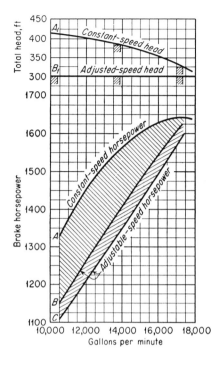

Fig. 24-5 Head and power curves with constant and adjustable speed.

This is simply a multistage centrifugal unit having a cone-shape nozzle fitted with a water-dispersion plate.   Liquid leaving the first stage of the pump enters the nozzle and leaves through an opening formed by the nozzle and plate.   Discharge of the liquid radially in all directions over

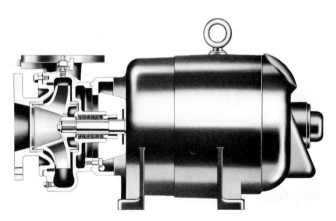

Fig. 24-6 Single-stage close-coupled general-purpose pump. (*Gorman-Rupp Co.*)

the edge of the plate accomplishes the conversion of velocity to pressure with small loss.

Eductors, which are devices resembling steam-jet ejectors, but which use water instead of steam as the motive fluid, find some use in a number of industries for pump priming and assistance. As a primer the eductor acts much like the devices in Fig. 7-37. When designed to assist a pump, the eductor functions only when the liquid level in the suction tank falls below the normal height. Then the eductor starts and discharges liquid into the pump suction line, assuring positive flow into the pump.

Index